Applications and Technologies of Renewable Energy

Applications and Technologies of Renewable Energy

Guest Editors

Ayman Al-Quraan
Ahmad M. A. Malkawi

MDPI

Basel • Beijing • Wuhan • Barcelona • Belgrade • Novi Sad • Cluj • Manchester

Guest Editors

Ayman Al-Quraan
Electrical Power Engineering
Department
Yarmouk University
Irbid
Jordan

Ahmad M. A. Malkawi
Mechatronics Engineering
Department
The University of Jordan
Amman
Jordan

Editorial Office
MDPI AG
Grosspeteranlage 5
4052 Basel, Switzerland

This is a reprint of the Special Issue, published open access by the journal *Sustainability* (ISSN 2071-1050), freely accessible at: www.mdpi.com/journal/sustainability/special_issues/96L861Z1P7.

For citation purposes, cite each article independently as indicated on the article page online and using the guide below:

Lastname, A.A.; Lastname, B.B. Article Title. *Journal Name* **Year**, *Volume Number*, Page Range.

ISBN 978-3-7258-2778-7 (Hbk)
ISBN 978-3-7258-2777-0 (PDF)
https://doi.org/10.3390/books978-3-7258-2777-0

Contents

About the Editors

Ayman Al-Quraan

Ayman A. Al-Quraan was born in Abu Dhabi, UAE, in 1986. He received his Ph.D. degree in Electrical and Computer Engineering from Concordia University, Montreal, in 2016. He is currently an Associate Professor of the Electric Power and Energy Department with Hijjawi Faculty for Engineering Technology, Yarmouk University (YU), Irbid, Jordan. He joined YU in January 2017 after a brief postdoctoral position at Concordia University. He is a founder of the power and energy research lab at Hijjawi Faculty for Engineering Technology, Yarmouk University, Jordan. He granted several research funds and projects in the field of power and energy. He is also the author and reviewer of many papers and books in this field. He holds the academic editor position with several international journals and publishers. Recently, he has been involved as a PI in a research project related to n-layer optimization problems applied to hybrid renewable energy systems (HRESs), funded by Yarmouk University. He currently serves as an Academic Editor to several international electric and energy journals with different publishers/ Wiley, Hindawi, and IET: International Journal of Energy Research, Journal of Electrical and Computer Engineering, and IET Electrical Systems in Transportation. He also served as a guest editor for the Special Issue, "Applications and Technologies of Renewable Energy" for *Sustainability* at MDPI. He is currently serving as a guest editor for the Special Issue, "Modelling Control and Optimization of Hybrid Energy Systems" for *Sustainability* at MDPI.

Ahmad M. A. Malkawi

Dr. Ahmad M. A. Malkawi joined The University of Jordan's Department of Mechatronics Engineering in 2019. He is currently an associate professor at the institute. Between 2019 and 2024, he served as an assistant professor in the Department of Mechatronics Engineering at the University of Jordan. From 2020 until 2022, he served as the Deanship of Scientific Research's Assistant Dean for Journals and Computing. Since 2022, he has served as the Deanship of Scientific Research's assistant dean for Computing and Management. His research interests are power electronics and drive, renewable energy, smart grid, microgrid/nanogrid, and hybrid energy storage systems.

Preface

Renewable energy technologies have diverse applications across various sectors, with the primary goal of reducing dependency on fossil fuels and minimizing environmental impacts. The most common types of renewable energy sources—i.e., solar, wind, hydroelectric, biomass, and geothermal—have applications in electricity generation, transportation, heating, and industrial processes. Solar energy is harnessed through photovoltaic cells to generate electricity or thermal energy for heating water and air. Wind energy technology provides another significant renewable energy source with wind turbines converting kinetic energy from the wind into mechanical power or electricity. Hydroelectric power plants utilize flowing water to spin turbines and generate electricity; meanwhile, biomass involves using organic materials like wood waste or agricultural residues as fuel for producing heat or electricity.

In addition to direct applications, renewable energy technologies have made substantial advancements in integration and optimization systems. Smart grids, equipped with advanced communication and control technologies, enable efficient distribution and management of renewable resources within the electricity network. Energy storage technologies such as batteries and pumped hydro storage are crucial in stabilizing grid operation by storing excess energy generated during peak production times for use during periods of low generation. Innovations in energy management systems facilitate better forecasting, monitoring, and control of renewable energy production which is essential for maximizing efficiency and minimizing costs.

The adoption of renewable energy technologies has led to significant environmental benefits by decreasing greenhouse gas emissions and pollution. Technological advancements have improved the efficiency and affordability of renewable systems, making them a viable option even in remote areas far from conventional power grids. In the transportation sector, electric vehicles powered by renewable sources such as solar or wind power provide a clean alternative to traditional gasoline-powered engines. Continued research efforts are concentrating on enhancing the performance and sustainability of renewable technologies, promising a future where cleaner energy solutions can meet increasing global demands while also supporting economic growth and energy security.

Ayman Al-Quraan and Ahmad M. A. Malkawi
Guest Editors

sustainability

MDPI

Article

Optimal Design of Grid-Connected Hybrid Renewable Energy System Considering Electric Vehicle Station Using Improved Multi-Objective Optimization: Techno-Economic Perspectives

Ameer A. Kareem Al-Sahlawi [1,2,*], Shahrin Md. Ayob [1,*], Chee Wei Tan [1], Hussein Mohammed Ridha [3,4] and Dhafer Manea Hachim [5]

[1] Division of Electrical Power Engineering, Faculty of Electrical Engineering, Universiti Teknologi Malaysia, UTM, Skudai 81310, Johor, Malaysia; cheewei@utm.my

[2] Department of Electrical Engineering, Faculty of Engineering, University of Kufa, Kufa 54001, Iraq

[3] Advanced Lightning, Power and Energy Research (ALPER), Department of Electrical and Electronics Engineering, Faculty of Engineering, Universiti Putra Malaysia, Serdang 43400, Selangor, Malaysia; hussain_mhammad@uomustansiriyah.edu.iq

[4] Department of Computer Engineering, Mustansiriyah University, Baghdad 14022, Iraq

[5] Engineering Technical College/Najaf, Al-Furat Al-Awsat Technical University, Najaf 31001, Iraq; coj.dfr@atu.edu.iq

* Correspondence: ameeralikareem451984@gmail.com or ameerkareim@graduate.utm.my or ameera.abbas@uokufa.edu.iq (A.K.A.-S.); shahrin@fke.utm.my (S.M.A.)

Abstract: Electric vehicle charging stations (EVCSs) and renewable energy sources (RESs) have been widely integrated into distribution systems. Electric vehicles (EVs) offer advantages for distribution systems, such as increasing reliability and efficiency, reducing pollutant emissions, and decreasing dependence on non-endogenous resources. In addition, vehicle-to-grid (V2G) technology has made EVs a potential form of portable energy storage, alleviating the random fluctuation of renewable energy power. This paper simulates the optimal design of a photovoltaic/wind/battery hybrid energy system as a power system combined with an electric vehicle charging station (EVCS) using V2G technology in a grid-connected system. The rule-based energy management strategy (RB-EMS) is used to control and observe the proposed system power flow. A multi-objective improved arithmetic optimization algorithm (MOIAOA) concept is proposed to analyze the optimal sizing of the proposed system components by calculating the optimal values of the three conflicting objectives: grid contribution factor (GCF), levelled cost of electricity (LCOE), and energy sold to the grid (E_{SOLD}). This research uses a collection of meteorological data such as solar radiation, temperature, and wind speed captured every ten minutes for one year for a government building in Al-Najaf Governorate, Iraq. The results indicated that the optimal configuration of the proposed system using the MOIAOA method consists of eight photovoltaic modules, two wind turbines, and thirty-three storage batteries, while the fitness value is equal to 0.1522, the LCOE is equal to 2.66×10^{-2} USD/kWh, the GCF is equal to 7.34×10^{-5} kWh, and the E_{SOLD} is equal to 0.8409 kWh. The integration of RESs with an EV-based grid-connected system is considered the best choice for real applications, owing to their remarkable performance and techno-economic development.

Keywords: renewable energy sources; grid-connected; V2G; multi-objective optimization; arithmetic optimization algorithm

Citation: Al-Sahlawi, A.A.K.; Ayob, S.M.; Tan, C.W.; Ridha, H.M.; Hachim, D.M. Optimal Design of Grid-Connected Hybrid Renewable Energy System Considering Electric Vehicle Station Using Improved Multi-Objective Optimization: Techno-Economic Perspectives. *Sustainability* **2024**, *16*, 2491. https://doi.org/10.3390/su16062491

Academic Editors: Ayman Al-Quraan and Ahmad M. A. Malkawi

Received: 16 January 2024
Revised: 25 February 2024
Accepted: 1 March 2024
Published: 17 March 2024

1. Introduction

The energy crisis resulting from the rapid depletion of fossil resources has raised public awareness of the need for environmental conservation. Thanks to the united efforts of scientists, significant progress has been accomplished during the past ten years. Distributed renewable energy sources (RESs) are integrated into the electrical grid to meet the energy demand [1], and these distributed generation (DG) systems have made considerable use of RESs and electric vehicle charging stations (EVCSs) [2]. The idea of multi-objective techno-economic optimization was put forward in [3] as a way to plan when to charge and discharge electric vehicles. For the first time, frequency regulation was provided while simultaneously modeling and optimizing end-user energy costs, battery degradation, grid interactions, and CO_2 emissions in the context of home microgrids. However, to increase reliability, optimize renewable energy sources, and lower overall costs, appropriate energy management and operation are necessary, along with an appropriate optimization technique based on techno-economic viewpoints.

In [4], the authors suggested a versatile multi-objective optimization method that considers the technological, financial, and environmental aspects while assessing and implementing V2G and grid-to-vehicle technologies. In addition, plug-in electric vehicle (PEV) users' driving habits, charging and discharging habits, and battery life cycles are considered. The firefly algorithm is applied inside a stochastic optimization framework to run simulations on a modified IEEE 69-bus radial distribution test system. The goal is to minimize two objective functions: CO_2 emissions and operational costs. The framework considers renewable generation, load usage, and the charging/discharging time of PEVs as ambiguous variables. The work in [5] offered hybrid renewable energy systems combined with mobile hydrogen vehicle storage and stationary batteries for a zero-energy community comprising office, residential, and academic buildings based on real-world energy consumption data and simulations. A time-of-use grid penalty cost model was presented to achieve electricity grid economy and flexibility, which evaluates grid export and import during on-peak and off-peak times. In the coupled platform of TRNSYS and jEplus + EA, multi-objective optimizations are carried out to size zero-energy buildings and the community while considering the self-consumption of renewable energy, on-site load coverage, and grid penalty cost. Methods for incorporating hydrogen energy technology into hybrid energy systems, focusing on hydrogen fuel cell power generation, were examined in [6]. Energy storage integration, sizing techniques, energy flow control, and the software implementation and optimization methods that go along with them were covered. Published case studies seldom address issues beyond technical ones. The authors talked about this fact in the context of accessible software packages. To meet the design objectives for the energy system, a four-dimensional multi-objective metaheuristic function was suggested, with weights assigned to environmental, economic, socio-political, and technical aspects.

Researchers in [7] looked into how responsive loads and the stochastic behavior of EVs (including their departure/arrival times and charge levels) could be used as demand-side management tools to improve the efficiency of a grid-connected microgrid that combines power, heating, and cooling systems. They suggested a multi-objective model considering responsive loads and electric cars for feeder reconfiguration, capacitor switching, and economical dispatching. The suggested model considers operating expenses, greenhouse gas emissions, the voltage stability index, and active power losses as objective functions. In addition to thermal and electrical energy storage devices, the microgrid based on a combined cooling, heating, and power system was outfitted with non-dispatchable distributed generators (photovoltaic (PV) cells and wind turbines (WTs)). Electric cars, thermal and electrical needs, and the stochastic behavior of non-dispatchable generators were considered for appropriate modeling. The max-geometric mean operator and fuzzy scaling were used in conjunction with a multi-objective hybrid big bang–big crunch algorithm to obtain the best answers. In [8], the multi-objective sand cat swarm optimization (MSCSO) algorithm was utilized to find a solution for the suggested model. Based on this, the daily stochastic economic scheduling of an electric thermal hydrogen integrated energy system (ETH-IES) was conducted to reduce operational expenses. The main concern of the authors of [9] was the economic and environmental aspects of microgrid (MG) functioning under different conditions. An analysis is conducted on an AC/DC hybrid MG with solar, diesel generator, lithium battery, and electric car charging stations. A constrained multi-objective optimization problem (CMOP) was constructed considering the operating restrictions of MG. The fuel cost, depreciation expenditure, and emission cost of distributed generators are the optimization goals of the proposed CMOP. A method for converting a multi-objective issue into a single-objective issue was introduced: the fuzzy comprehensive evaluation. Then, the solutions of distributed generator outputs are solved using the comprehensive learning particle swarm optimization (CLPSO). The optimization outcomes in grid-connected and islanded modes demonstrate the efficacy of the suggested models, techniques, and algorithm.

A multi-objective optimization approach utilizing the Normalized Normal Constraint (NNC) was utilized to evaluate two competing objectives: minimizing the operating costs of the Active Distribution System (ADS) and minimizing the power losses in the ADS [10]. In the interim, variable wind patterns, solar systems, and electric car arrival and departure timings are considered. The suggested model is a multi-objective problem with two stochastic phases run on a modified IEEE 18-bus test system in a General Algebraic Modeling System (GAMS) environment. In [11], to optimize the environmental and economic performance of an intelligent parking lot (IPL) with electric vehicles (EVs) while utilizing time-of-use (TOU) rates for demand response programs (DRPs), a bi-objective optimization model has been presented. ε-constraint and fuzzy decision-making strategies are applied to tackle this kind of problem, and the outcomes, which show the efficacy and efficiency of the methods used, are displayed for comparison. The IPL linked to the upstream net, renewable and non-renewable resources, and a hydrogen storage system make up the examined example model in that research. The bi-objective issue in question has been modelled using a MIP model, which is then simulated using GAMS. To facilitate the electrification of green transport, the authors in [12] suggested a multi-objective planning framework for electric vehicle (EV) charging stations in developing power networks. The effects of EV integration on financial and environmental criteria are examined in four examples. The suggested model was designed to integrate the planning models of transmission lines, energy storage systems (ESSs), renewable energy systems, and thyristor-controlled series compensators into the EV-based planning problem to make the construction of EVCSs easier. The second objective, on the other hand, was focused on decreasing the carbon dioxide emissions from fossil fuel-based power units to help the environment. The first objective function seeks to maximize the penetration of EVs by increasing the networks' capacity to provide charging stations continuously throughout the day. Reducing the initial outlay and ongoing expenses for the installed equipment is the third goal, which aims to satisfy the financial needs. The multi-objective variant of the Gazelle optimization algorithm (MGOA) was used to find a solution for the suggested model, which was expressed as a multi-objective optimization problem. The suggested issue and a set of four benchmark test functions were solved to gauge the MGOA's effectiveness.

A multi-objective optimization model was developed in [13] to reduce the gearbox losses, operational expenses, and carbon emissions of many microgrid systems. First, a brand-new technique based on a back propagation neural network enhanced by long short-term memory deep learning was put forth to anticipate the charging loads of EVs. A double-layer solution algorithm was proposed based on the forecast data. At the multiple-microgrid layer, it comprises an adaptive multi-objective evolutionary algorithm based on decomposition and differential evolution. At the single-microgrid layer, it consists of a modified consistency algorithm for rapid economic scheduling. In the end, a case study consisting of four interconnected IEEE microgrids was used to simulate the model system, and the suggested algorithm's performance was contrasted with that of traditional multi-objective evolutionary algorithms based on decomposition. A methodology for optimization bound by dependability was introduced in [14] to determine the quantity and dimensions of microgrid (MG) system components. To accomplish this, issue reliability indicators for lost load anticipation and anticipated energy not delivered are introduced. The Monte Carlo sampling technique was used to represent the uncertainties related to load forecasting, modeling of all MG units, and random outage of all units. The suggested paper's major objective was to determine the ideal MG size that would minimize operating, emission, and MG investment costs. Additionally, under the usage of time-of-use (TOU) rates of demand response programs (DRPs), a bi-objective optimization model has been presented for the best possible environmental performance and economic operation of MGs, including EVs. Fuzzy decision-making techniques and the ε-constraint are applied to alleviate this issue. Long-term planning was tackled as an optimization issue using the Tabu search method.

In [15], the authors compared the output of vehicle-to-home (V2H) and stationary battery (SB). They devised a multi-objective optimization approach for the household of EV owners utilizing energy expenses, comprising investment and CO_2 emissions, as indices. As a case study, the authors used an imagined average Japanese detached house to assess the environmental and economic effects of solar electricity self-consumption utilizing SB or V2H. The findings indicated that, by 2030, non-commuting EV owners should consider investing in V2H if the cost of a bidirectional charger is one-third that of an affordable SB. In [16], for regional integrated energy systems (RIESs), a multi-objective optimization that takes electric cars (EVs) and renewable energy uncertainty into account was suggested. The RIES can balance the system's environmental friendliness and economy. First, an orderly model for charging and discharging EVs with the following driving rules is built. It considers the impact of elements like disorderly access and EV charging/discharging on system functioning. Then, to address the uncertainty of renewable energy generation, a robust optimization model with a polyhedral uncertainty set was built. Additionally, a multi-objective function is constructed to minimize both operation costs and carbon emissions. A carbon emission penalty component is implemented to reduce the multi-objective solution to a single-objective solution. Ultimately, an actual RISE performs the validation.

The authors of [17] examined how a commercial PV charging station with ten electric car chargers should be commissioned. The charging station may purchase and sell power to the grid as it is linked to the main distribution network. A multi-objective optimization technique has been devised that minimizes the expenses associated with power losses in the distribution grid and the operational costs of the charging station. The owner of the charging station and the distribution system operator have their interests taken into account in the suggested solution. The minimal charge when the car exits the charging station, user comfort restrictions, and grid technical limits were also considered. A single day with a fifteen-minute resolution is the analysis period. In [18], the authors employed multi-objective optimization to find the best combination of energy and transportation technologies while maximizing the positive effects on the economy and environment.

In contrast to continuous multi-objective linear programming with average cost intervals, the authors showed the extra benefit of using multi-objective mixed integer linear programming (MOMILP) while considering economies of scale. The authors solved MOMILPs precisely using an enhanced version. The effect of policies on the Pareto frontier is evaluated to distinguish between optimum solutions with and without subsidies. The writers distinguished between the need for investments (bounded rationality) and minimizing economic life cycle costs (full rationality). An electrical and transportation-related Belgian corporation serves as an example of the methodology. Transportation technologies include internal combustion engine cars, grid-powered battery electric vehicles (BEVs), and solar-powered BEVs; electricity technologies include solar photovoltaics and the grid. Grid-powered BEVs have a limited ability to reduce greenhouse gas emissions, but they are less expensive to use than solar panels. It was discovered that current policy initiatives appropriately target rational investors who take life cycle costs into account, but private (possibly constrained rational) investors frequently concentrate primarily on needed investments.

To account for uncertainties arising from wind speed, solar irradiance, the conventional load, and PEV load demand, the authors in [19] proposed a multi-objective optimization methodology for the siting and sizing of solar distributed generations (SDGs), wind distributed generations (WDGs), and capacitor banks (CBs) inside the system of power. The primary goals are the overall cost, greenhouse gas emissions, and the voltage stability index.

The associated uncertainties are handled using an unconventional point estimate method (PEM), while a chance-constrained programming technique handles the smooth constraints. Using the greatest entropy approach, the associated probability distribution functions of the output variables are computed. Moreover, Monte Carlo simulation was used for robustness analysis (MCS). The suggested approach was implemented on a standard radial distribution network. The fuzzy satisfactory technique chooses the Pareto front's optimal trade-off solution.

An interval optimization strategy was suggested by [20] to represent the unpredictability of upstream grid prices. By changing the uncertainty-based profit function to a deterministic multi-objective problem and treating average and deviation profits as competing objective functions—average profit should reach the maximum, while deviation profit should reach the minimum—the authors presented a novel solution to the problem of electric vehicle aggregator uncertainty. The two-dimensional problem was also solved using the ε-constraint approach to obtain optimum Pareto solutions. Finally, the fuzzy satisfactory strategy was used to select a trade-off solution among Pareto solutions as a target to show the examined technique's capabilities; it was also contrasted with the deterministic strategy and the proposed interval optimization approach.

Reference [21] describes the design of an islanded hybrid system (IHS) that includes a diesel generator, solar system, wind turbine (WT), and energy storage systems (ESSs) that are both mobile (electric cars) and stationary (battery). The suggested approach uses two distinct goal functions in a multi-objective optimization to reduce the overall cost of building, maintaining, and operating the sources and ESSs within the IHS and the system's emission level. A single-objective optimization problem is created for the suggested design using the Pareto optimization methodology based on the ε-constraint method. In [22], the authors used a multi-objective framework to consider two objectives. A multi-objective mixed binary linear programming was provided to minimize the overall cost of energy consumption and peak load in communal residential structures. This programming considers the scheduling of electric car charging and discharging and battery energy storage systems. Then, the Pareto front solutions of the provided multi-objective model are obtained using the Pascoletti–Serafini scalarization technique. In the end, the suggested model's performance was examined and documented using model simulations in two distinct scenarios.

Plug-in Electric Vehicles (PEVs) are used as storage units in a multi-objective power dispatching issue that the authors in [23] defined. The authors minimized three objectives, analyzed three criteria, and framed the energy storage planning as a Mixed-Integer Linear Programming (MILP) problem while adhering to PEV constraints. Two cost-to-variability measures based on the Sharpe Ratio are presented to analyze the energy storage schedules' volatility. Energy storage planning was optimized by adding these extra parameters to minimize the difference between two Sharpe Ratio indices, maximum peak load, PEV battery utilization, total Microgrid (MG) expenses, and maximum peak load. Pareto fronts are examined and discussed about various scenarios for energy storage. The most significant outcome of the study would be that schedules that lower the system's total cost could potentially be less dependable since they increase the maximum peak load and its unpredictability under different circumstances. In [24], a multi-objective framework was suggested for the day-to-day management of a smart grid (SG) with a high level of sensitive load penetration. To provide highly dependable power for sensitive loads, the Virtual Power Player (VPP) oversees the day-ahead scheduling of energy resources in the smart grid, considering the extensive usage of Distributed Generation (DG) and V2G. The collection of non-dominated solutions is identified by applying a Pareto front technique. To account for the dependability needs of sensitive and vulnerable loads, the mathematical formulation incorporates the maximization of the minimum available reserve in addition to the cost reduction.

By calculating the available capacity of EV aggregators, a preliminary investigation of the multi-objective optimum dispatch of the smart grid was suggested in [25]. A statistical model was used to determine the maximum possible capacity of EVs by describing their behavior. Subsequently, the multi-objective optimum dispatch was defined together with its constraints. The high-dimensional multi-objective enhancement issue was solved using the multi-objective genetic particle swarm optimizer to find the Pareto front. In [26], a mixed-integer linear programming (MILP) framework is created to measure flexibility in a sizable business park with limited access to historical time series data. The suggested mathematical model considers renewable energy sources, including solar energy, electric vehicle (EV) charging stations, heat pumps, and centralized energy storage systems. The quantification of flexibility was formulated as a bi-objective optimization problem, which was solved by utilizing the epsilon-constraint approach to approximate the set of Pareto-efficient solutions. The authors' goal [27] was to optimize power loss, voltage deviation, and voltage imbalance factor—three significant objective functions—by concurrently allocating EVCSs and smart photovoltaic inverters in distribution networks. A unique hybrid fuzzy Pareto dominance idea with a differential evolution technique was suggested to address such a multi-objective optimization issue. A scenario-based approach was also employed to incorporate the uncertainties of the model that includes loads, PV generation, and EVCS demand. Next, under various case scenarios, the efficacy of the stochastic multi-objective strategy was investigated and confirmed on an imbalanced 37-bus network.

In [28], a residential microgrid with plug-in hybrid electric cars, PV units, battery energy storage systems, combined cooling, heating and power, and other components was modeled to determine the best scheduling state for each unit while accounting for the uncertainty of distributed energy resources. This was accomplished by modeling the uncertainties of solar irradiance, electrical and heat demand, and electrical market pricing using a scenario-based approach that uses the Normal, Weibull, and Beta probability distribution functions, respectively. Scenario reduction strategies are employed to choose representative situations generated using the scenario tree. The suggested issue was a mixed-integer nonlinear programming problem to minimize emissions and operation costs. The optimal solution on the Pareto front set is identified using a fuzzy approach, and the augmented ε-constraint method was utilized to solve this multi-objective problem. In [29], considering the time-of-use rates of a demand response program, a multi-objective optimization approach was proposed for the cost-effective operation and environmental performance of intelligent parking lots (IPLs). It was advised to address this problem using the multi-objective grasshopper optimization technique, since such a model is related to several practical bounds. The results show how well the compared methods using fuzzy decision-making strategies worked. To improve this approach and advance searching operators, chaos theory was utilized. Furthermore, the suggested multi-objective approach is a model developed utilizing non-dominated sorting theory, variable detection, fuzzy theory, and strategy selection-based memory to choose the best Pareto among a range of reliable options for handling the abovementioned difficulties.

Optimal nature-inspired metaheuristics algorithms tend to experience premature convergence and, in general, swiftly obtain both the local and almost global optimal states. The No Free Lunch (NFL) theorem [30] for finding the best algorithm to tackle all optimization problems states that not all algorithms are suited to address all problems. This is because no method is optimal for solving all optimization problems. On the other hand, each of the previously discussed approaches has its unique way of delivering the optimal solution in terms of performance when it comes to resolving and optimizing power issues.

A metaheuristic approach based on simulation was created [31] to identify the ideal size of a hybrid renewable energy system for residential buildings. The development of a dynamic multi-objective particle swarm optimization method was needed to solve this multi-objective optimization problem. The approach should maximize the buildings' renewable energy ratio while minimizing the overall net present cost and carbon dioxide emissions for any necessary system modifications. The standard of the Pareto front generated by the proposed technique was assessed using three established performance indicators. A multi-objective optimization model that considers a multi-energy system and minimizes the overall cost and the life cycle emissions of vehicles and buildings was developed [32]. The model chooses the sorts of vehicle powertrains and the storage and conversion technologies the community utilizes to produce heat and electricity. The Pareto solutions that emerge depend on the shift from internal combustion engine cars to battery electric cars, and to a much lesser extent, plug-in hybrid cars. The heating energy is decarbonized by switching from gas boilers to heat pumps. In [33], the authors developed a multi-objective hydro–thermal–wind with EV scheduling (MOHTWES) issue by combining large-scale electric vehicles (EVs) with wind power generation. Furthermore, a better multi-objective particle swarm optimization (IMOPSO) approach was suggested to solve the aforementioned issue under several restrictions. The IMOPSO may provide good and well-distributed Pareto optimum solutions in objective space by providing a novel dual-population evolution mechanism and a hierarchical elitism-preserving method based on crowding entropy.

In [34], the issue of stochastic dynamic pricing and energy management policy for providers of EV charging services was examined. EV charging service providers face numerous uncertainties when energy storage systems and renewable energy integration are present. These include fluctuations in wholesale electricity prices, inherent intermittency in renewable energy generation, and volatility in charging demand. The goal was to provide guidelines to charging service providers to manage electricity and set appropriate charging prices while balancing the competing goals of increasing customer satisfaction, boosting profitability, and lessening their impact on the power grid despite these uncertainties. The authors devised a new metric to assess the impact on the power system without having to solve the full power flow equations. The approach incorporates a safeguard of profit to insulate service providers from extreme financial losses. Two algorithms are used to determine the price and electricity procurement policy: the greedy algorithm (benchmark algorithm) and the stochastic dynamic programming (SDP) algorithm. The multi-objective optimization's Pareto front was determined. In [2], the authors suggested a multi-objective planning approach for allocating EVCSs and RESs in the best possible way. In particular, voltage variations, energy losses, and EV owners' discontent are considered three sub-objectives to be minimized in the proposed RES and EVCS planning framework.

Furthermore, considering the various operating constraints of the grid, RESs, and EVCSs, active power curtailment of RESs is not an option. The suggested framework takes into consideration enhanced control systems for linking RES inverters, as well as grid-to-vehicle (G2V) and V2G schemes, to yield additional benefits. To address this holistic framework with conflicting sub-roles and find the Pareto-optimal solutions, a two-level method was proposed.

In this research, the proposed system comprises a PV–Wind–Battery system combined with EVCS using the vehicle-to-grid (V2G) technique. The thrilling Arithmetic Optimization Algorithm (AOA) has been improved to overcome its drawbacks, such as being trapped in a local search (stagnation in local minima), premature convergence, and neither the addition (A) nor the multiplication (M) operators being obtainable for the exploitation or exploration phases. In addition, in the AOA, the rudimentary mathematical models obtained in both the exploration and exploitation phases (There can never be a perfect balance between exploration and exploitation). Furthermore, most of the methods in the previous literature considered only a single objective during the optimization process for the proposed system (either an economic or technical objective). In this study, a Multi-Objective Improved Arithmetic Optimization Algorithm (MOIAOA) based on the Non-Scale multiple-run Pareto Front concept has been proposed to analyze the optimal sizing design of the proposed system components by calculating the optimal values of the three conflicting objectives, Grid Contribution Factor (GCF), Levelized Cost of Electricity (LCOE), and Energy sold to the grid (E_{SOLD}). These three constraint objectives are used as renewability, economic, and technical criteria. The RB-EMS is used for controlling and monitoring the power flow of the proposed system. The results are performed to analyze the optimal sizing of the proposed system by using an optimal sizing method based on the MOIAOA Non-Scale multiple-run Pareto Front concept.

The following categories apply to the remaining sections. A description of the proposed system's modeling components is given in Section 2. The criteria for the renewability, economic, and technical assessments are given in Section 3. Section 4 outlines the suggested methodology for sizing the PV/WT/battery system in conjunction with EVCS utilizing a vehicle-to-grid (V2G) technique in a grid-connected system. In contrast, Section 5 presents the findings and discussion. Section 6 wraps up the analysis and suggests next steps.

2. Mathematical Modeling of the Grid-Connected PV/WT/Battery System Combined with EVCS Using Vehicle-to-Grid (V2G) Technique

A mathematical equation is used to model the two different RESs stated earlier, namely the PV and WT, with additional components. That leads to determining the output power under different climate data of a government building located in Al-Najaf Governorate in Iraq.

The PV array, WT, EMS control, storage battery, unidirectional converter, bidirectional converter, grid, building load, and EVCS are the main components of the proposed system, as seen in Figure 1. These could differ greatly depending on several factors, including the availability of meteorological data, renewability–economic–technical parameters, and the intended power demand. The technical and economical specifications of the PV module used in the proposed system are given in Table 1.

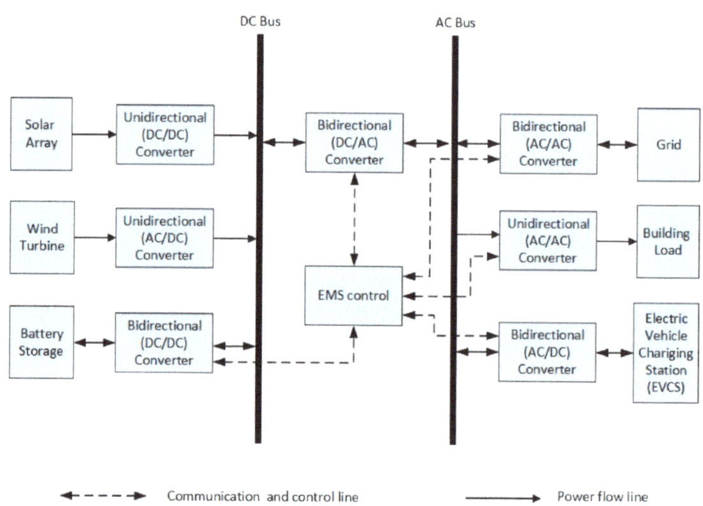

Figure 1. The proposed system.

Table 1. Economical and technical data of the proposed system components.

Components	Parameter	Value	Unit
Wind Turbine (WT)	Rated Power of Wind Turbine (P_r)	1	kW
	Cut-in speed (Vcin)	3	m/s
	Cut-out speed (Vco)	20	m/s
	Rated wind speed (Vrat)	11	m/s
	Capital cost (per kW)	2300	USD
	Replacement cost (per kW)	1500	USD
	O & M cost (per kW) [operation + maintenance]	2	USD/yr
	Hub height	50	M
	Overall efficiency	26	%
	Lifetime	20	years

Table 1. *Cont.*

Components	Parameter	Value	Unit
Solar (PV)	Rated power (Ps r)	325	W
	Derating factor (f loss)	88	%
	Capital cost (per kW)	1200	USD
	Replacement cost (per kW)	1200	USD
	O & M cost (per kW)	4	USD/yr
	Lifetime	20	years
Battery	kVAh or kWh capacity	6	kWh
	Minimum state of charge (SOCmin)	30	%
	Maximum state of charge (SOCmax)	100	%
	Round trip efficiency (gbatt)	92	%
	Capital cost (per unit battery)	167	USD
	Replacement cost (per unit)	67	%
	M & O cost (per unit)	1.67	USD/yr
	Lifetime	5	years
	Nominal battery capacity	41	Ah
	Battery capacity	75	Ah
Rectifier (grec) and inverter (ginv)	Efficiency		97%
	Installation and capital cost (per kW)	127	USD/yr
	O & M cost (per kW)	1	USD/yr
	Lifetime	20	years
General Requirement	Interest rate	6%	
	Project life (N)	20	years
	EVs Capacity in kWh	20	kWh
	Utility prices:		
	Power export price to utility (selling)	0.015	USD/kWh
	Power import price from utility (purchasing)	0.013	USD/kWh
Optimization of lower and upper bounds	Solar	1200	1
	Wind	1000	15
	Battery	1000	1

2.1. Photovoltaic Panel Mathematical Modeling

PV is the most widely used RES for generating. In this research, polycrystalline solar panels (KD325GX-LFB) are taken into consideration. The panel manufacturers' specifications and solar parameters are reported in [35–37]. The panels are inclined with an angle of 30° to the direction of the south. The modeled equation for the output power produced from the PV system is given in Equation (1) and reported in [37–39].

$$P_{pv_{out}}(t) = P_{(PV_{rated})} \times \frac{G_{(t)}}{1000} \times \left[1 + \alpha_t\left(\left(T_{amb} + (0.03125 \times G_t)\right) - T_{C_{STC}}\right)\right] \quad (1)$$

where $P_{(PV_{rated})}$ indicates the rated power for PV (in watts), α_t is the temperature coefficient (-3.7×10^{-3}) 1/C, $T_{C_{STC}}$ is the cell temperature (in °C) under standard test condition (STC), and T_{amb} is the ambient temperature (in °C), respectively. $G_{(t)}$ refers to solar irradiance (in W/m^2), 1000 W/m^2 is the reference irradiance, and $P_{pv_{out}}(t)$ is the PV output power (in watts). Equation (2) can be used to obtain the $T_{C(STC)}$ [40]. *NOCT* is the nominal operating cell temperature in °C that the manufacturer can model.

$$T_{C(STC)} = T_{amb} + G_{(t)} \times \left(\frac{NOCT - 20}{800}\right) \quad (2)$$

Additionally, the value 0.03125 °C was obtained by subtracting the value of Nominal Operation Cell Temperature (NOCT), which is 45 °C in this study, from air temperature (20 °C) based on the PV module that the manufacturer has specified; the acquired result was divided by the irradiance on the cell surface (800 W/m^2) to obtain 0.03125 °C [35]. The technical and economical specifications of the PV module used in the proposed system are given in Table 1.

2.2. Wind Turbine Mathematical Modeling

Vertical and horizontal axis wind turbines are the products of axial categorization of wind turbines [41]. The horizontal axis wind turbine is the most widely utilized type of wind turbine for various reasons, including its capacity to capture the most wind energy, adaptability to low-wind conditions, and pitch angle adjustment capability to prevent strong windstorms [40]. Therefore, a wind turbine comprises three basic components: the generator, which has a gearbox and controls, and the rotor, which houses the blade and the frame. Equation (3) [38] presents the model equation for the output power produced by the WT. The economical and technical specifications of the WT used in the proposed system are given in Table 1.

$$P_{WT} = \begin{cases} 0 & v(t) \leq v_{cut-in} \text{ or } v(t) \geq v_{cut-out} \\ P_r \frac{v(t)-v_{cut-in}}{v_r - v_{cut-out}} & v_{cut-in} < v < v_r \\ P_r & v_r < v(t) < v_{cut-out} \end{cases} \quad (3)$$

The variables v_{cut-in} and $v_{cut-out}$ represent the cut-in speed and cut-out speed, respectively. P_r stands for rated power, v_r for rated wind speed, and P_{WT} is the produced output power of the WT, as indicated by the manufacturer [40]. Equation (4) illustrates how taking hub height into account can yield output power from WT with improved precision.

$$V_2 = V_1 * \left(\frac{h}{h_{ref}}\right)^{\alpha} \quad (4)$$

where h is hub height, h_{ref} is the reference height anemometer, and α is the power-law exponential, known as wind gradient, Hellmann exponent, or friction coefficient, which equals 1/7 [42]. The wind speed (m/s) is represented by V_2 and V_1. It is evident that the cut-in wind speed v_{cut-in}, cut-out wind speed $v_{cut-out}$, and rated wind speed v_r, in that order, determine the output power generated from WT [43].

2.3. Battery Mathematical Modeling

Systems that store and release energy from renewable energy sources (RESs), such as solar, wind, and hydropower, are known as battery energy storage systems [44]. However, these RESs are known to have high intermittency. Energy is stored in a battery storage system (BSS) that can be used during a grid outage to reduce intermittency and boost system reliability and efficiency. Equation (5) [45] provides the mathematical formula to determine the nominal battery capacity.

$$C_B = \frac{E_L * AD}{DOD * \eta_{inv} * \eta_b} \quad (5)$$

where C_B is the battery's nominal capacity, E_L is its daily average load demand, DoD is the suggested depth of drain (80%), and autonomy days (usually 3–5 days) are represented, while the inverter's efficiency, η_{inv}, equals 95% and the battery's efficiency, η_b, is 85%. The battery's energy storage capacity is known as its state of charge (SoC) and its energy consumption is known as its depth of discharge (DoD) [42,43,45,46]. Additionally, the minimum depth of discharge and the value of DOD, which is set at 80%, may be computed using Equation (6) [47]. $SoC_{min} \leq SoC_{batt} \leq SoC_{max}$ represents the SoC's border. Furthermore, the battery's power output has a mathematical expression, as shown in Equation (7).

$$SOC_{BT\_MIN} = (1 - DOD) \times C_B \quad (6)$$

$$P_b(t) = (P_{pv}(t) + P_{WT}(t)) - \frac{P_l(t)}{\eta_{inv}} \quad (7)$$

where η_{inv} is the inverter efficiency (95%) and $P_b(t)$ is the total power delivered from the battery, $P_{PV}(t)$ is the total power produced from PV, $P_{WT}(t)$ is the total power produced from WT, and $P_l(t)$ is the total energy demand [38]. One crucial factor that indicates the battery's performance is its state of charge (SoC) [48]. Equations (8) and (9) are used to determine the state of charge (SoC) of the battery when charging or discharging. According to the mathematical computation provided by Equation (8), the battery is in a charging condition when the total generated output power from PV and WT exceed the load.

$$SoC(t) = SoC(t-1).(1-\sigma) + \left((P_{pv}(t) + P_{wt}(t)) - \frac{P_l(t) + P_{EV_{dem}}}{\eta_{inv}} \right) * \eta_b \quad (8)$$

The output power produced by the PV and WT, respectively, is expressed as $P_{PV}(t)$ and $P_{WT}(t)$. The battery's self-discharge rate, or σ, equals 0.007%/h [49]. The overall energy demand is represented by $P_{PV}(t)$ *Pl(t)*, the state of charge of the battery at a time (t) is indicated by *SoC (t)*, the inverter efficiency is represented by η_{inv}, and the battery efficiency is marked by η_b, which is equal to 85% [38]. The EV battery specification is also considered to obtain the SoC, charging decision, and energy demand. If the total generated output power from PV and WT is less than the load demand, as determined by Equation (9), the battery's state of charge (SoC) will be in a discharging situation.

$$SoC(t) = SoC(t-1).(1-\sigma) + \left(\frac{P_l(t) + P_{EV_{dem}}}{\eta_{inv}} - (P_{pv}(t) + P_{wt}(t)) \right) * \eta_b \quad (9)$$

Equation (10), which represents the battery power during discharging when the SoC exceeds the RESs, may be derived from the previously mentioned facts.

$$P_{BATT}(t) = [P_l(t) - P_{wt}(t)] * \eta_{inv} - P_{PV}(t) \quad (10)$$

The technical and economical specifications of the BT used in the proposed system are given in Table 1.

2.4. Converter Mathematical Modeling

Power converters, such as DC/AC and AC/DC, are required when a system consists of both AC and DC components; Table describes the converter. In this analysis, batteries that generate DC output, solar PV panels (DC), and household (AC) demands are taken into consideration. Peak load demand (P_L^m) at a time (t) and inverter efficiency (η_{inv}) are combined to estimate the converter size and Equation (11) [50] is used to determine the inverter rating ($P_{inv}(t)$).

$$P_{inv}(t) = \frac{P_L^m(t)}{\eta_{inv}} \quad (11)$$

2.5. The Grid Mathematical Modeling

The grid can supplement the energy deficit if the RESs and battery bank are unable to meet the load needs [51]. Equation (12) can be used to determine the money received from energy sales to the utility grid.

$$R_{grid} = \sum_{t=1}^{8760} rate_{feed-in} \times E_{grid(selling)} \quad (12)$$

where $E_{grid(selling)}$ represents the selling energy price (USD 0.015/kWh) and $rate_{feed-in}$ refers to the feed-in tariff rate, which is USD 0.02/kWh. Moreover, Equation (13) is used to compute the cost of power from the grid [52]. On the other hand, 8760 is the amount of hours in a year. The following is the purchasing price of the grid-purchased power:

$$C_{grid} = C_p \times \sum_{t=1}^{8760} E_{grid(purchased)} \tag{13}$$

where $\sum_{t=1}^{8760} E_{grid(purchased)}$ is the hourly total of yearly grid power purchases for a year [52], and C_p is the cost of purchasing electricity in Iraq, which equals USD 0.013/kWh.

2.6. Mathematical Modeling of Electric Vehicle Charging Station

The battery of electric vehicles (EVs) is used to overcome several supply constraints to improve security and financial sustainability. One of the fundamental needs for modeling an electric vehicle charging station is knowing its rated capacity. One can compute the rated capacity using the method shown in Equation (14).

$$S_{rated} = \frac{k_{load} * N_{slot} * P_{EV}}{COS \varnothing} \tag{14}$$

where N_{slot} is the number of charging slots for each EV, which equals 3, k_{load} is the overload factor for cover overloading in transients, which is 1.1, P_{EV} is the maximum power rate of each EV, which is 90 kW, and S_{rated}, the station's rated capacity, equals 850.97 VAr [53].

3. Date Collection and Renewability–Economic–Technical Assessments

In this research, real meteorological data have been used to model the proposed system. These meteorological data are recorded every 10 min for an entire year (from 1 January to 31 December of 2018) in a government building (Engineering Technical College) located in Al-Najaf Governorate in Iraq [54], which is located at the coordinates 31° north latitude and 44° east longitude. These meteorological data consist of solar radiation, ambient temperature, and wind speed. The meteorological data were collected for an entire year.

3.1. The Study Site and Load Profile

This study investigates the efficacy and potency of the recommended strategies for the ideal sizing of the suggested system in Al-Najaf Governorate in Iraq.

3.1.1. Al-Najaf Governorate in Iraq

Al-Najaf Governorate is a city in central Iraq about 160 km (99 mi) south of Baghdad; see Figures 2 and 3. This study uses the climatology data and load demand to implement the mathematical equations to calculate the total amount of power generated during an entire year. Real meteorological data have been used in the modeling of the proposed system. These meteorological data are recorded every 10 min for an entire year (from 1 January to 31 December of 2018) of a government building (Engineering Technical College) located in Al-Najaf Governorate in Iraq [54], which is located at coordinates 31° north latitude and 44° east longitude. These data were gathered from a weather station installed ten meters above the ground, as shown in Figure 4. These meteorological data consist of solar radiation, ambient temperature, a wind speed. The energy demand profile data were assumed for an entire year.

Figure 2. Map of Iraq.

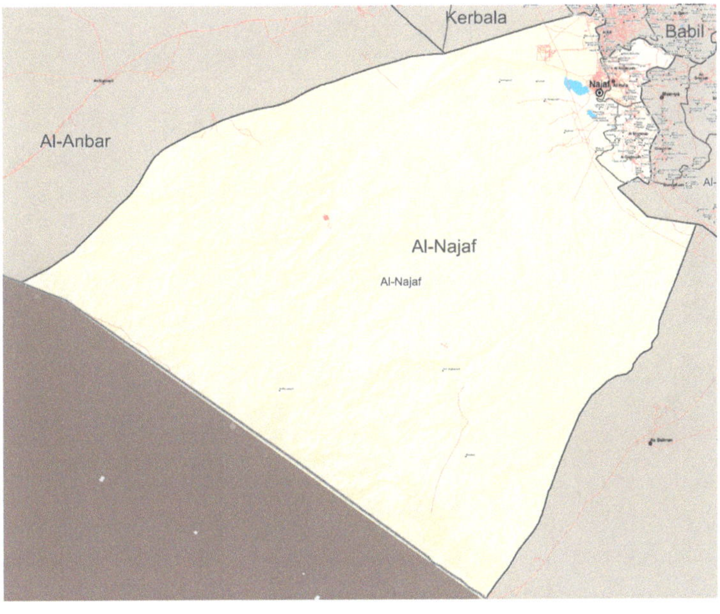

Figure 3. Map of Al-Najaf Governorate.

(a) (b)

Figure 4. Weather station: (**a**) Davis weather station; (**b**) Vantage Pro2.

3.1.2. Load Profile

The data of the load demand for the entire year of the building where the study was conducted have a maximum value of 4.9687 kW and a minimum value of 0.6417 kW, recorded every 10 min.

3.2. *Objective Function Formulation*

3.2.1. Levelized Cost of Electricity (LCOE)

The LCOE for a power system is the ratio between the total costs of the system and its total electricity production over its economic lifetime [55]. It is regarded as the minimum cost at which electricity must be sold to break even over the project's lifetime. LCOE is measured in USD/kWh. The first objective can be presented as shown in Equation (15):

$$\begin{aligned} objective_1 &= min\ (LCOE) \\ &= min \begin{pmatrix} The\ reation\ betweenhe\ the\ total\ costs\ of\ the\ system\ and \\ its\ total\ electricity\ production\ over\ its\ economic\ lifetime \end{pmatrix} \end{aligned} \quad (15)$$

3.2.2. Grid Contribution Factor (GCF)

Grid Contribution Factor (GCF) can minimize the maximized Renewable Energy Fraction (REF). GCF is given in Equation (16) as follows:

$$GCF = 1 - RFE \quad (16)$$

The GCF is the amount the grid contributes to meeting the necessary energy demand by minimizing REF [52]. The GCF is measured in kWh. The second objective can be presented as shown in Equation (17):

$$objective_2 = min\ (GCF) = min\ (1 - REF) \quad (17)$$

3.2.3. Energy Sold to the Grid (E_{SOLD})

E_{SOLD} is the annual energy sold to the grid and not self-consumed by the charging station [56]. The E_{SOLD} is defined as the quantity of electricity sold from any part of the system (such as PV and/or wind and/or battery and/or EV battery) into the main grid. One of the objective functions of this research is to maximize the value of E_{SOLD}, which means decreasing the dependency on the grid; this is achieved by increasing the dependency on renewable energy and/or battery and/or EV batteries. The E_{SOLD} is measured in kWh. The third objective can be presented as shown in Equation (18):

$$objective_3 = max\ (E_{SOLD}) \tag{18}$$

4. The Proposed Methodology for Sizing of the Grid-Connected PV/WT/Battery/EVCS System

The proposed system comprises a grid-connected PV/WT/battery combined with EVCS. The proposed system's energy management strategy and scenarios are provided in phase 1 of this part, which is separated into three sections. Phase 2 clarifies the Arithmetic Optimization Algorithm (AOA). Phase 3 presents the proposed Improved Arithmetic Algorithm (IAOA) with benchmark algorithms such as AOA, Ant Lion Optimizer (ALO), and particle swarm optimization (PSO). Finally, the proposed Multi-Objective Improved Arithmetic Optimization Algorithm (MOIAOA) is explained.

4.1. Energy Management Strategy and Its Scenarios in the Proposed System

Information management included in such a system is known as an energy management system (EMS); it provides the ability to guarantee that energy is supplied through generation, transmission, and distribution at the lowest feasible cost. EMS is believed to use several methods to supply the load needed, as described in [57] and [58]. Additionally, according to the literature, it can categorized into three groups: Rule-Based (RB), Learning-Based (LB), and Optimization-Based (OB), each of which has a subclassification [59]. In addition, it is resource-dependent, balances BT SoC power, and lowers system running costs [60]. There will be difficulties when integrating RESs with the grid, such as overloading [61]. To get around this integration limitation, EMSs can be used to monitor and control the energy systems of RESs in situations where the data obtained from strategies of controlling are inaccurate because the design variable is not taken into account as a crucial feature by taking advantage of sizing algorithms [62]. The optimization algorithms are integrated with EMS to ensure a steady power flow into the suggested system [63]. The system's energy management is configured to meet load needs while considering the dynamic energy flow among all system components [64].

In this research, the system's integrated RB-EMS considered the following four operating modes for three EVs. The proposed system's energy management is crucial to manage the power flow during the optimization process. The operating modes of the RB-EMS are used for controlling and observing the power flow of the proposed system. RB-EMS-based operating modes (scenarios) and their working rules can be illustrated as follows:

1. Operating Mode 1: Renewable energy sources (photovoltaic and wind power) supply power for running the system and charging the battery and the electric vehicle.
2. Operating Mode 2: The battery supplies power for load and electric vehicle charging if there is no grid and insufficient RESs.
3. Operating Mode 3: The main grid supplying power for electric vehicle charging (Buying–Charging–G2V) when batteries and RESs are not available and grid demand is required. The power flow will be unidirectional.
4. Operating Mode 4: The electric vehicle supplying power for the grid (V2G–Sell–Discharging) when grid demand is high and batteries and RESs are unavailable. The flow of power will be bidirectional. The proposed operation modes of RB-EMS for the proposed system are listed in Table 2.

Table 2. The rule-based EMS scenarios for the proposed system.

Rule No.	Modes	IF	THEN
1	RESs	$(P_{pv}(t) + P_{WT}(t)) > P_l(t)$	$(P_{pv}(t) + P_{WT}(t))$ *to* $P_l(t)$ *and* $EV(t)$
2	BT	$P_b(t) > [P_l(t) - P_{WT}(t)] - P_{PV}(t) * \eta_{inv}$	$P_b(t) > [P_l(t) - P_{WT}(t)] - P_{PV}(t) * \eta_{inv}$ *to* $P_l(t)$ *and* $EV(t)$
3	Charge (G2V)	$E_{grid} < EV_{demand}$	$E_{grid} < EV_{demand}$ *to* EV (G2V)
4	Discharge (V2G)	$E_{grid} > EV_{demand}$	$E_{grid} > EV_{demand}$ *to* EV (G2V)

RB-EMS is used in this study because of its advantages, which include its ability to precisely solve problems and make quick judgments to fulfill load demand while minimizing operating costs. The flowchart in Figure 5 illustrates how the metaheuristic technique and system configuration sizing (PV-WT-BT) are being considered to meet the study's objective functions. The (if, otherwise, and then) statement governs the primary mechanism of rule-based strategy [37]. The if-then conditions for the charging and discharging function with the previously mentioned modes are shown in Figure 6. The proposed system's RB-EMS, as seen in Figure 6, presents the power flow via the system's components as a flowchart.

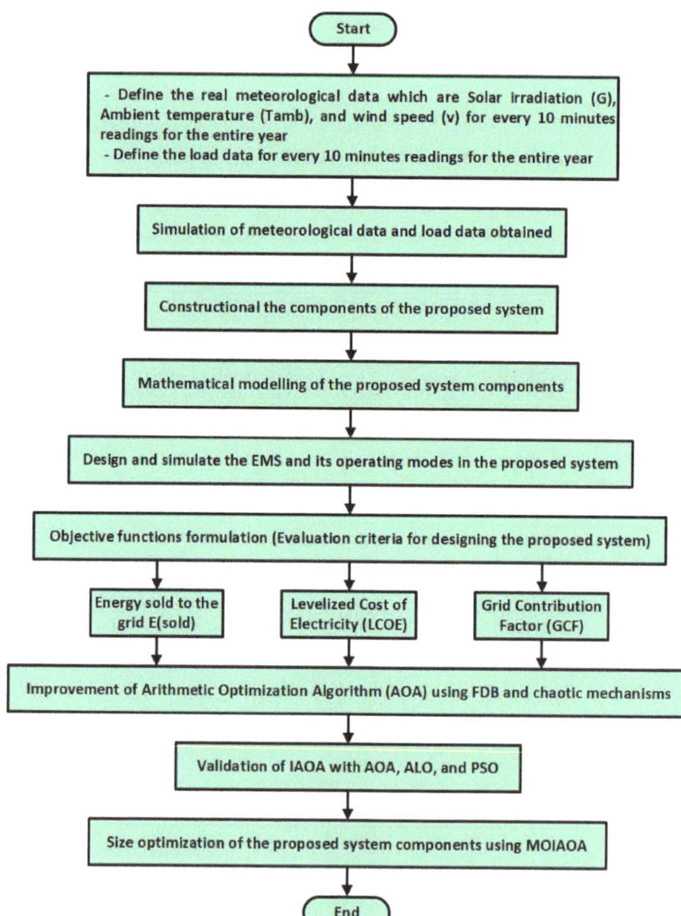

Figure 5. Research methodology flowchart.

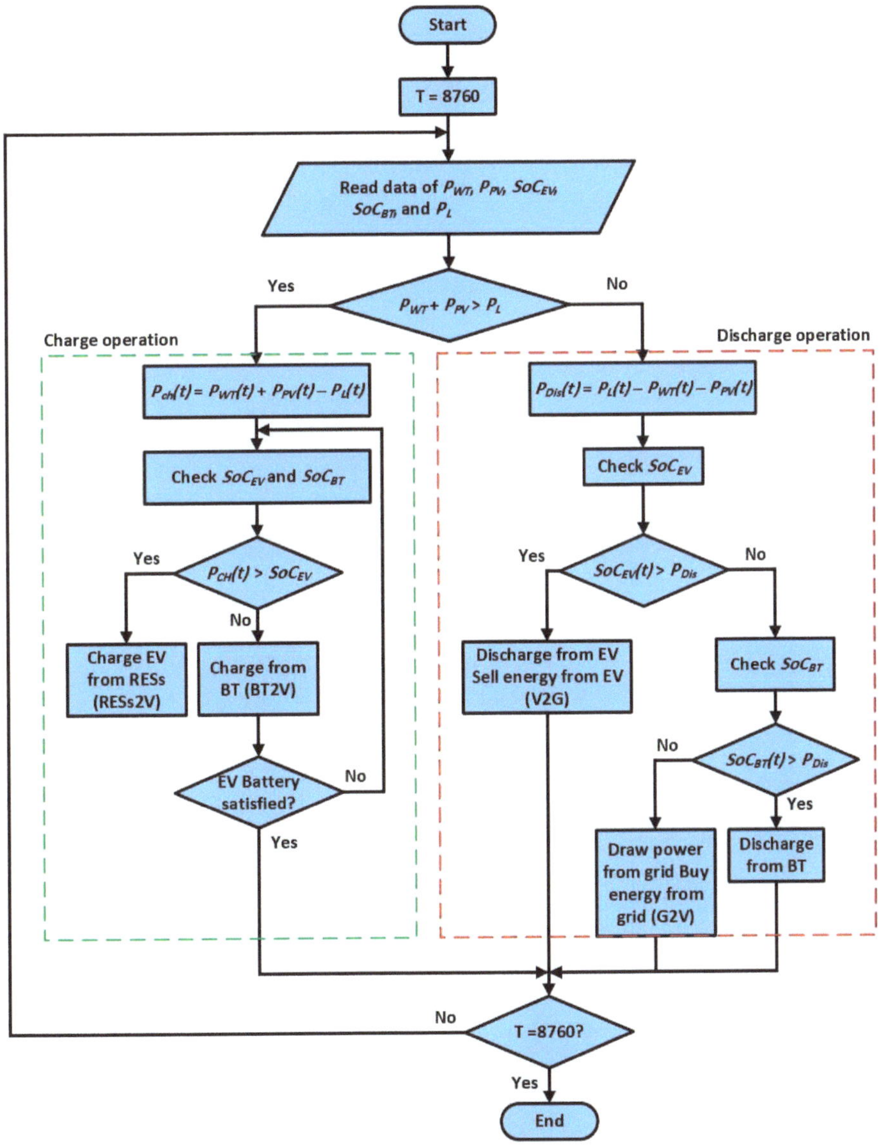

Figure 6. If-then condition operations along with the previously described scenarios.

4.2. Arithmetic Optimization Algorithm (AOA)

Comparable to other population-based optimization techniques, the Arithmetic Optimization Algorithm (AOA) was released in 2021 by Abualigah et al. [65]. The variety and exploitative stages in AOA were produced by the mathematical operators addition (A "+"), division (D "÷"), multiplication (M "×"), and subtraction (S "−").

4.2.1. Inspiration

Arithmetic is a sufficient yet necessary prerequisite for algebra, number theory, analysis, geometry, and modern mathematics. Therefore, these four simple operators might be used to find the best solutions while preserving between the exploitation and exploration periods.

4.2.2. Initialization Phase

A list of potential solutions (X) is established in the first phase. For the most optimal solutions thus far, the best solution from all iterations is kept. As given in Equation (19):

$$X = X_{LB} + rand(X_{UB} - X_{LB}) \tag{19}$$

X_{UB} and X_{LB} establish the upper and lower boundaries of the problem, where X is a collection of initialized solutions and rand is a random variable in the range [0, 1]. The Math Optimizer Accelerated (MOA) function is employed to discern the exploration and exploitation stages. It is calculated in the manner described in Equation (20) below:

$$MOA(C_{iter}) = Min + C_{iter} \times \frac{Max - Min}{M_{iter}} \tag{20}$$

where C_{iter} specifies the current iteration and ranges between 1 and the maximum number of iterations (M_{iter}), and MOA (C_{iter}) defines the value at the t th iteration. Max and Min represent the highest and lowest values of the accelerated function. The steps of exploration and exploitation will be covered in detail in the following sections.

4.2.3. Exploration Phase

Since the the D and M operators have widely distributed values in the design space, they are used during the exploratory stage. The MOA function limits the exploration phase; the D and M operators are utilized if r1 > MOA is found; otherwise, the A and S operators are kept in place. The following equations can be used to express the exploring portion:

$$x_{ij}(C_{iter}+1) = \begin{cases} best(X_j) \div (MOP + \varepsilon) \times ((UB_j - LB_j) \times \mu + LB_j) & r2 > 0.5 \\ best(X_j) \times MOP \times ((UB_j - LB_j) \times \mu + LB_j) & otherwise \end{cases} \tag{21}$$

where the second integer, *r2*, is conditioned between the D and M operations and is generated at random. ε is a small integer value, while a control variable called μ is set to 0.5 to change the search process. UB_j and LB_j stand for the bottom and upper limits, respectively.

$$MOP(C_{iter}) = 1 - \frac{C_{iter}^{\frac{1}{\alpha}}}{M_{iter}^{\frac{1}{\alpha}}} \tag{22}$$

In this work, Math Optimizer Probability is a coefficient represented by the symbol MOP. The sensitivity control parameter indicates the accuracy of the exploitation throughout the iterations α, which is set to 5.

4.2.4. Exploitation Phase

Despite the large density of the A and S operators, their small dispersion makes them easily accessible. The following can be used to represent the S and A operators:

$$x_{ij}(C_{iter}+1) = \begin{cases} best(X_j) - MOP \times ((UB_j - LB_j) \times \mu + LB_j) & r3 > 0.5 \\ best(X_j) \times MOP \times ((UB_j - LB_j) \times \mu + LB_j) & otherwise \end{cases} \tag{23}$$

where the third number, *r3*, is a randomly generated number that represents the A and S operators.

4.3. *The Proposed Improved Arithmetic Optimization Algorithm (IAOA)*

There are two improvements key to overcoming the limitations of the original AOA by employing the fitness distance balance (FDB) integrated with chaotic map strategies [66,67]. Firstly, the FDB is implemented to ensure the best new solutions are chosen to import high-quality solutions into the new generation. The selection mechanism is determined by computing the fitness function values of all particles and their distance from the best solution position. Secondly, the FDB is integrated with a chaotic map tactic for local minima avoidance. Therefore, the newly developed IAOA can enhance the exploitation by utilizing the selection of the FDB technique. At the same time, the new search areas are discovered by obtaining a chaotic map tactic, where the balance between exploitation and exploration tendencies is achieved. The proposed IAOA is proposed to handle three conflicting objectives: GCF, E_{SOLD}, and LCOE. The correlation between these is an inverse relationship, where their values are varied based on the given weights for each one. PF solutions will be executed after executing all weights' statuses.

These objectives are transformed into a single objective after performing a normalization operation [68]. The improvements of IAOA are described by the following:

- FDB Strategy

At first, the distance of particles from the best solution P_{best} is computed using Equation (24) below:

$$\begin{aligned} &\forall_{i=1}^{n}, P_i \neq P_{\text{best}}, D_{P_i} \\ &= \sqrt{\left(x_{1P_i} - x_{1P_{\text{best}}}\right)^2 + \left(x_{2P_i} - x_{2P_{\text{best}}}\right)^2 + \cdots + \left(x_{mP_i} - x_{mP_{\text{best}}}\right)^2} \end{aligned} \quad (24)$$

The distance matrix (D_P) is generated for particle candidates, as given in Equation (25) below:

$$D_P \equiv \begin{bmatrix} d_1 \\ \vdots \\ d_n \end{bmatrix}_{n \times 1} \quad (25)$$

Secondly, the scores of the particles are determined according to the distance and fitness values, as seen in Equation (23). These two variables, $normF$ and $normD_x$, are normalized with a range of [0, 1] to avoid one dominating the other. Then, the scores of particles (Sx_i) are determined with $normF$ and $normD_x$ according to the following expression:

$$\forall_{i=1}^{n} P_i, S_{FDB^1 P_i} = \text{normF } F_{P_i} + \text{ norm } D_{P_i}$$

Finally, the score vector (S_x) can be presented by the following equation:

$$S_x \equiv \begin{bmatrix} s_{x,1} \\ \cdot \\ \cdot \\ s_{x,n} \end{bmatrix}_{nx1} \quad (26)$$

According to our new strategy, the S_x vector is implemented with a chaotic map tactic to boost the convergence and prevent the premature convergence during the optimization process.

- Chaotic map tactic

It is described as follows:

$$X_{new} = X - Xnew(S_x) * (m_c - 1) \quad (27)$$

where m_c is a vector and it is computed as follows:

$$m = rand;$$

$$m_c = 4 \cdot m \cdot (1 - m) \quad (28)$$

The main benefit of chaos is to explore new search areas and information about the candidate particles in the FDB strategy, which can concurrently and perfectly enrich the population with high-quality solutions (exploitation) while exploring new promising zones in the search space.

- Handling upper and lower boundaries

Because several random techniques were used throughout the optimization process, some newly created particles exceeded the upper and lower bounds of the optimization problem. Most of the optimization methods produce simple upper and lower boundaries, which may delay the convergence speed to optimal solutions. To address this issue, we suggested a new method that makes use of the following mathematical framework to transform particles from predetermined upper and lower boundaries to places that are close to optimal areas:

$$x_{i,j} = best(x_j) + \epsilon \times (rand \times (UB_j - LB_j)) \times rand \times LB_j \tag{29}$$

The aforementioned equations improve the diversity of the best optimal solutions discovered thus far. This means that particles are not just moved from locality to optimal regions, but also the quality of the solution is increased by obtaining information from the best particle's neighborhood. The operation process of the proposed IAOA is demonstrated in Figure 7.

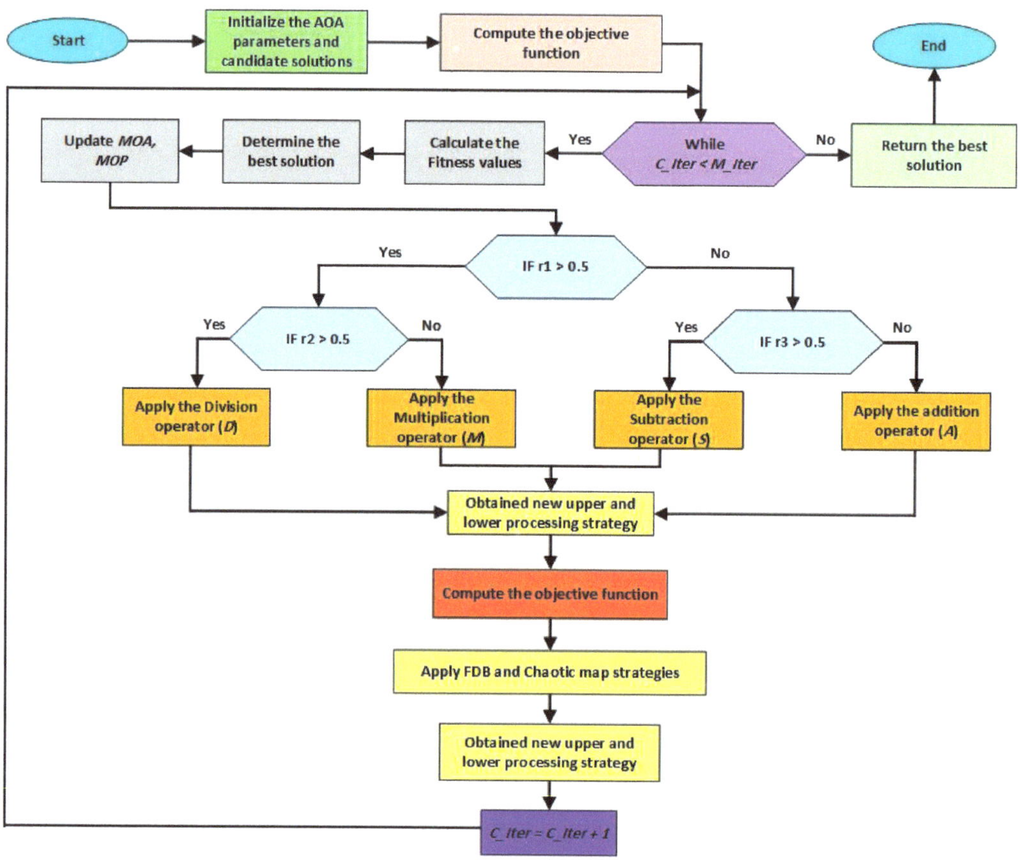

Figure 7. Flowchart of the proposed IAOA.

4.4. The Proposed MOIAOA Method

In developing a set of PF solutions, the multi-objective optimization (MOO) methods present several difficulties regarding efficiency, convergence, and diversity. In the case of large goal optimization problems, most MOO methods produce unsatisfactory optimal PF solutions by simultaneously attempting to enhance diversity and convergence. Furthermore, most of the methods in the previous literature considered only a single objective during the optimization process for the proposed system (either an economic or technical objective).

The proposed system in this work is to define the scope of REMS in terms of system size by calculating the minimum (optimal) number of PV modules, wind turbines, and ESU batteries. This will be achieved by calculating the LCOE, GCF, and E_{SOLD} for the proposed system. It will ensure that the REMS operates the charging station without economic losses for a specific number of EVs per day (three EVs). The proposed system, displayed in Figure 1, can be considered a typical grid-connected system for supplying electricity with the assistance of V2G technology and is presented as a test-case system to verify the effectiveness of the devised optimal design method. In this research, an AOA has been proposed and improved to become IAOA. In addition, the optimal design for the proposed system's components has been determined using a multi-objective improved arithmetic optimization algorithm (MOIAOA).

Multi-objective IAOA (MOIAOA) based on the Non-Scale multiple-run Pareto Front concept has been used to calculate the optimal values of the three conflicting objectives, which are Grid Contribution Factor (GCF), Levelized Cost of Electricity (LCOE), and Energy sold to the grid (E_{SOLD}). This method is classified as a Non-Scale (NS) multiple-run Pareto Front method and deals with multi-objective optimization problems. In this paper, for the LCOE, GCF, and E_{SOLD}, the aggregation function transforms objectives into a mono-objective problem, where the aggregation function treats the multi-objective optimization problems as a mono-objective problem, as described below:

$$f_i\,(t) = \sum_{i=1}^{k} w_i \times f_i\,(t) \tag{30}$$

where k is an aggregated function that belongs to the individual objective function number and x is the decision variable vector related to the search space. The range of weight coefficients is $0 > wi < 1$, denoting the relative importance of the k objective function of the problem, and it is assumed as follows:

$$\sum_{i=1}^{k} w_i = 1 \tag{31}$$

On the other hand, the three objective functions are not scalable. Normalizing and implementing the goal function described below is essential [50,51].

$$f_i\,(t) = \frac{f_i\,(t) - f_i^{min}\,(t)}{f_i^{max}\,(t) - f_i^{min}\,(t)} \tag{32}$$

where the upper and lower bounds of the ith individual objective function are denoted as $f_i^{max}(t)$ and $f_i^{min}(t)$, respectively.

5. Results and Discussion

In previous sections, the concept and operation of RB-EMS and optimal design have been established for the proposed system. The modeling of the components and power flow control strategy of the proposed system has been presented. This section will present the results obtained from the data collection. The results of the optimal design of the proposed system will be presented.

The proposed system consists of a grid-connected PV-WT battery including EVCS. The EVCS is combined within the proposed system by using V2G technology. In summary, the RB-EMS aims to operate the charging station while keeping the charging cost lower than the average grid electricity price (without economic losses) and reducing the grid burden and system economic losses. In addition, the RB-EMS is embedded in the central controller of the proposed system for real-time decision making without the physical presence of a human operator, so the results of EV charging are derived using the proposed system operated under the control of RB-EMS. The optimal design of the proposed system is provided in this study to show how to meet the load requirement for a government building in the most efficient manner.

The meteorological data of solar insolation (G), temperature (Ta), and wind speed (v) have been used in this research in Al-Najaf Governorate, Iraq. The data obtained are used throughout the simulation process in MATLAB simulations. Data from 1 January 2018 to 31 December 2018 were recorded for one year and collected every ten minutes in a government building. The topographical location of the study region is identified as 31° north latitude and 44° east longitude. Figures 8–10 depict the G, Ta, and v plots, respectively.

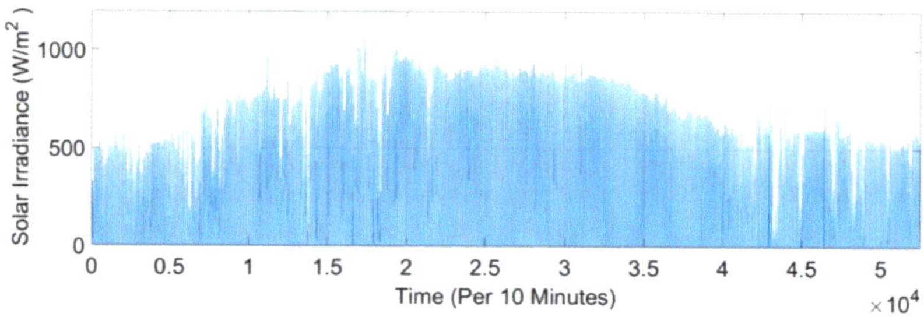

Figure 8. Ten minutes of data readings of solar irradiance during the entire year.

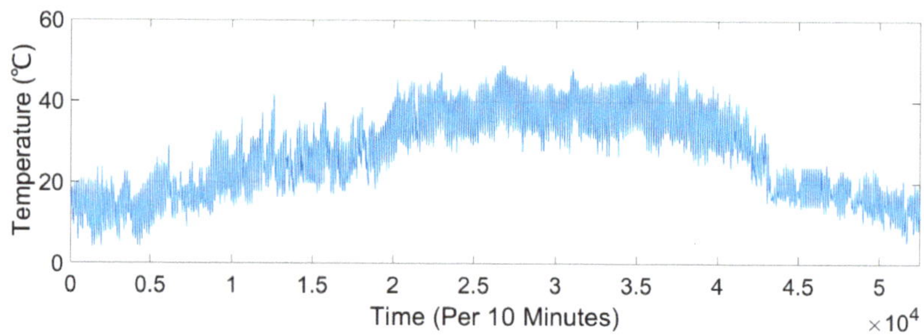

Figure 9. Ten minutes of data readings of temperature during the entire year.

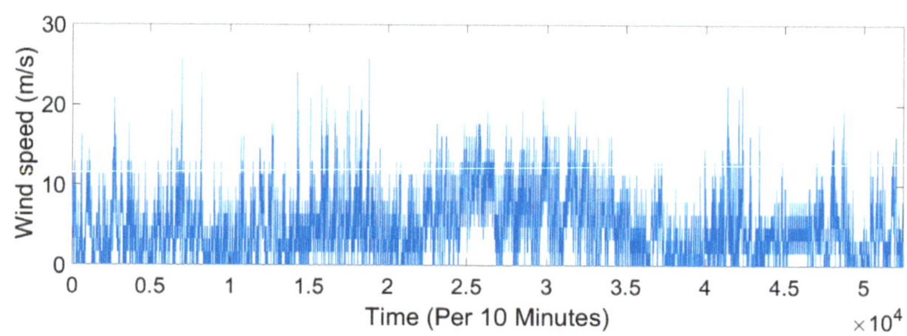

Figure 10. Ten minutes of data readings of wind speed during the entire year.

Data analysis significantly improves our understanding of consumer energy needs from the available Renewable Energy Sources (RESs). This is crucial for handling difficult situations like days without sunlight or wind. The area under study is fortunate to have abundant solar energy and wind throughout the year, with the highest levels observed in July. Most solar radiation, which holds immense potential for generating electricity using photovoltaic (PV) systems, is experienced during summer, followed by spring, autumn and winter. The wind speed is at its maximum in spring, followed by summer, winter, and autumn, offering substantial potential for harnessing wind turbines for electricity generation. Air conditioning units are primarily used during the summer, while consumers predominantly use heaters in winter.

The data of the load demand for the entire year of the building that the study was conducted in show a maximum value of 4.9687 kW and a minimum value of 0.6417 kW every 10 min. A critical stage in optimizing the energy system is accurately estimating the energy demand to be fulfilled to avoid oversizing or under-sizing the system. In this research, the load demand profile for the chosen building is considered for one year with a minimum value of 0.6417 kW and a maximum value of 4.9687 kW. These load data are considered for every ten minutes for the entire year. The load data are graphically presented in Figure 11. Energy demand can be categorized into domestic loads, including appliances in the selected building. Given the case study area's two distinct seasonal variations (cold and hot), the energy demand profile data reveal that energy consumption is high during the hot season, in contrast to the cold season.

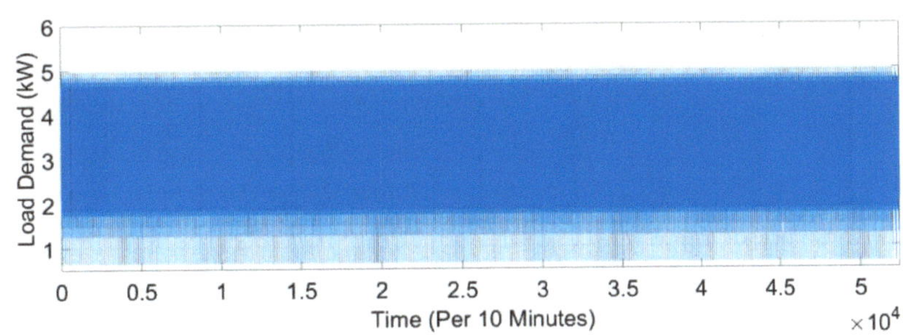

Figure 11. Ten-minute data values of load profile of the studied area during the entire year.

The plots of the annual PV output power (P_{PV}) and wind turbine output power (P_{WT}) for the optimal configuration achieved by the proposed system are displayed in Figures 12 and 13, respectively. The proposed system is primarily designed to take advantage of the RESs in the location and interchange power via V2G technology between the EVCS and the utility grid. One of the RESs taken into account in the proposed system is PV. The solar irradiance (G) and ambient temperature (T_{am}) are the primary climatological factors that affect the output power produced by the PV. Figure 12 shows the output power produced by PV in the proposed system. The wind turbine is the second RES regarded in the proposed system and the output power generated from the wind turbine in the proposed system is illustrated in Figure 13.

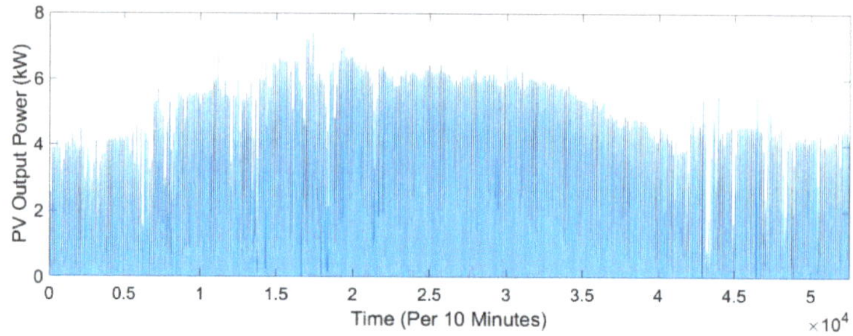

Figure 12. PV output power in the proposed system.

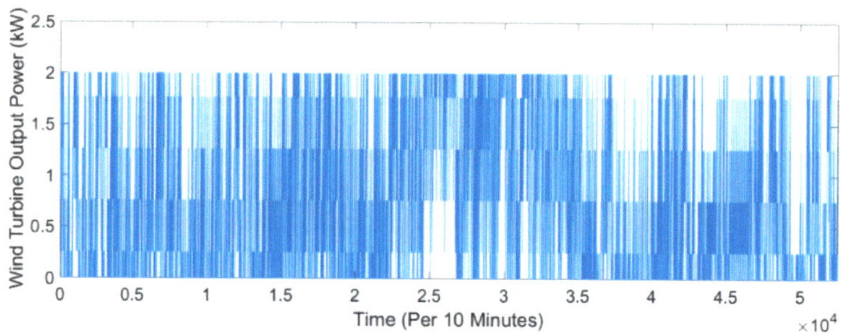

Figure 13. The wind turbine output power in the proposed system.

5.1. Performance Comparison between the Proposed IAOA and AOA, PSO, ALO

This study used four standard benchmark cases to confirm and verify the original AOA's stability [69] and performance. These benchmarks can reasonably approximate the capability of exploitation and exploration, respectively, for the pending algorithm. Numerous experiments were conducted to evaluate the IAOA's properties, for instance, the differences between IAOA and the original AOA were examined by utilizing several benchmarks, as indicated in Table 3.

Table 3. Details of benchmark functions.

Benchmark Function	Dim	Range	Optimal Value
$f_1(x) = \sum_{i=1}^{d} x_i^2$	10	[−100, 100]	0
$f_2(x) = \sum_{i=1}^{d} \lvert x_i \rvert + \prod_{i=1}^{d} \lvert x_i \rvert$	10	[−100, 100]	0
$f_3(x) = \max_i\{\lvert x_i \rvert, 1 \leq i \leq t\}$	10	[−10, 10]	0
$f_4(x) = \sum_{i=1}^{d} (x_i^2 - 10 * \cos(2\pi x_i) + 10d)$	10	[−5.12, 5.12]	0

The detailed experimental findings produced by AOA, ALO, PSO, and IAOA on these benchmarks are displayed in Table 4 concerning best value, worst value, average value and STD (standard deviation) value. This table shows that the IAOA can retrieve the best values by obtaining the minimum values of best value, worst value, average value and STD value. Hence, IAOA outperforms AOA, ALO, and PSO, indicating that algorithm stability can be guaranteed. Furthermore, to further illustrate descript convergence, the evolution curves of each approach on most of the benchmarks in this work are shown in Figures 14–17. These figures show that, on these benchmarks, the proposed IAOA has satisfied quicker convergence than the AOA, PSO, and ALO methods.

Table 4. Results of IAOA compared to several peers on benchmark functions.

Function	Algorithm	Best Value	Worst Value	Average Value	STD
$f_1(x)$	ALO	1.6291×10^{-9}	8.8636×10^{-9}	4.0018×10^{-9}	1.9334×10^{-9}
	PSO	8.2671×10^{-121}	2.5752×10^{-48}	1.2876×10^{-49}	5.7582×10^{-49}
	AOA	0	0	0	0
	IAOA	0	0	0	0
$f_2(x)$	ALO	1.1730×10^{-5}	0.4850	0.0254	0.1082
	PSO	1.2952×10^{-14}	9.0712×10^{-6}	1.0976×10^{-6}	2.5864×10^{-6}
	AOA	0	0	0	0
	IAOA	0	0	0	0
$f_3(x)$	ALO	1.1878×10^{-6}	0.0025	1.8484×10^{-4}	5.4351×10^{-4}
	PSO	1.4774×10^{-25}	4.0204×10^{-13}	2.0227×10^{-14}	8.9870×10^{-14}
	AOA	0	0	0	0
	IAOA	0	0	0	0
$f_4(x)$	ALO	5.1427×10^{-5}	0.0027	4.3189×10^{-4}	5.8547×10^{-4}
	PSO	2.0618×10^{-19}	2.5902×10^{-13}	2.3924×10^{-14}	6.2083×10^{-14}
	AOA	0	$\mathbf{1.1591 \times 10^{-186}}$	$\mathbf{5.7954 \times 10^{-188}}$	0
	IAOA	0	$\mathbf{1.7994 \times 10^{-250}}$	$\mathbf{8.9970 \times 10^{-252}}$	0

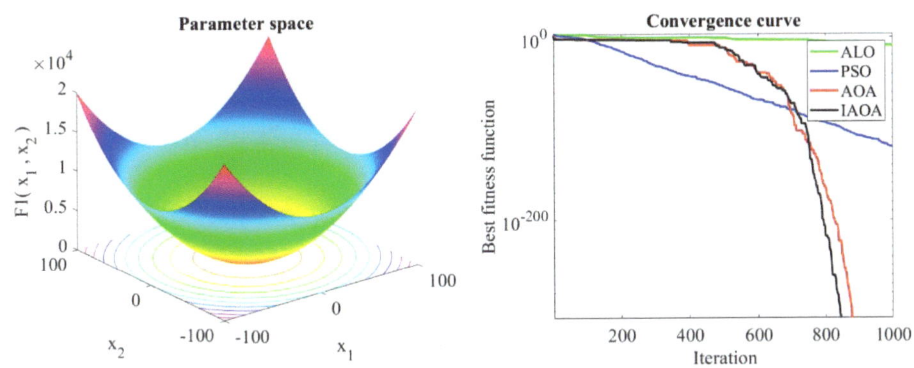

Figure 14. Parameter space and convergence curve of the IAOA, ALO, AOA, and PSO on the first test function.

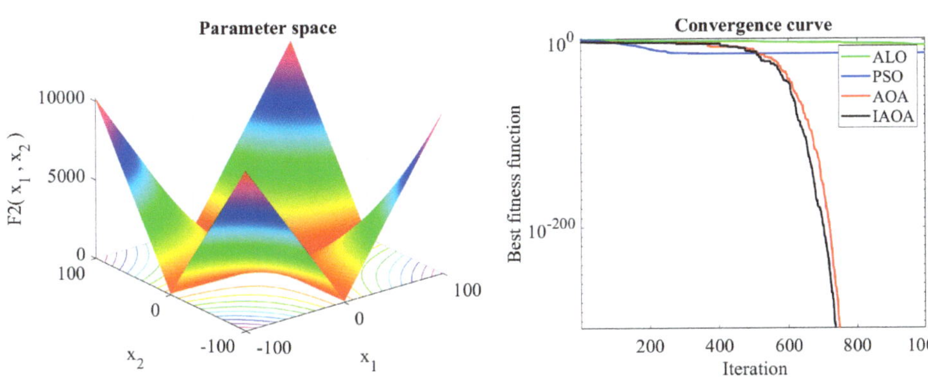

Figure 15. Parameter space and convergence curve of the IAOA, ALO, AOA, and PSO on the second test function.

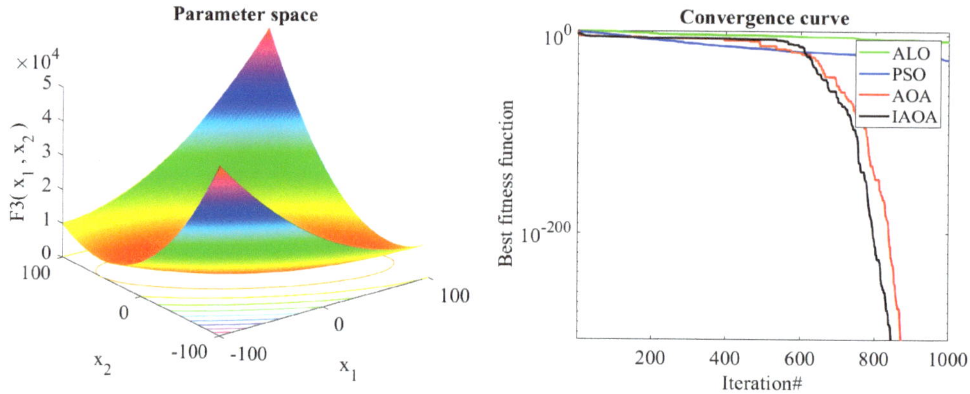

Figure 16. Parameter space and convergence curve of the IAOA, ALO, AOA, and PSO on the third test function.

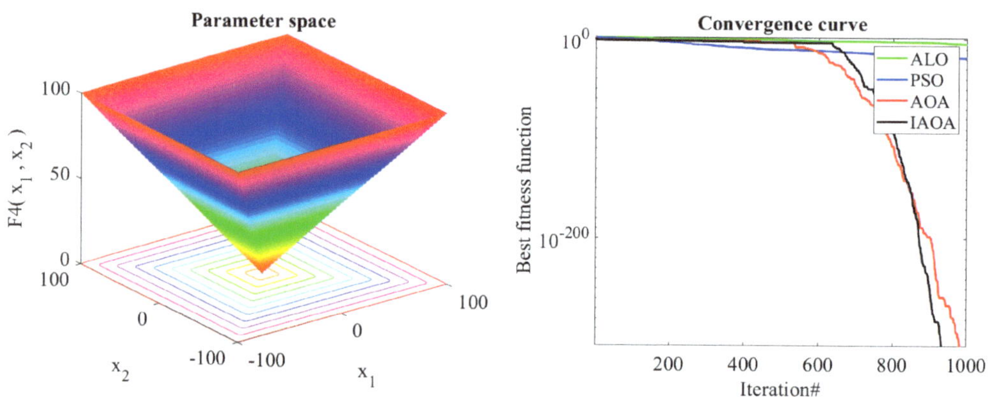

Figure 17. Parameter space and convergence curve of the IAOA, ALO, AOA, and PSO on the fourth test function.

Exploiting and exploring individual algorithms are very common in gauging the individual algorithm search capacity for all metaheuristic optimization strategies. The first step in each algorithm is thoroughly exploring the promising areas of the given solution search space. Depending on the optimization technique, optimizers can support this phase by using some stochastic operators to search the given space globally and randomly. The exploitation step, however, is a local search in which the optimizers look near the most promising regions identified thus far in the exploration phase. There is always a challenge in the optimization period to effectively balance these two stages, which can be performed using the controlling parameter. These parameters were carefully selected and tested on the employed standard benchmark test functions in this research.

Fitness–distance balance (FDB) and chaotic map mechanisms have been applied to improve the AOA. The AOA, PSO, and ALO benchmark algorithms are selected to perform the comparative analysis. A comparative analysis between IAOA, AOA, PSO, and ALO has been performed to test the efficiency and reliability of the algorithms, as shown in Figures 14–17. Four popular standard mathematical benchmark functions comprising of the Unimodal and Multimodal functions have been used for comparison implementation.

Figures 14–17 display the convergence curve for the IAOA, AOA, PSO, and ALO methods. The convergence curve shows how the algorithm converges to the best solution.

So, the convergence curve indicates how fast the fitness value converges towards an optimal solution through iterations. The final value of the fitness value shows the best solution, while the nature/slope of this curve shows how fast the algorithm converges to the final solution. AOA has been improved by enhancing the exploration and exploitation phases. Fitness–distance balance (FDB) mechanisms have enhanced the exploitation phase. At the same time, a chaotic map mechanism has been used to enhance the exploration phase.

IAOA has been compared with the AOA, ALO, and PSO on four benchmark cases, as shown in Section 5.1. The experimental results, which are average values and evolution curves, are vividly recorded in Table 4 and Figures 14–17, respectively. These records demonstrate that the efficiency of IAOA was observed according to the enhanced evolutionary convergence in competition with other counterparts. The curves in Figures 14–17 demonstrate an accelerated drift for the proposed IAOA.

5.2. Results of the MIAOA

In this section, an optimization of the proposed system using a Multi-Objective Improved Arithmetic Optimization Algorithm has been implemented. The outcomes of the optimal design method based on RB-EMS-MOIAOA based on the Non-Scale multiple-run Pareto Front concept are shown in this section. The suggested optimization method aims to identify the best layout for the proposed system that would supply building demand at a desired value of LCOE, GCF, and E_{SOLD} to guarantee that the proposed RB-EMS-MOAOA is reliable and valid when calculating the optimal system size required to meet the required demand of the chosen building.

In this research, the proposed system comprises a grid-connected PV–Wind–Battery system combined with EVCS via using the V2G technique. Multi-Objective Improved Arithmetic Optimization Algorithm (MOIAOA) based on the Non-Scale multiple-run Pareto Front concept has been used to improve the optimal design of the proposed system components. Multi-objective IAOA (MOIAOA) Non-Scale multiple-run Pareto Front concept has been used to calculate the optimal values of the three conflicting objectives, which are Grid Contribution Factor (GCF), Levelized Cost of Electricity (LCOE), and Energy sold to the grid (E_{SOLD}). This method is classified as a Non-Scale (NS) multiple-run Pareto Front method, which deals with multi-objective optimization problems. Rule-Based Energy Management Strategy (RB-EMS) controls and monitors the proposed system's power flow. Three constraint objectives are used; the technical, economic, and renewability criteria are all weighted, normalized, and combined using a mono-objective function.

Table 5 shows the optimal configurations of the proposed system with sets of weights using the proposed MOIAOA based on the Non-Scale multiple-run Pareto Front concept. For the proposed system, Table 5 indicates the initializing weights (W1, W2, W3), number of wind turbines, number of PV modules, number of batteries, fitness value, LCOE, GCF, E_{SOLD}, and CPU execution time. The range of weight sets is [1, 36] with a step size of 0.1. In Table 5, the optimal weight set and configurations of the proposed system using the MOIAOA Non-Scale multiple-run Pareto Front concept have been tabulated. The maximum fitness function (f) value recorded is 0.1649 at the set of weights [0.5, 0.3, 0.2]. In contrast, the minimum fitness function (f) value recorded is 0.0611 at the set of weights [0.1, 0.1, 0.8]. The given weight value changes the value of the individual objective. An accurate selection of the effective W1, W2, and W3 weights from the design space is required to find the best trade-off between the technical, economic, and renewability objectives. The trade-off between the defined level of renewability, economic, and technical criteria is required to choose an optimal configuration of the proposed system.

Table 5. Optimal weight sets and configurations of the proposed system using MOIAOA based on Non-Scale multiple-run Pareto Front concept.

W1	W2	W3	WT	PV	Bat	f	LCOE	GCF	E_{SOLD}	Elapsed Time (Seconds)
0.1	0.8	0.1	1	31	138	0.0762	2.37×10^{-2}	0.0171	0.8596	4956.680489
0.1	0.7	0.2	1	32	108	0.1149	2.37×10^{-2}	0.0171	0.8596	5536.762598
0.1	0.6	0.3	4	18	100	0.1344	2.33×10^{-2}	0.0274	0.8337	4827.420494
0.1	0.5	0.4	1	18	93	0.145	0.022	0.8377	0	3438.61148
0.1	0.4	0.5	1	14	73	0.1343	0.0261	0.8376	0	2909.015921
0.1	0.3	0.6	1	13	63	0.1144	0.0304	0.8375	0	2731.221521
0.1	0.2	0.7	1	11	49	0.0942	0.0531	0.8371	0	2678.625856
0.1	0.1	0.8	1	12	106	0.0611	0.022	0.8377	0	5521.751529
0.2	0.7	0.1	6	27	70	0.0943	0.022	0.8377	0	2740.512867
0.2	0.6	0.2	4	22	147	0.1306	0.0237	0.0171	0.8596	4810.072815
0.2	0.5	0.3	1	20	91	0.1522	0.0237	0.0171	0.8596	3293.332732
0.2	0.4	0.4	2	15	63	0.1487	0.0253	0.0012	0.8475	3465.500523
0.2	0.3	0.5	1	15	83	0.1359	0.0531	0.8371	0	2780.494461
0.2	0.2	0.6	1	14	63	0.1095	0.022	0.8377	0	2766.968238
0.2	0.1	0.7	1	12	56	0.0779	0.0261	0.8376	0	3786.031941
0.3	0.6	0.1	5	31	59	0.1154	0.0216	0.1136	0.7579	3414.903356
0.3	0.5	0.2	3	24	68	0.1419	0.0233	0.0274	0.8337	2633.810036
0.3	0.4	0.3	1	19	103	0.1646	0.0237	0.0171	0.8596	5378.376291
0.3	0.3	0.4	1	15	56	0.1481	0.0246	0.0035	0.8554	2637.907928
0.3	0.2	0.5	1	12	63	0.13	0.0304	0.8375	0	2654.862484
0.3	0.1	0.6	1	12	56	0.0975	0.0261	0.8376	0	5068.761916
0.4	0.5	0.1	5	31	59	0.132	0.0243	0.0069	0.8475	3235.824878
0.4	0.4	0.2	4	12	56	0.1573	0.0239	0.0738	0.7026	4983.926465
0.4	0.3	0.3	1	16	63	0.161	0.0253	0.0012	0.8475	2682.447224
0.4	**0.2**	**0.4**	**2**	**8**	**33**	**0.1522**	$\mathbf{2.66 \times 10^{-2}}$	$\mathbf{7.34 \times 10^{-5}}$	**0.8409**	**3272.623188**
0.4	0.1	0.5	3	1	2	0.1156	0.0261	0.8376	0	2731.882521
0.5	0.4	0.1	3	24	68	0.1425	0.0262	0.0119	0.7838	2762.613745
0.5	0.3	0.2	1	19	68	0.1649	0.0239	0.0738	0.7026	5103.018621
0.5	0.2	0.3	1	15	52	0.1595	0.0246	0.0035	0.8554	5587.216803
0.5	0.1	0.4	3	1	1	0.1284	0.0246	0.0035	0.8554	2951.953591
0.6	0.3	0.1	5	17	42	0.1534	0.0262	0.0119	0.7838	2522.240268
0.6	0.2	0.2	5	7	25	0.1668	0.0266	7.34×10^{-5}	0.8409	2629.085911
0.6	0.1	0.3	3	1	2	0.1418	2.46×10^{-2}	3.50×10^{-3}	0.8554	2904.915789
0.7	0.2	0.1	8	6	17	0.1542	0.0262	0.0119	0.7838	2771.440828
0.7	0.1	0.2	4	3	11	0.1518	0.0266	7.34×10^{-5}	0.8409	5176.083577
0.8	0.1	0.1	5	10	4	0.1387	0.0246	3.50×10^{-3}	0.8554	5424.871393

In Table 5, the first range of weights used in the MATLAB simulation of this work are W1 = 0.1, W2 = 0.8, and W3 = 0.1, and that leads to the results of the number of PV modules being equal to 31, the number of wind turbines equal to 1, number of batteries equal to 138, fitness value equal to 0.0762, LCOE equal to 2.37×10^{-2} USD/kWh, GCF equal to 0.0171 kWh, and E_{SOLD} equal to 0.8596 kWh. Figure 18 presents the distribution of the solutions using the NS multiple-run Pareto Front method, where the aggregation of the three objectives is based on a predetermined set of weights.

Based on Table 5, the balance can be achieved when optimal weights W1, W2, and W3 are 0.4, 0.2, and 0.4, respectively. The weights of 0.4, 0.2, and 0.4 are optimal. Therefore, at the optimal weights and by comparison with other optimal weights, it can be observed that the value of LCOE is small, and the value of GCF is small, too. At the same time, the value of E_{SOLD} is high. The results indicated that by employing the proposed MOIAOA Non-Scale multiple-run Pareto Front concept, the optimal configurations of the proposed system are as follows: number of PV modules equal to 8, number of wind turbines equal to 2, number of batteries equal to 33, fitness value equal to 0.1522, LCOE equal to 2.66^{-2} USD/kWh, GCF equal to 7.34^{-5} kWh, and E_{SOLD} equal to 0.8409 kWh.

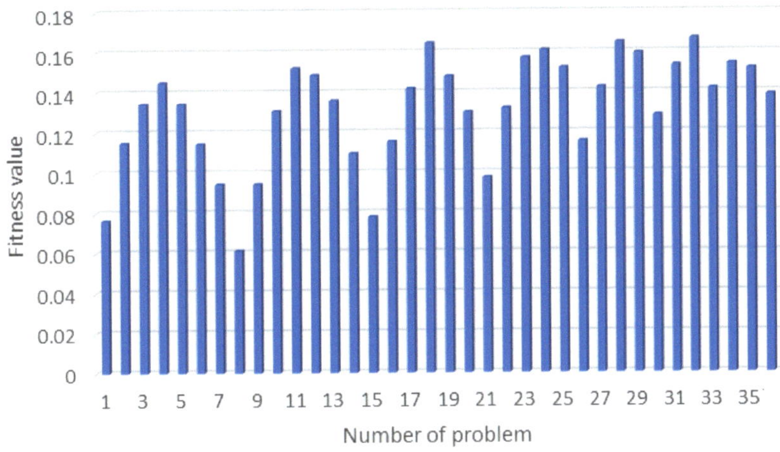

Figure 18. Development of the aggregation function based on the NS multiple-run Pareto Front method.

Figure 19 displays the evolution of the mono-objective function (f) (fitness function) based on optimal solutions using MOIAOA based on the Non-Scale multiple-run Pareto Front concept. Figure 19 depicts the aggregation function's development for the proposed sizing algorithm at the optimal weights. Figure 19 demonstrates the development of the evaluation of the aggregation function with the maximum iteration of the proposed method to obtain the optimal size of the components of the proposed system.

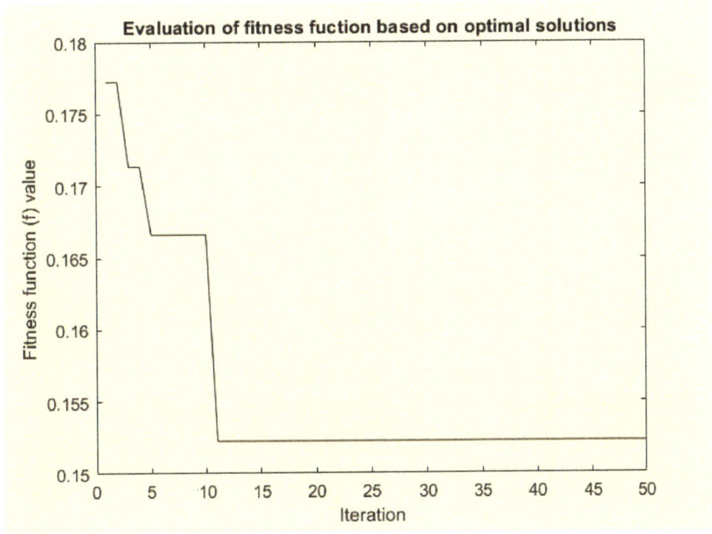

Figure 19. Evaluation of fitness function based on optimal solutions using MOIAOA based on Non-Scale multiple-run Pareto Front concept.

Figure 20 exhibits the optimal weight sets based on three objectives using MOIAOA Non-Scale multiple-run Pareto Front concept; this figure presents the optimal configuration based on MOIAOA Non-Scale multiple-run Pareto Front concept, with the three objectives of E_{SOLD}, LCOE, and GCF.

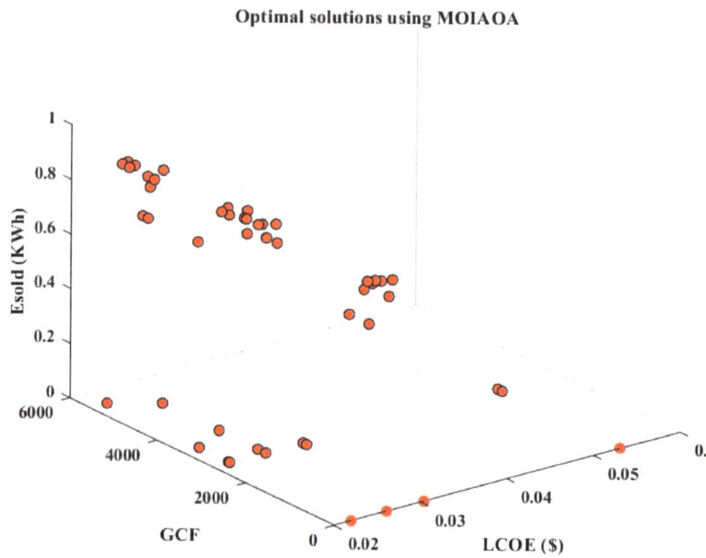

Figure 20. Evaluation of E_{SOLD} (kWh), GCF (kWh), and LCOE (USD) values with different weight sets using MOIAOA based on Non-Scale Pareto Front concept.

The results section shows that the control strategy can effectively schedule the generator at all times and meet the load demand. The simulation results demonstrate the superiority and the rapid convergence performance of the proposed improved (MOIAOA) Non-Scale multiple-run Pareto Front concept algorithm.

6. Conclusions and Future Direction

In this work, the IAOA is proposed and compared with AOA, PSO, and ALO methods to prove that the proposed algorithm is superior (to justify the superiority of the proposed algorithm). Then, a new MOIAOA is proposed for finding the optimal design of the proposed system that includes a PV–WT–battery system combined with EVCS using V2G technology in Al-Najaf Governorate in Iraq. The most desirable configurations for the proposed system are defined based on renewability–economic–technical criteria using ten-minute data readings of real meteorological data during the entire year. LCOE, GCF, and E_{SOLD} are utilized as economic, renewability, and technical criteria, respectively. The MOIAOA based on a Non-Scale multiple-run Pareto Front concept was proposed to choose an optimal design for the proposed system. The FDB mechanism was employed to enhance the proposed system's exploitation phase, and the chaotic map mechanism was employed to enhance the exploration phase of the proposed system. By using Non-Scale multiple-run PF, LCOE, GCF, and E_{SOLD} are used as three constraint objective functions, which are aggregated after normalization and weighting by the mono-objective function. LCOE, GCF, and E_{SOLD} are utilized as economic criteria, renewability criteria, and technical criteria, respectively. LCOE needed to be minimized, GCF needed to be minimized, and E_{SOLD} needed to be minimized. The Rule-Based Energy Management Strategy (RB-EMS) was used to control and observe the power flow in the proposed system. The Pareto Front method was employed to obtain the optimal values of LCOE, GCF, and E_{SOLD}; these three constraint objectives are utilized as renewability criteria, economic criteria, and technical criteria. MATLAB R2020b is used for simulations in this research. For performance comparison, it is noted that the proposed IAOA is more efficient than AOA, ALO, and PSO because IAOA converges to the optimal solution in fewer iterations. The results indicated that by employing the proposed MOIAOA Non-Scale multiple-run Pareto Front concept, the optimal configurations of the proposed system are as follows: number of PV modules equal to 8, number of wind turbines equal to 2, number of batteries equal to 33, fitness value equal to 0.1522, LCOE equal to 2.66–2 USD/kWh, GCF equal to 7.34–2 kWh, and E_{SOLD} equal to 0.8409 kWh.

This paper also describes integrating EVs with photovoltaics (PV) and wind turbines as Renewable Energy Sources (RESs) to address the problems associated with fossil fuels. Alternative energy sources can be used to handle the situation where fossil fuels have started to dwindle, leading to various power and environmental difficulties. This work is fully satisfied and meets the grid-connected system's load demand. The constraints in electricity and environmental systems are resolved by integrating RESs with other sources. To handle the complexity of PV–wind hybrid systems, a metaheuristic optimization approach (MOIAOA) was combined with RB-EMS to achieve the objective functions. For the future direction of this work, this work highly recommends using other metaheuristic algorithms to investigate the renewability–economic–technical criteria and additional development of energy management strategies. Additionally, more examinations may be undertaken in the utilization of commercial load needs in the Al-Najaf Governorate in Iraq.

Author Contributions: Conceptualization, A.A.K.A.-S.; data curation, A.A.K.A.-S. and D.M.H.; formal analysis, A.A.K.A.-S., H.M.R. and D.M.H.; investigation, S.M.A., C.W.T., H.M.R. and D.M.H.; methodology, A.A.K.A.-S.; project administration, A.A.K.A.-S.; resources, A.A.K.A.-S.; software, A.A.K.A.-S.; validation, A.A.K.A.-S. and H.M.R.; visualization, A.A.K.A.-S. and S.M.A.; supervision, S.M.A. and C.W.T.; writing—original draft preparation, A.A.K.A.-S.; writing—review and editing, A.A.K.A.-S., S.M.A. and H.M.R.; funding acquisition, A.A.K.A.-S. and S.M.A. All authors have read and agreed to the published version of the manuscript.

Funding: This research was funded by University of Kufa (17348), Universiti Teknologi Malaysia: UTMPGIS[UTM.J.08.02.01/13.14/1/1 JId.2 (47)].

Institutional Review Board Statement: Not applicable.

Informed Consent Statement: Not applicable.

Data Availability Statement: The data presented in this study are available on request from the corresponding author.

Acknowledgments: The first author would like to thank the Iraqi government for supporting this study with a scholarship awarded through the University of Kufa by a vote of 17,348. The authors would like to thank Universiti Teknologi Malaysia (UTM) for the library facility and scholarship awarded through UTMPGIS by the vote of UTM.J.08.02.01/13.14/1/1 JId.2 (47).

Conflicts of Interest: The authors declare no conflicts of interest.

Abbreviations

E_{SOLD}	The annual energy sold to the grid
WT	Wind turbine
EV	Electric Vehicle
EVCS	electric vehicle charging station
NS	Non-Scale
PSO	Particle swarm optimization
ALO	Ant Lion Optimizer
EMS	Energy Management Strategy
AOA	Arithmetic optimization algorithm
HRES	Hybrid renewable energy system
MOO	Multi-objective optimization
STC	Standard Test Condition
$P_{pv_{out}}(t)$	The output power generated from PV
$G_{(t)}$	Solar irradiance
$P_{(PV_{rated})}$	Rated power for PV
NOCT	The nominal operating cell temperature

v_{cut-in}	cut-in speed of the WT
$v_{cut-out}$	cut-out speed of the WT
P_r	Rated power of the WT
v_r	Rated wind speed of the WT
P_{WT}	The generated output power of the WT
BSS	Battery Storage System
DOD	The depth of discharge
S_{rated}	The station rated capacity
$\cos\varnothing$	The power factor
N_{slot}	The amount of charging slots for each EV
k_{load}	The overload factor for cover overloading in transients
P_{EV}	The maximum power rate of each EV
$P_{inv}(t)$	The inverter rating
$P_L^m(t)$	The peak load demand
BT	Battery
X	a collection of initialized solutions
Rand	a random variable in the range [0, 1]
X_{UB} and X_{LB}	the upper and lower limits of the problem
MOA (C_{iter})	the value at the t th iteration
M_{iter}	the maximum number of iterations
r_2	randomly generated number that is conditioned between the D and M operations
UB_j and LB_j	the upper and lower limits
$\varepsilon\varepsilon$	a tiny integer value
r_3	a randomly generated number that serves as a denotation for the A and S operators
FF	Fitness Function
PV	photovoltaic
RESs	Renewable energy sources
LCOE	Levelized Cost of Energy
V2G	Vehicle-to-grid
STC	Standard Test Conditions
RB-EMS	Rule-Based Energy Management Strategy
GCF	Grid Contribution Factor
REF	Renewable Energy Fraction
NPC	Economic criterion of net present cost
IAOA	Improved arithmetic optimization algorithm
MOIAOA	Multi-objective improved arithmetic optimization algorithm
α_t	Temperature coefficient
$T_{C_{STC}}$	The cell temperature as reference temperature
T_{amb}	The ambient temperature
C_B	Capacity of the battery
E_L	The daily average load demand
AD	the autonomy days
v1, v2	The wind speed
h	hub height
h_{ref}	The reference height anemometer
α	The power-law exponential known as wind gradient, Hellmann exponent, or friction coefficient
SOC	State of Charge
$P_b(t)$	The battery's output of electricity
$P_{pv}(t)$	The total power generated by PV
$P_{WT}(t)$	The total power generated by WT
$P_l(t)$	The total energy demand
η_{inv}	The inverter efficiency
σ	The self-discharge rate of the battery

η_b	Battery efficiency
$rate_{feed-in}$	The feed-in tariff rate
$E_{grid(selling)}$	The cost of selling energy
C_p	The cost of buying electricity from the grid
$\sum_{t=1}^{8760} E_{grid(purchaed)}$	The per hour summation of annually buying electricity from the grid for one year
MOA	Math Optimizer Accelerated
C_{iter}	the current iteration
Max & Min	The accelerated function's maximum and lowest values are denoted by Max and Min (Maximum and minimum values of the MOA function)
μ	a control variable set
MOP	Math Optimizer Probability
$\propto$	a sensitive control parameter set
FDB	Fitness–distance balance

References

1. Ali, S.; Zheng, Z.; Aillerie, M.; Sawicki, J.-P.; Péra, M.-C.; Hissel, D. A Review of DC Microgrid Energy Management Systems Dedicated to Residential Applications. *Energies* **2021**, *14*, 4308. [CrossRef]
2. Ali, A.; Shaaban, M.F.; Awad, A.S.A.; Azzouz, M.A.; Lehtonen, M.; Mahmoud, K. Multi-Objective Allocation of EV Charging Stations and RESs in Distribution Systems Considering Advanced Control Schemes. *IEEE Trans. Veh. Technol.* **2023**, *72*, 3146–3160. [CrossRef]
3. Das, R.; Wang, Y.; Putrus, G.; Kotter, R.; Marzband, M.; Herteleer, B.; Warmerdam, J. Multi-objective techno-economic-environmental optimization of electric vehicle for energy services. *Appl. Energy* **2020**, *257*, 113965. [CrossRef]
4. Ahmadi, S.E.; Kazemi-Razi, S.M.; Marzband, M.; Ikpehai, A.; Abusorrah, A. Multi-objective stochastic techno-economic-environmental optimization of distribution networks with G2V and V2G systems. *Electr. Power Syst. Res.* **2023**, *218*, 109195. [CrossRef]
5. Liu, J.; Chen, X.; Yang, H.; Shan, K. Hybrid renewable energy applications in zero-energy buildings and communities integrating battery and hydrogen vehicle storage. *Appl. Energy* **2021**, *290*, 116733. [CrossRef]
6. Eriksson, E.; Gray, E.M. Optimization and integration of hybrid renewable energy hydrogen fuel cell energy systems—A critical review. *Appl. Energy* **2017**, *202*, 348–364. [CrossRef]
7. Sedighizadeh, M.; Fazlhashemi, S.S.; Javadi, H.; Taghvaei, M. Multi-objective day-ahead energy management of a microgrid considering responsive loads and uncertainty of the electric vehicles. *J. Clean. Prod.* **2020**, *267*, 121562. [CrossRef]
8. Jia, S.; Kang, X.; Cui, J.; Tian, B.; Xiao, S. Hierarchical stochastic optimal scheduling of electric thermal hydrogen integrated energy system considering electric vehicles. *Energies* **2022**, *15*, 5509. [CrossRef]
9. Zhang, Q.; Ding, J.; Li, G. Optimal operation of PV-diesel-battery MG based on fuzzy comprehensive evaluation. In Proceedings of the 2018 Thirteenth International Conference on Ecological Vehicles and Renewable Energies (EVER), Monte Carlo, Monaco, 10–12 April 2018; pp. 1–5.
10. Saffari, M.; Misaghian, M.S.; Flynn, D.; Kia, M.; Vahidinasab, V.; Lotfi, M.; Catalão, J.P.; Shafie-Khah, M. Multi-objective optimization of an active distribution system using normalised normal constraint method. In Proceedings of the 2019 IEEE Milan PowerTech, Milan, Italy, 23–27 June 2019; pp. 1–6.
11. Jannati, J.; Nazarpour, D. Multi-objective scheduling of electric vehicles intelligent parking lot in the presence of hydrogen storage system under peak load management. *Energy* **2018**, *163*, 338–350. [CrossRef]
12. Ahmed, H.Y.; Ali, Z.M.; Refaat, M.M.; Aleem, S.H.A. A Multi-Objective Planning Strategy for Electric Vehicle Charging Stations towards Low Carbon-Oriented Modern Power Systems. *Sustainability* **2023**, *15*, 2819. [CrossRef]
13. Tan, B.; Chen, H. Multi-objective energy management of multiple microgrids under random electric vehicle charging. *Energy* **2020**, *208*, 118360. [CrossRef]
14. Hosseinnia, H.; Nazarpour, D.; Talavat, V. Multi-objective optimization framework for optimal planning of the microgrid (MG) under employing demand response program (DRP). *J. Ambient Intell. Humaniz. Comput.* **2019**, *10*, 2709–2730. [CrossRef]
15. Kataoka, R.; Shichi, A.; Yamada, H.; Iwafune, Y.; Ogimoto, K. Comparison of the economic and environmental performance of V2H and residential stationary battery: Development of a multi-objective optimization method for homes of EV owners. *World Electr. Veh. J.* **2019**, *10*, 78. [CrossRef]
16. Wu, G.; Yi, C.; Xiao, H.; Wu, Q.; Zeng, L.; Yan, Q.; Zhang, M. Multi-objective Optimization of Integrated Energy Systems Considering Renewable Energy Uncertainty and Electric Vehicles. *IEEE Trans. Smart Grid* **2023**, *14*, 4322–4332. [CrossRef]
17. Stojkovic, J. Multi-objective optimal charging control of electric vehicles in pv charging station. In Proceedings of the 2019 16th International Conference on the European Energy Market (EEM), Ljubljana, Slovenia, 18–20 September 2019; pp. 1–5.
18. De Schepper, E.; Van Passel, S.; Lizin, S.; Vincent, T.; Martin, B.; Gandibleux, X. Economic and environmental multi-objective optimization to evaluate the impact of Belgian policy on solar power and electric vehicles. *J. Environ. Econ. Policy* **2016**, *5*, 1–27. [CrossRef]

19. Zeynali, S.; Rostami, N.; Feyzi, M. Multi-objective optimal short-term planning of renewable distributed generations and capacitor banks in power system considering different uncertainties including plug-in electric vehicles. *Int. J. Electr. Power Energy Syst.* **2020**, *119*, 105885. [CrossRef]
20. Ahmadi-Nezamabad, H.; Zand, M.; Alizadeh, A.; Vosoogh, M.; Nojavan, S. Multi-objective optimization based robust scheduling of electric vehicles aggregator. *Sustain. Cities Soc.* **2019**, *47*, 101494. [CrossRef]
21. Yang, Z.; Ghadamyari, M.; Khorramdel, H.; Alizadeh, S.M.S.; Pirouzi, S.; Milani, M.; Banihashemi, F.; Ghadimi, N. Robust multi-objective optimal design of islanded hybrid system with renewable and diesel sources/stationary and mobile energy storage systems. *Renew. Sustain. Energy Rev.* **2021**, *148*, 111295. [CrossRef]
22. Foroozandeh, Z.; Ramos, S.; Soares, J.; Vale, Z. Energy management in smart building by a multi-objective optimization model and pascoletti-serafini scalarization approach. *Processes* **2021**, *9*, 257. [CrossRef]
23. Coelho, V.N.; Coelho, I.M.; Coelho, B.N.; Cohen, M.W.; Reis, A.J.; Silva, S.M.; Souza, M.J.; Fleming, P.J.; Guimaraes, F.G. Multi-objective energy storage power dispatching using plug-in vehicles in a smart-microgrid. *Renew. Energy* **2016**, *89*, 730–742. [CrossRef]
24. Soares, J.; Ghazvini, M.A.F.; Vale, Z.; de Moura Oliveira, P. A multi-objective model for the day-ahead energy resource scheduling of a smart grid with high penetration of sensitive loads. *Appl. Energy* **2016**, *162*, 1074–1088. [CrossRef]
25. Han, W.; Zhang, J.; Chen, L. Multi-Objective Optimal Dispatch of Smart Grid by Estimating Available Capacity of EV Aggregators. In Proceedings of the 2021 IEEE Sustainable Power and Energy Conference (iSPEC), Nanjing, China, 23–25 December 2021; pp. 1299–1303.
26. Panda, N.K.; Paterakis, N.G. A multi-objective optimization model for the quantification of flexibility in a large business park. In Proceedings of the 2021 International Conference on Smart Energy Systems and Technologies (SEST), Vaasa, Finland, 6–8 September 2021; pp. 1–6.
27. Gholami, K.; Karimi, S.; Anvari-Moghaddam, A. Multi-objective stochastic planning of electric vehicle charging stations in unbalanced distribution networks supported by smart photovoltaic inverters. *Sustain. Cities Soc.* **2022**, *84*, 104029. [CrossRef]
28. Sedighizadeh, M.; Esmaili, M.; Mohammadkhani, N. Stochastic multi-objective energy management in residential microgrids with combined cooling, heating, and power units considering battery energy storage systems and plug-in hybrid electric vehicles. *J. Clean. Prod.* **2018**, *195*, 301–317. [CrossRef]
29. Zhu, J.; Jin, Y.; Zhu, W.; Lee, D.-K.; Bohlooli, N. Multi-objective planning of micro-grid system considering renewable energy and hydrogen storage systems with demand response. *Int. J. Hydrog. Energy* **2023**, *48*, 15626–15645. [CrossRef]
30. Sharafi, M.; ElMekkawy, T.Y.; Bibeau, E.L. Optimal design of hybrid renewable energy systems in buildings with low to high renewable energy ratio. *Renew. Energy* **2015**, *83*, 1026–1042. [CrossRef]
31. Murray, P.; Carmeliet, J.; Orehounig, K. Multi-objective optimization of power-to-mobility in decentralised multi-energy systems. *Energy* **2020**, *205*, 117792. [CrossRef]
32. Zhang, Y.; Le, J.; Liao, X.; Zheng, F.; Liu, K.; An, X. Multi-objective hydro-thermal-wind coordination scheduling integrated with large-scale electric vehicles using IMOPSO. *Renew. Energy* **2018**, *128*, 91–107. [CrossRef]
33. Luo, C.; Huang, Y.-F.; Gupta, V. Stochastic dynamic pricing for EV charging stations with renewable integration and energy storage. *IEEE Trans. Smart Grid* **2017**, *9*, 1494–1505. [CrossRef]
34. Wolpert, D.H.; Macready, W.G. No free lunch theorems for optimization. *IEEE Trans. Evol. Comput.* **1997**, *1*, 67–82. [CrossRef]
35. "Kyocera KD325GX-LFB (325 W) Solar Panel", SolarDesignTool. Available online: http://www.solardesigntool.com/components/module-panel-solar/Kyocera/2788/KD325GX-LFB/specification-data-sheet.html (accessed on 4 June 2023).
36. Mouli, G.C.; Bauer, P.; Zeman, M. System design for a solar powered electric vehicle charging station for workplaces. *Appl. Energy* **2016**, *168*, 434–443. [CrossRef]
37. Bhatti, A.R.; Salam, Z. A rule-based energy management scheme for uninterrupted electric vehicles charging at constant price using photovoltaic-grid system. *Renew. Energy* **2018**, *125*, 384–400. [CrossRef]
38. Bukar, A.L.; Tan, C.W.; Lau, K.Y. Optimal sizing of an autonomous photovoltaic/wind/battery/diesel generator microgrid using grasshopper optimization algorithm. *Sol. Energy* **2019**, *188*, 685–696. [CrossRef]
39. Diaf, S.; Diaf, D.; Belhamel, M.; Haddadi, M.; Louche, A. A methodology for optimal sizing of autonomous hybrid PV/wind system. *Energy Policy* **2007**, *35*, 5708–5718. [CrossRef]
40. Bhandari, B.; Poudel, S.R.; Lee, K.-T.; Ahn, S.-H. Mathematical modeling of hybrid renewable energy system: A review on small hydro-solar-wind power generation. *Int. J. Precis. Eng. Manuf. -Green Technol.* **2014**, *1*, 157–173. [CrossRef]
41. Al-Kharbosy, M.E.H. Enhancement Protection and Operation of the Doubly Fed Induction Generator during Grid Fault. Master's Thesis, South Valley University-Qena-Egypt, Qena, Egypt, 2012.
42. Borhanazad, H.; Mekhilef, S.; Ganapathy, V.G.; Modiri-Delshad, M.; Mirtaheri, A. Optimization of micro-grid system using MOPSO. *Renew. Energy* **2014**, *71*, 295–306. [CrossRef]
43. Poompavai, T.; Kowsalya, M. Control and energy management strategies applied for solar photovoltaic and wind energy fed water pumping system: A review. *Renew. Sustain. Energy Rev.* **2019**, *107*, 108–122. [CrossRef]
44. Leone, C.; Longo, M.; Fernández-Ramírez, L.M.; García-Triviño, P. Multi-Objective Optimization of PV and Energy Storage Systems for Ultra-Fast Charging Stations. *IEEE Access* **2022**, *10*, 14208–14224. [CrossRef]

45. Bukar, A.L.; Tan, C.W.; Yiew, L.K.; Ayop, R.; Tan, W.-S. A rule-based energy management scheme for long-term optimal capacity planning of grid-independent microgrid optimized by multi-objective grasshopper optimization algorithm. *Energy Convers. Manag.* **2020**, *221*, 113161. [CrossRef] [PubMed]
46. Maleki, A.; Khajeh, M.G.; Ameri, M. Optimal sizing of a grid independent hybrid renewable energy system incorporating resource uncertainty, and load uncertainty. *Int. J. Electr. Power Energy Syst.* **2016**, *83*, 514–524. [CrossRef]
47. Kaabeche, A.; Ibtiouen, R. Techno-economic optimization of hybrid photovoltaic/wind/diesel/battery generation in a stand-alone power system. *Sol. Energy* **2014**, *103*, 171–182. [CrossRef]
48. Chang, W.-Y. The state of charge estimating methods for battery: A review. *Int. Sch. Res. Not.* **2013**, *2013*, 953792. [CrossRef]
49. RELiON Batteries, Rechargable Deep Cycle Lithium Batteries Rb75 12.8. 2020. Available online: https://relionbattery.com/products/lithium/application/electric-vehicles (accessed on 14 May 2023).
50. Singh, S.; Singh, M.; Kaushik, S.C. Feasibility study of an islanded microgrid in rural area consisting of PV, wind, biomass and battery energy storage system. *Energy Convers. Manag.* **2016**, *128*, 178–190. [CrossRef]
51. Natesan, C.; Ajithan, S.K.; Chozhavendhan, S.; Devendiran, A. Power management strategies in microgrid: A survey. *Int. J. Renew. Energy Res.* **2015**, *5*, 334–340.
52. Barakat, S.; Ibrahim, H.; Elbaset, A.A. Multi-objective optimization of grid-connected PV-wind hybrid system considering reliability, cost, and environmental aspects. *Sustain. Cities Soc.* **2020**, *60*, 102178. [CrossRef]
53. Arancibia, A.; Strunz, K. Modeling of an electric vehicle charging station for fast DC charging. In Proceedings of the 2012 IEEE International Electric Vehicle Conference, Greenville, SC, USA, 4–8 March 2012; pp. 1–6.
54. AL-Hasnawi, D.M.H. *Experimental Study the Effect of Ambient Temperature on the Performance of Photovoltaic System Work with MPPT Charge Controller in Technical Engineering College of Najaf*; Southern Technical University: Basra Governorate, Iraq, 2018.
55. Wirth, H.; Schneider, K. *Recent Facts about Photovoltaics in Germany*; Fraunhofer Ise: Freiburg im Breisgau, Germany, 2015; p. 92.
56. Bartolucci, L.; Cordiner, S.; Mulone, V.; Santarelli, M.; Ortenzi, F.; Pasquali, M. PV assisted electric vehicle charging station considering the integration of stationary first-or second-life battery storage. *J. Clean. Prod.* **2023**, *383*, 135426. [CrossRef]
57. Duman, A.C.; Erden, H.S.; Gönül, Ö.; Güler, Ö. A home energy management system with an integrated smart thermostat for demand response in smart grids. *Sustain. Cities Soc.* **2021**, *65*, 102639. [CrossRef]
58. Alsharif, A.; Tan, C.W.; Ayop, R.; Dobi, A.; Lau, K.Y. A comprehensive review of energy management strategy in Vehicle-to-Grid technology integrated with renewable energy sources. *Sustain. Energy Technol. Assess.* **2021**, *47*, 101439. [CrossRef]
59. Tran, D.-D.; Vafaeipour, M.; El Baghdadi, M.; Barrero, R.; Van Mierlo, J.; Hegazy, O. Thorough state-of-the-art analysis of electric and hybrid vehicle powertrains: Topologies and integrated energy management strategies. *Renew. Sustain. Energy Rev.* **2020**, *119*, 109596. [CrossRef]
60. Wu, X.; Hu, X.; Moura, S.; Yin, X.; Pickert, V. Stochastic control of smart home energy management with plug-in electric vehicle battery energy storage and photovoltaic array. *J. Power Sources* **2016**, *333*, 203–212. [CrossRef]
61. Hussain, M.T.; Sulaiman, D.N.B.; Hussain, M.S.; Jabir, M. Optimal Management strategies to solve issues of grid having Electric Vehicles (EV): A review. *J. Energy Storage* **2021**, *33*, 102114. [CrossRef]
62. Lorestani, A.; Gharehpetian, G.B.; Nazari, M.H. Optimal sizing and techno-economic analysis of energy- and cost-efficient standalone multi-carrier microgrid. *Energy* **2019**, *178*, 751–764. [CrossRef]
63. Bouchekara, H.R.E.-H.; Javaid, M.S.; Shaaban, Y.A.; Shahriar, M.S.; Ramli, M.A.M.; Latreche, Y. Decomposition based multiobjective evolutionary algorithm for PV/Wind/Diesel Hybrid Microgrid System design considering load uncertainty. *Energy Rep.* **2021**, *7*, 52–69. [CrossRef]
64. Ridha, H.M.; Hizam, H.; Mirjalili, S.; Othman, M.L.; Ya'acob, M.E.; Ahmadipour, M. Innovative hybridization of the two-archive and PROMETHEE-II triple-objective and multi-criterion decision making for optimum configuration of the hybrid renewable energy system. *Appl. Energy* **2023**, *341*, 121117. [CrossRef]
65. Abualigah, L.; Diabat, A.; Mirjalili, S.; Abd Elaziz, M.; Gandomi, A.H. The Arithmetic Optimization Algorithm. *Comput. Methods Appl. Mech. Eng.* **2021**, *376*, 113609. [CrossRef]
66. Duman, S.; Kahraman, H.T.; Sonmez, Y.; Guvenc, U.; Kati, M.; Aras, S. A powerful meta-heuristic search algorithm for solving global optimization and real-world solar photovoltaic parameter estimation problems. *Eng. Appl. Artif. Intell.* **2022**, *111*, 104763. [CrossRef]
67. Kahraman, H.T.; Aras, S.; Gedikli, E. Fitness-distance balance (FDB): A new selection method for meta-heuristic search algorithms. *Knowl. -Based Syst.* **2020**, *190*, 105169. [CrossRef]
68. Ridha, H.M.; Gomes, C.; Hizam, H.; Ahmadipour, M.; Heidari, A.A.; Chen, H. Multi-objective optimization and multi-criteria decision-making methods for optimal design of standalone photovoltaic system: A comprehensive review. *Renew. Sustain. Energy Rev.* **2021**, *135*, 110202. [CrossRef]
69. Zeng, H.B.; Liu, X.G.; Wang, W.; Xiao, S.P. New results on stability analysis of systems with time-varying delays using a generalized free-matrix-based inequality. *J. Frankl. Inst.* **2019**, *356*, 7312–7321. [CrossRef]

sustainability

MDPI

Article

Velocity Augmentation Model for an Empty Concentrator-Diffuser-Augmented Wind Turbine and Optimisation of Geometrical Parameters Using Surface Response Methodology

Ngwarai Shambira *, Golden Makaka and Patrick Mukumba

Physics Department, Faculty of Science & Agriculture, University of Fort Hare, Alice 5700, South Africa; medzai9@gmail.com (G.M.); pmukumba@ufh.ac.za (P.M.)
* Correspondence: shamby000@gmail.com

Abstract: Wind energy, renowned for cost-effectiveness and eco-friendliness, addresses global energy needs amid fossil fuel scarcity and environmental concerns. In low-wind speed regions, optimising wind turbine performance becomes vital and achievable by augmenting wind velocity at the turbine rotor using augmentation systems such as concentrators and diffusers. This study focuses on developing a velocity augmentation model that correctly predicts the throat velocity in an empty concentrator-diffuser-augmented wind turbine (CDaugWT) design and determines optimal geometrical parameters. Utilising response surface methodology (RSM) in Design Expert 13 and computational fluid dynamics (CFD) in ANSYS Fluent, 86 runs were analysed, optimising parameters such as diffuser and concentrator angles and lengths, throat length, and flange height. The ANOVA analysis confirmed the model's significance ($p < 0.05$). Notably, the interaction between the concentrator's length and the diffuser's length had the highest impact on the throat velocity. The model showed a strong correlation ($R^2 = 0.9581$) and adequate precision (ratio value of 49.655). A low coefficient of variation (C.V.% = 0.1149) highlighted the model's reliability. The findings revealed a 1.953-fold increase in inlet wind speed at the throat position. Optimal geometrical parameters for the CDaugWT included a diffuser angle of 10°, concentrator angle of 20°, concentrator length of 375 mm ($0.62R_{th}$), diffuser length of 975 mm ($1.61R_{th}$), throat length of 70 mm ($0.12R_{th}$), and flange height of 100 mm ($0.17R_{th}$) where R_{th} is the throat radius. A desirability value of 0.9, close to 1, showed a successful optimisation. CFD simulations and RSM reduced calculation cost and time when determining optimal geometrical parameters for the CDaugWT design.

Keywords: concentrator-diffuser-augmented wind turbine; response surface methodology; computational fluid dynamics; optimisation; wind energy; Betz limit

Citation: Shambira, N.; Makaka, G.; Mukumba, P. Velocity Augmentation Model for an Empty Concentrator-Diffuser-Augmented Wind Turbine and Optimisation of Geometrical Parameters Using Surface Response Methodology. *Sustainability* **2024**, *16*, 1707. https://doi.org/10.3390/su16041707

Academic Editors: Ayman Al-Quraan and Ahmad M. A. Malkawi

Received: 27 December 2023
Revised: 15 February 2024
Accepted: 16 February 2024
Published: 19 February 2024

1. Introduction

Many without electricity worldwide reside in rural areas, and South Africa is not an exception, with an estimated 3.5 million households not connected to the national electricity grid [1]. The Electricity Supply Commission (ESKOM), the main energy supplier, notes that expanding the grid to these rural communities is expensive. Given their remote nature, low population density, and low-income levels, it becomes challenging for ESKOM to recover capital and operating costs solely from tariffs [1,2]. The potential expansion of wind energy into these areas could drive technological advances, contributing to a cleaner energy mix. As a clean and consistently renewable resource, wind energy plays a crucial role in reducing carbon dioxide emissions and mitigating the effects of climate change [3,4]. Although wind turbine technology may be an alternative source of electricity in such areas, its effectiveness is often hindered by low wind speeds. In addition, most commercially available wind turbines were designed for high wind speeds; thus, they do not work efficiently in areas of low wind speeds. Therefore, shrouded wind turbines have been suggested in the literature to augment the power output of wind turbines in low-wind speed regions in order to address this issue [5,6]. Shrouded wind turbines enhance wind energy capture by increasing wind velocity as it passes through the throat region of ducts while retaining the original energy extraction method employed by bare wind turbines [7,8]. Numerous studies suggest that shrouded wind turbines can outperform their bare counterparts, producing more wind power at lower cut-in wind speeds and allowing for extended operation throughout the year [7,9–11].

Additionally, shrouded turbines may surpass the Betz limit, signifying peak efficiency and mitigating energy losses related to tip vortices [10–13]. Shrouded wind turbines offer additional advantages, including noise reduction, reduced risk of blade failure, improved bird safety, and enhanced performance in turbulent environments [14]. Despite being promising for built environments and small-scale applications [14], hub flow separation can impact shrouded turbine performance, hindering the shroud's ability to capture mass [12]. The concentrator-diffuser-augmented wind turbine (CDaugWT) consists of a concentrator, a cylindrical section (throat section) housing the wind turbine, and a diffuser with a flange, as shown in Figure 1.

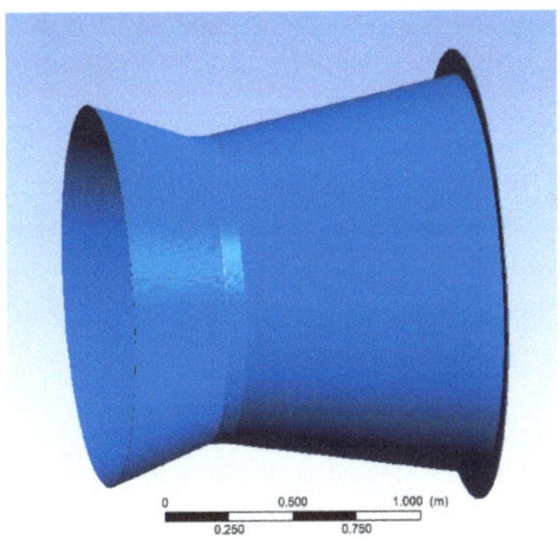

Figure 1. The CDaugWT design.

The CDaugWT significantly enhances wind speed at the rotor or throat section [8,15]. Commonly referred to as a wind lens, the CDaugWT, when equipped with a flange, can generate 2–5 times more power than conventional wind turbines of the same rotor diameter and upstream wind speed [16]. The CDaugWT is advocated for use in low-wind-speed areas and urban environments due to its ability to reduce drag forces and enhance power coefficients.

Several research studies have contributed to the understanding and optimisation of the CDaugWT. In [14], high-fidelity numerical modelling techniques were utilised to investigate six geometric parameters of the wind lens and assess its responsiveness to wind direction. The optimised configuration of the wind lens attained a power coefficient (Cp) of 0.702 at a tip speed ratio of 1, surpassing the Betz limit. This optimisation also narrowed the range of the torque ripple factor, establishing it as a practical and economical design for power generation, especially in the built environment. A wind tunnel study by [17] examined the influence of turbulence on a wind turbine equipped with a wind lens under yaw conditions. They explored the wind lens turbine's performance across varying turbulence intensity levels (10% and 15%) and yaw angles (0° to 30°). The findings indicated that an increase in yaw angle correlated with a decrease in the power coefficient for the turbine equipped with the wind lens. Additionally, the wind lens turbine demonstrated superior performance compared to the turbine without the flow concentrator in turbulent flow and yaw angles of 20° or less.

The research conducted in [18] investigated the impact of a brimmed wind lens on power augmentation in wind turbines. The numerical analysis focused on the low-pressure region and its influence on vortices formed by the brim attached to the diffuser. Comparative numerical predictions were employed to optimise torque augmentation. The results demonstrated a substantial increase in wake formation and vortex strength when incorporating the brimming effect into the diffuser. Furthermore, Ref. [19] utilised computational fluid dynamics (CFD) to analyse a diffuser-augmented hydrokinetic turbine. They predicted performance and flow speeds for various shrouds encasing the turbine, including nozzle, diffuser, and combined nozzle-diffuser turbines. The CFD analysis provided insights for optimising the turbine design to enhance hydrokinetic energy capture. Modified nozzle-diffuser-augmented turbines demonstrated improved efficiency, with a 36.73% increase in the performance of the modified combined model, albeit with effects on torque and power output due to pressure drop and rotor rotation.

Additionally, Ref. [20] conducted a feasibility study on a nozzle-diffuser duct as an energy harvester for Kuroshio currents. It employed computational fluid dynamics simulations in ANSYS Fluent to calculate the drag and mass coefficients of the duct anchored to the seabed. The results indicated that the duct could remain stable 25 m below the sea surface under normal wave conditions and achieve a peak power take-off of 15 kW.

In [21], computational analysis was carried out on three diffuser duct configurations: the straight wind lens, curved wind lens, and vortex generators (VGs) wind lens. Their study investigated the impact of these designs on wind turbine performance. The findings revealed that the curved wind lens and VGs-assisted wind lens setups generated increased turbulence behind the wind turbine, forming low-pressure areas. Notably, the curved diffuser wind turbine achieved the highest power generation in this analysis.

Research efforts have also focused on enhancing the design and geometry of shrouded wind turbines. Another study [22] employed wind lens technology to increase wind velocity at the rotor. It conducted a numerical simulation for a wind turbine model, both with and without a diffuser, optimising the diffuser's geometry for maximum efficiency. The study concluded that incorporating a diffuser with an 8° divergence angle around a wind turbine operated at an optimal wind velocity of 2 m/s resulted in a power coefficient of 0.6, surpassing the Betz limit of 59.3% for horizontal axis wind turbines. The investigation in [23] found that converging–diverging ducts for low-speed wind electricity generation can increase wind velocity by 1.32-fold at a 15° outlet angle and an initial wind speed of 2.45 m/s. However, it identified external backflow zones that hinder internal flow. The authors recommended counteracting this by creating a negative pressure zone outside the duct, enhancing internal flow, and avoiding excessive outlet angles.

Furthermore, Ref. [24] developed a compact wind turbine featuring a variable nozzle-diffuser duct. This innovative design allowed control over the duct's geometry, optimising its performance in various wind conditions. Using numerical simulations and experimental testing, it achieved a significant average augmentation ratio of 39.75% and a rotor speed-up ratio of 53% in low-wind speed tunnel experiments.

Geometric characteristics are key in governing the aerodynamic performance of shrouded wind turbines [14,25]. Therefore, an in-depth exploration of the impact of each parameter is essential [26]. It has been established that the ability of the concentrator to increase velocity is determined by both the angle and length of the concentrator [9,27–30]. Concurrently, the augmentation of velocity for diffusers is significantly influenced by the length of the diffuser and its opening angle [31–34]. Moreover, adding a flange to the diffuser is thought to further improve velocity augmentation [11,35–37]. While the cylindrical section primarily serves as a housing for the turbine, its throat length is crucial as it can influence airflow, thereby affecting velocity augmentation. It is important to recognise that interactions among these parameters impact the overall airflow within the structure of the CDaugWT.

As emphasised by [14], the geometric parameters of the CDaugWT design exhibit diverse impacts on downstream vortices. The concentrator angle stands out as a pivotal factor influencing wind turbine performance, impacting wind velocity enhancement, flow disruption prevention, and pressure recovery through the diffuser. Abdelrazek et al. [30] emphasised the significance of a well-designed concentrator angle, highlighting its role in significantly increasing wind velocity by guiding, collecting, and concentrating airflow within the nozzle. Furthermore, their study reported that wind velocity experiences an increase as the concentrator length increases from $0.15D$ to $0.2D$, with fluctuations beyond $0.2D$, where D represents the throat diameter. In a similar finding, Ref. [32] noted that wind velocity tends to increase with the concentrator length up to a certain point, beyond which it fluctuates.

Regarding the diffuser angle, Ref. [33] indicated that the optimal opening angle falls within the range of 6° for intermediate and long diffusers, while 8° is ideal for short diffusers. The authors of [37] suggested that increasing diffuser length while preventing flow separation within the shroud results in enhanced velocity at the rotor. However, Ref. [36] cautioned against excessive diffuser length, as it may lead to decreased performance due to potential flow instability. The work in [32] highlighted the critical role of throat section length in shrouded wind turbine systems, emphasising its significance in preventing non-uniform velocity distribution before the turbine and potential flow separation behind the rotor, particularly at the diffuser's outlet. Adding a flange to the diffuser exit creates a low-pressure region, drawing more mass flow and causing wind turbines to rotate and automatically align with changes in wind direction [11,26].

Response surface methodology (RSM) is a versatile tool in engineering and optimising wind energy systems. Researchers such as the authors of [38] underscore its crucial role in elucidating parameter-response relationships and facilitating the identification of optimal combinations. Rahmatian et al. [9] illustrate the prowess of RSM in optimising various aspects of wind energy systems, including turbine energy recovery and control parameter optimisation. The authors of [39] recognise RSM's broad applicability, but they also acknowledge its limitations when dealing with nonlinear outputs. The work described in [40] introduces response surface optimisation (RSO) through the use of surrogate models, providing insights into the intricate impact of design parameters on vertical axis wind turbine performance. Stressing the robustness of RSM, Ref. [41] positions it as a benchmark against which machine learning methods in optimisation can be evaluated. Collectively, these perspectives underscore the enduring significance of RSM in the ever-evolving landscape of wind energy optimisation.

A review of studies suggests that the surface response methodology can be effectively combined with other methods in the context of shrouded wind turbines. The authors of [9] employed the response surface methodology (RSM) and genetic algorithms (GA) for the optimisation of a convergent–divergent shroud in wind turbines. In their study, seven parameters are simultaneously investigated: the throat diameter and the length and angle of the duct components (nozzle, diffuser, and flange). The findings indicated a substantial increase in wind speed of up to 2.18 times and a remarkable 3.94-fold improvement in the wind turbine power coefficient. Additionally, the optimised duct configuration led to reduced vortices behind the turbine, resulting in decreased noise levels and dynamic forces. The analysis presented in [15] utilised a multi-objective genetic algorithm to optimise a nozzle-divergent duct for wind turbines. The simultaneous optimisation of turbine and duct components resulted in a significant 12.3% increase in turbine power. This approach led to reduced drag and thrust coefficients, highlighting improved aerodynamic efficiency and structural performance.

Ramayee and Supradeepan [42] conducted an optimisation study on a shrouded wind turbine enclosure using numerical simulations and the design of experiments (DOE) approach. The study identified optimal parameters, including a shroud length-to-diameter (L/D) ratio of 0.4, shroud angle of 9°, flap L/D ratio of 0.2, flap angle of 16°, and a radial distance of $0.2R$. This optimisation resulted in a higher acceleration factor, reduced material volume, and a shorter enclosure length. The authors of [38] optimised a wind turbine with a concentrator and flap using the response surface method (RSM) and Box Behnken experimental design. They conducted a 2D computational fluid dynamics (CFD) analysis and employed the actuator porous disc model. The combined approach achieved a 1.2-fold increase in flow speed at the turbine zone. Bouvant et al. [43] optimised an Archimedes screw turbine using response surface methodology and central composite design (CcompD). The study achieved a predicted Cp of 0.5137.

A research study in [7] utilised response surface methodology (RSM) and computational fluid dynamics (CFD) to optimise duct performance. Employing central composite design (CcompD) for 27 runs, the analysis revealed that optimal design parameters, such as a duct throat diameter of 0.16 m, contraction ratio of 2, and length-to-throat diameter ratio of 1.5, could increase power output up to six times. Numerical data emphasised the substantial impact of well-selected parameters on wind-enhanced power and speed, confirming the efficiency of RSM-enabled CFD simulations in optimising duct performance. Netto et al. [44] conducted an optimisation study on an H-Darrieus vertical axis wind turbine using surrogate-based optimisation with three models. The extreme learning machine surrogate outperformed with the smallest root mean square error at 11.24%, followed by kriging at 17.64% and response surface at 22.17%. The authors of [11] optimised the Archimedean-spiral type wind turbine (ASWT) for off-grid power generation using an evolutionary algorithm with a kriging model and computational fluid dynamics simulations. The optimised shroud design achieved a remarkable Cp of 0.502, a 2.58-fold increase over the bare ASWT's Cp of 0.195 at a tip speed ratio of 2.5. The authors of [41] optimised a U-type Darrieus wind turbine (UDWT) using machine learning (ML) with a back-propagation neural network and three optimisation algorithms (genetic algorithm, particle swarm optimisation, and simulated annealing). The optimised UDWT achieved significant power coefficient improvements, surpassing bare turbines in aerodynamic and structural performance. Kaseb and Montazeri [45] utilised metamodels to optimise converging–diverging ducted openings in high-rise buildings. They demonstrated that such ducts could enhance wind power, with various metamodels yielding power estimates showing up to 153% variation.

While considerable research has explored enhancing wind turbine power output through CDaugWTs, further investigation is required to design a CDaugWT with optimal geometrical parameters. For economic and structural reasons, the CDaugWT structure must have an optimal length with increased velocity augmentation at the wind turbine rotor. Many previous studies overlooked interactions that occur among the parameters affecting throat velocity. This study addresses this gap by employing the surface response methodology, an optimisation approach that considers the interactions between geometrical parameters, thus enhancing the precision of the optimisation process. The study also seeks to develop a velocity augmentation model that considers the relationship between the throat velocity and six geometrical parameters and to identify optimal geometrical parameters for CDaugWT design, utilising computational fluid dynamics simulations carried out in ANSYS Fluent workbench and Design Expert 13 software. This is a novel approach because, according to the authors, to their knowledge, this has never been done. The model is important because it assists in the design process of the CDaugWT by accurately predicting the throat velocity and is also utilised as an objective function in evolutionary optimisation methods when finding optimal geometrical parameters.

The study, however, shares similarities with the work conducted in [7,9]. A contrast arises in selecting geometric parameters, particularly lengths, as outlined in [9]. At the same time, the authors of [7], in their mathematical model, focused on two response outputs (velocity and power output), three parameters, and a design structure, which differ from those of the present study. The present study also seeks to promote using surrogate models and computational fluid dynamics simulations in shroud optimisation, a practice not well-established in current wind energy applications.

Section 2 outlines the methodology employed in the research, introducing the geometry of the CDaugWT structure and outlining the surface response methodology and numerical methodology. The main findings of the study are subsequently discussed in Section 3, and finally, the conclusion is drawn in Section 4.

2. Methodology

The methodology phase of the research focused on the development of the velocity augmentation model and identifying the optimal geometrical parameters of the concentrator-diffuser design that maximised the weighted average throat velocity of a CDaugWT. This involved the application of response surface methodology using the central composite design and numerical methodology to establish the velocity augmentation model. In numerical methodology, computational fluid dynamics (CFD) was employed, utilising ANSYS Fluent, Reynolds averaged Naiver–Stokes equations, and the shear stress transport k-ω turbulence model. These phases are presented in each subsection.

2.1. Surface Response Methodology

The response surface methodology (RSM) was utilised in the present study. RSM is a collection of mathematical and statistical techniques that serve the purpose of designing and constructing empirical models [46–48]. It analyses the impact of various factors and determines the optimal conditions [37,49,50]. An outline of the steps of the response surface methodology is given in Figure 2.

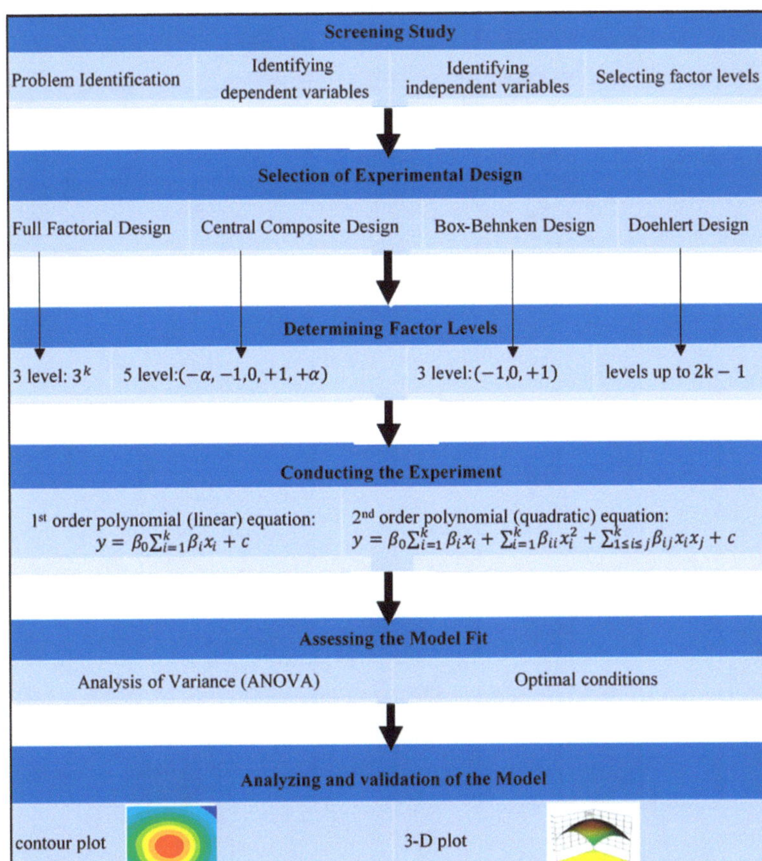

Figure 2. Surface response methodology procedures [48].

In the present study, six geometrical parameters were identified from the literature: diffuser angle (θ_d), concentrator angle (θ_c), concentrator length (L_c), diffuser length (L_d), length of the throat (L_{th}), and flange height (H_f) [51–54]. Figure 3 depicts the geometry of the empty CDaugWT with the design parameters shown.

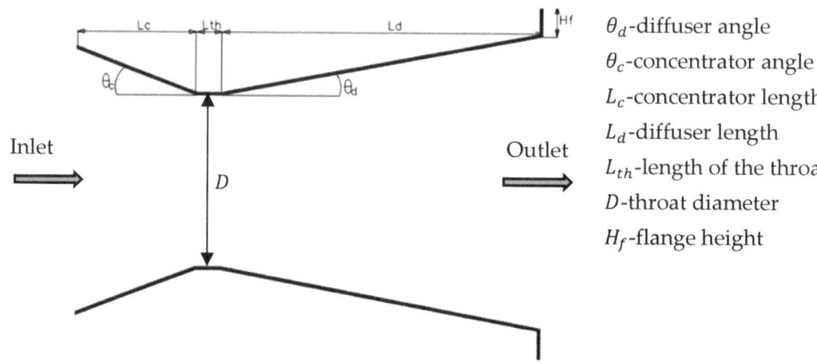

Figure 3. Geometry of the empty CDaugWT with the design parameters.

Research studies have indicated that these parameters affect the velocity augmentation of the diffuser-augmented wind turbine (DaugWT), the concentrator-augmented wind turbine (CaugWT), and the CDaugWT [26,29,55,56]. However, most of these studies focused on analysing each parameter's influence independently, while other parameters were held constant [42,57,58]. The present study investigated the interaction between these parameters and their contribution towards velocity augmentation. The geometrical parameter range chosen based on previous literature studies is shown in Table 1.

Table 1. The parameter ranges.

Parameter	Name	Units	Low	High
A	Diffuser angle (θ_d)	degrees	9.5	10
B	Concentrator angle (θ_c)	degrees	19.5	20
C	Concentrator length (L_c)	mm	350	375
D	Diffuser length (L_d)	mm	950	975
E	length of the throat (L_{th})	mm	65	70
F	Flange height (H_f)	mm	95	100

This study utilised the central composite design (CcompD) with n number of independent variables ($X_1, X_2, \ldots, X_n$), encompassing three segments: the standard 2^n factorial points with their origin at the centre, an axial section with ($2n$) points, and repeated runs at the centre (n_c) to prevent errors. The axial (star) points ($2n$) served for screening analysis, maintaining constant variance of model prediction at points equidistant from the design centre. The parameter n_c represents the number of central points providing test replicates at the centre, crucial for the independent estimation of the experimental error [59]. The number of experiments needed in a CcompD to quantify the effects of six factors is given by Equation (1) [60,61].

$$N = 2^n + 2n + n_c = 2^6 + 2 \times 6 + 10 = 86 \quad (1)$$

In experiments involving six factors, the CcompD included 64 factorial points, 12 axial points, and 10 replicates at central points, resulting in a total of 86 tests (N), as indicated in Equation (1). Once the desired range of variable values was defined, they were coded at five levels to ± 1 for the factorial points, 0 for centre points, and $\pm\alpha$ for the axial points [37,59,62]. The experimental sample schedule used in CFD analysis is given in Table 2.

Table 2. Central composite experimental design (CcompD) samples.

Run	A: θ_d[°]	B: θ_c[°]	C: [L_c/mm]	D: [L_d/mm]	E: [L_{th}/mm]	F: [H_f/mm]
1	10	19.5	350	975	70	100
2	9.8	19.8	362.5	962.5	67.5	97.5
3	9.5	20	350	950	65	100
4	9.5	19.5	350	950	70	95
5	9.5	20	375	950	70	95
6	9.8	19.8	362.5	962.5	67.5	97.5
7	10	20	350	975	65	95
8	9	19.8	362.5	962.5	67.5	97.5
9	9.8	19.8	362.5	927.1	67.5	97.5
10	10	20	350	950	65	95
11	9.8	19.8	362.5	997.9	67.5	97.5
12	9.5	20	375	950	70	100
13	10	20	350	950	65	100
14	9.5	19.5	350	950	65	100
15	9.8	19.8	362.5	962.5	74.6	97.5
16	9.8	19.8	327.1	962.5	67.5	97.5
17	10	19.5	375	975	70	95
18	9.8	19	362.5	962.5	67.5	97.5
19	10	19.5	350	975	65	95
20	9.8	19.8	362.5	962.5	67.5	97.5
21	9.5	19.5	350	975	65	95
22	9.5	19.5	350	975	70	95
23	10	20	350	975	70	100
24	9.5	20	350	950	65	95
25	9.5	19.5	375	950	65	100
26	9.8	19.8	362.5	962.5	67.5	97.5
27	9.5	20	350	950	70	95
28	10	19.5	350	975	70	95
29	9.5	19.5	350	975	70	100
30	9.8	19.8	362.5	962.5	67.5	97.5
31	9.5	20	350	975	70	100
32	9.5	19.5	350	975	65	100
33	9.5	19.5	375	975	65	100
34	9.5	20	350	950	70	100
35	10	20	350	950	70	100
36	9.5	20	375	975	65	95
37	10	20	375	950	70	95
38	10	19.5	375	950	70	95
39	9.5	19.5	375	975	70	100
40	10	19.5	375	975	65	95
41	10	19.5	375	950	65	100

Table 2. *Cont.*

Run	A: θ_d[°]	B: θ_c[°]	C: [L_c/mm]	D: [L_d/mm]	E: [L_{th}/mm]	F: [H_f/mm]
42	10.5	19.8	362.5	962.5	67.5	97.5
43	9.8	19.8	362.5	962.5	60.4	97.5
44	9.8	19.8	362.5	962.5	67.5	97.5
45	9.8	19.8	362.5	962.5	67.5	97.5
46	10	19.5	375	950	65	95
47	9.5	19.5	350	950	70	100
48	9.5	19.5	375	950	70	95
49	9.8	19.8	362.5	962.5	67.5	104.6
50	9.5	20	375	950	65	100
51	10	20	375	950	70	100
52	9.5	19.5	350	950	65	95
53	9.8	19.8	362.5	962.5	67.5	97.5
54	10	19.5	375	975	70	100
55	9.5	20	375	975	70	95
56	9.5	20	375	975	70	100
57	10	19.5	350	950	70	100
58	10	19.5	350	975	65	100
59	9.8	19.8	362.5	962.5	67.5	97.5
60	10	20	375	975	70	95
61	10	19.5	375	950	70	100
62	9.8	19.8	362.5	962.5	67.5	97.5
63	10	20	350	975	65	100
64	10	20	350	975	70	95
65	9.5	19.5	375	975	70	95
66	9.5	19.5	375	950	70	100
67	9.8	20.5	362.5	962.5	67.5	97.5
68	9.5	19.5	375	975	65	95
69	10	20	350	950	70	95
70	10	20	375	975	65	100
71	10	20	375	975	70	100
72	9.5	20	350	975	65	95
73	9.5	20	350	975	65	100
74	9.5	20	350	975	70	95
75	9.8	19.8	397.9	962.5	67.5	97.5
76	10	19.5	350	950	70	95
77	9.5	19.5	375	950	65	95
78	9.8	19.8	362.5	962.5	67.5	90.4
79	10	19.5	375	975	65	100
80	10	20	375	975	65	95
81	10	20	375	950	65	95
82	10	19.5	350	950	65	95

Table 2. *Cont.*

Run	A: θ_d[°]	B: θ_c[°]	C: [L_c/mm]	D: [L_d/mm]	E: [L_{th}/mm]	F: [H_f/mm]
83	9.5	20	375	950	65	95
84	10	19.5	350	950	65	100
85	9.5	20	375	975	65	100
86	10	20	375	950	65	100

The 86 prepared empty concentrator-diffuser geometries (experimental samples), as depicted in Table 2, were subsequently designed and simulated using ANSYS Fluent workbench for 2D CFD analysis. This analysis aimed to obtain throat velocity values (V_{th}), which were rendered dimensionless by dividing them by the inlet free stream velocity (V_∞) to yield the throat velocity augmentation ratio $\left(\frac{V_{th}}{V_\infty}\right)$. The CFD experimental results of $\left(\frac{V_{th}}{V_\infty}\right)$ were subsequently used as input data in Design Expert 13.0 software to establish the velocity augmentation model. This model was further analysed and optimised to obtain the optimal geometrical parameters for the CDaugWT design. The present study, guided by insights from prior research [7,41,43,63] and the fact that the first-order multiple linear model is limited to a 2-level factorial fit due to the strong curvature of the response surface [48], led to the adoption of the more flexible and versatile second-order multiple linear model, or quadratic model. Equation (2) provides this quadratic regression equation for the wind speed augmentation model [7,38,43,64].

$$y = \alpha_0 + \sum_{i=1}^{n} \alpha_i x_i + \sum_{i=1}^{n-1} \sum_{j>1}^{n} \alpha_{ij} x_i x_j + \sum_{i=1}^{n} \alpha_{ii} x_i^2 + \varepsilon_0 \quad (2)$$

where y represents the predicted response variable or objective function (throat wind speed augmentation ratio $\frac{V_{th}}{V_\infty}$, α_0 is the offset term, α_i is the linear effect, α_{ij} is the squared effect, and α_{ii} is the interaction effect. ε_0 represents the total error; this value is usually assumed to follow a normal distribution with a mean of zero.

To assess the validity and significance of the wind speed augmentation model developed using response surface methodology (RSM), this study employed analysis of variance (ANOVA) and a lack-of-fit test [65]. The model's fitness was crucial for optimising the geometrical parameters of the CDaugWT and the velocity augmentation ratio. Various metrics, including F-value, correlation coefficient (R^2), adjusted determination coefficient (adjusted R^2), the p-value for regression parameters, and adequate precision were used to check the model adequacies. The model was considered adequate when the p-value was <0.05, lack-of-fit p-value was >0.05, $R^2 > 0.9$, and adequate precision > 4 [50,66]. The proposed model's 95% confidence level evaluated the significance of independent geometrical parameters on the velocity augmentation ratio. R^2, ranging from 0 to 1, indicated how well the model fit the data, with a value closer to 1 signifying a better fit. Additionally, p-values reflected the significance of parameters on the objective function, with lower p-values indicating a more significant role. The response surface optimiser in Design Expert 13 was used to obtain the optimum design parameters, and a reduced velocity augmentation model was obtained. The response surface (contours and 3D plots) was plotted and analysed at the optimised condition. The obtained values of geometrical parameters were verified with CFD analysis in ANSYS Fluent workbench.

2.2. Numerical Methodology

2.2.1. Governing Equations

The current study involved two-dimensional, incompressible, steady-state simulations. In the present study, only the continuity and momentum equations were considered. The conservation of energy equation was excluded from the analysis due to the negligible heat transfer and the assumption of incompressible flow. The governing equations considered were as follows:

Continuity equation

$$\frac{\partial(\rho u_i)}{\partial x_i} = 0 \tag{3}$$

Momentum equations

$$\frac{\partial(\rho u_i u_j)}{\partial x_j} = -\frac{\partial p}{\partial x_i} + \frac{\partial}{\partial x_j}\left[\rho v\left(\frac{\partial u_i}{\partial x_j} + \frac{\partial u_j}{\partial x_i}\right) - \rho\overline{u_i' u_j'}\right] \tag{4}$$

In Equation (4) $\left(-\rho\overline{u_i' u_j'}\right)$ denotes Reynolds stresses, p, u_i, u_i', and v, respectively, denote mean static pressure, mean velocity, turbulent fluctuation, and kinematic viscosity.

2.2.2. Turbulence Model Equations

The study employed the widely used shear stress transport (SST) k-ω turbulence model, which combines the accuracy of the k-ω model near the wall with the independence of the k-ε model in the far field [67–69]. This model effectively investigates boundary layers, considering factors such as free-stream flow turbulence, pressure gradient, and heat transfer influences, particularly in shrouded wind turbines [15,37,70,71]. The SST k-ω model transitions from near-wall regions to far-from-wall regions, accurately predicting flow separation and shear flows [11]. Notably, it is suitable for adverse pressure gradients, insensitive to free-stream turbulence in the far wake region, and eliminates the need for damping functions [72]. The model choice was based on extensive literature recommendations for aerodynamic purposes, especially in shrouded wind turbines [9,11,12,37,40,42,67–69,72–78].

The transport equations for the SST k-ω turbulence model were as follows:

Turbulent kinetic energy (k)

$$\frac{\partial}{\partial t}(\rho k) + \frac{\partial}{\partial x_i}(\rho k u_i) = \frac{\partial}{\partial x_j}\left[\Gamma_k \frac{\partial k}{\partial x_j}\right] + G_k - Y_k + S_k \tag{5}$$

Specific dissipation rate (ω)

$$\frac{\partial(\rho\omega)}{\partial t} + \frac{\partial(\rho\omega u_i)}{\partial x_i} = \frac{\partial}{\partial x_j}\left(\Gamma_\omega \frac{\partial \omega}{\partial x_j}\right) + G_\omega - Y_\omega + D_\omega + S_\omega \tag{6}$$

where G_k, denotes the generation of turbulence kinetic energy resulting from the mean velocity gradient, which can be computed using the following Equation (7):

$$G_k = -\rho \frac{u_i u_j}{\partial} \frac{\partial u_j}{\partial x_i} \tag{7}$$

The generation of G_ω is given by (8)

$$G_\omega = \alpha \frac{\omega}{k} G_k \tag{8}$$

The value of coefficient α depends on k and ω, and it is determined in such a way that it tends towards unity in the far field regions of the flow.

The terms Γ_k and Γ_ω, represent the effective diffusivity of k and ω, respectively, shown as:

$$\Gamma_k = \mu + \frac{\mu_t}{\sigma_k} \tag{9}$$

$$\Gamma_\omega = \mu + \frac{\mu_t}{\sigma_\omega} \tag{10}$$

where σ_k and σ_ω, are the turbulent Prandtl numbers for k and ω, respectively. μ_t is the turbulent viscosity; Y_k and Y_ω, describe turbulence-induced k and ω dissipation. The cross-diffusion term is controlled by D_ω, while user-defined source terms are stated as S_k and S_ω.

The SST k-ω model is classified as a low Reynolds turbulence model, which means that it is designed for regions where boundary layer effects are significant. To meet this criterion, the height of the first cell within the boundary layer should fall within the viscous sublayer. A recommended guideline is to ensure that the y^+ value of the mesh on the wall surfaces is approximately 1 [12,68].

This dimensionless wall distance, y^+, is given by the following equation:

$$y^+ = \frac{u^* y}{v} \tag{11}$$

where u^* stands for the friction velocity, y for the distance between the first node and the wall, and v for the kinematic viscosity [12,77].

2.2.3. Computational Domain and Boundary Conditions

The CDaugWT shroud was modelled using the design modeller in ANSYS Fluent. In order to reduce the computational time, a 2D axisymmetric domain was employed for the study [79]. The dimensions of the computational domain are illustrated in Figure 4 as a function of the throat diameter (D). The domain surrounding the CDaugWT shroud had a rectangular shape with 15D length and 5D height. The inlet and outlet of the domain were positioned at a distance of 5D upstream and 10D downstream of the CDaugWT shroud, respectively.

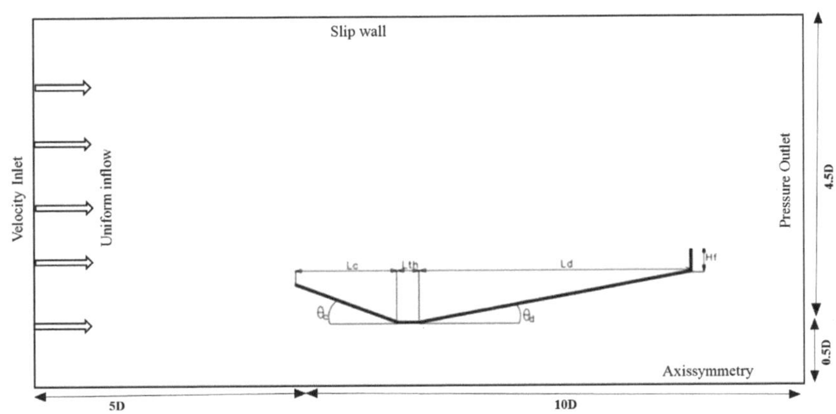

Figure 4. Computational domain dimensions.

2.2.4. Solver Settings and Boundary Conditions

Before conducting the fluent analysis, the 2D computational mesh was prepared using the pressure-based solver and the absolute velocity formulation. A single solver processor was utilised. The fluid medium employed was air, with a density of 1.225 kg/m^3 and a viscosity of 1.7894×10^{-5}kg/ms. Furthermore, the boundary conditions were defined with the inlet area designated as a velocity inlet, set at a velocity of 2 m/s, and the outlet area configured as a pressure outlet set at 0 Pa [80]. The wall boundary was established as a slip condition with a zero shear boundary condition, and the CDaugWT shroud was specified as a no-slip boundary [81]. The pressure-velocity coupling was achieved using a coupling scheme. The governing equations were solved using the second-order upwind method [68]. The chosen turbulence model was the SST k-ω, described by Equations (5) and (6). This model combines elements of the *k*-ε and *k*-ε turbulence models, as previously mentioned. The current study employed a root mean square (RMS) residual value of 10^{-6} as the stopping criterion for continuity, momentum, velocity components, and turbulence equations [64,68,76]. The solution converged after approximately 121 iterations.

2.2.5. Mesh Generation, Quality, and Grid Independence Test

The study employed the ANSYS Fluent 2022R1 software meshing tool for model creation and mesh generation. An unstructured quadrilateral mesh consisting of 84,388 cells and 85,096 nodes was utilised. To satisfy the required condition of $y^+ = 1$ for the SST k-ω turbulence model, the height of the first layer was set to 1.1 mm. Specifically, the wall boundaries of the shroud were equipped with ten inflation layers, each with a growth rate of 1.2, as illustrated in Figure 5. The mesh size was gradually increased to achieve solution convergence while avoiding abrupt increases in mesh size, because they can hinder solution convergence [82].

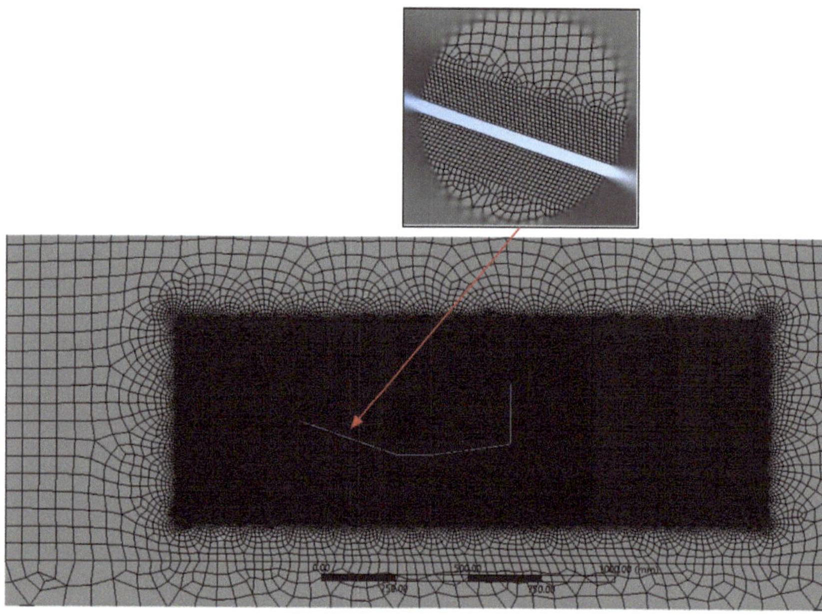

Figure 5. Computational meshing and inflation layers on the shroud.

Furthermore, the assessment of mesh quality considered aspects such as aspect ratio, orthogonal quality, and skewness [68,83]. To ensure solution convergence, strict criteria were applied, maintaining a maximum aspect ratio of 25 or less and a maximum skewness of 0.9, even for intricate geometries [82]. ANSYS recommends a maximum skewness ratio below 0.95 for optimal mesh quality, categorising average skewness ratios as excellent (0–0.25), very good (0.25–0.50), good (0.50–0.80), acceptable (0.80–0.94), poor (0.95–0.97), and unacceptable (0.98–1.00) [84]. This study's maximum skewness ratio was 0.7392, well below the recommended threshold, and the average skewness ratio was 0.0827, indicating excellent mesh quality. The low skewness enhanced simulation accuracy and stability, confirming the mesh's perfection. Orthogonal quality, another crucial parameter, is deemed excellent between 0.95 and 1.00, very good between 0.70 and 0.95, good between 0.20 and 0.69, acceptable between 0.10 and 0.20, poor between 0.001 and 0.10, and unacceptable below 0.001 [84]. Meeting the recommended minimum orthogonal quality rate above 0.1, the simulation's mesh exhibited a minimum orthogonal quality rate above 0.167. Furthermore, the average orthogonal quality rate, indicating perfect quality, was 0.9805 in this simulation. Table 3 shows mesh parameters.

Table 3. Details of the mesh parameters.

Mesh Metric	Element Quality	Aspect Ratio	Skewness	Orthogonal Quality
Min	0.2719	1	1.3508×10^{-10}	0.50298
Max	0.9995	5.1535	0.7392	1
Average	0.9246	1.0822	0.0827	0.9805
Standard Deviation	0.1089	0.1422	0.1163	0.0487

Conducting a mesh independence study is crucial to ensure that the obtained solution remains unaffected by mesh resolution or cell size [68]. In this study, a mesh independence analysis aimed to enhance accuracy by minimising the impact of cell size on computational results while reducing costs. For the grid independence test, eight different grid sizes were employed. The number of grid elements was gradually increased from 30,159 to 198,365 until minimal changes were observed in the results (Table 4 and Figure 6). The findings revealed that from grid number 5, comprising 84,388 elements, throat velocity deviated by less than 0.29%. Considering mesh independence and computational efficiency, grid number 5 was determined as the optimal grid size.

Table 4. Study of mesh independency.

Grid No	Number of Elements	Average Throat Velocity (V_{th})	Relative Difference (%)
1	30,159	2.7854613	-
2	40,238	2.7895242	0.146
3	49,663	2.7918505	0.083
4	66,047	2.7939905	0.077
5	84,388	2.7858975	0.290
6	120,840	2.7858098	0.003
7	192,859	2.7864958	0.025
8	198,365	2.7863879	0.004

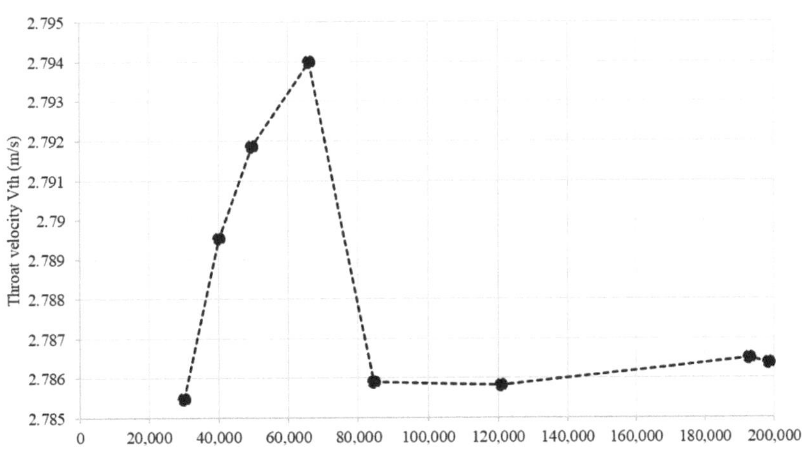

Figure 6. Mesh independence study for shroud (2D mode).

3. Results and Discussion

3.1. CFD Analysis

The outcomes and discussions concerning the performance of a CDaugWT structure under six geometrical parameters are presented and explored. The design's performance was influenced by the diffuser angle, concentrator angle, length of the concentrator, length of diffuser, length of throat, and flange height. Utilising ANSYS Fluent workbench, 86 randomly designed runs based on response surface methodology (RSM) were executed to optimise the area-averaged throat velocity and maximum velocity inside the flanged CDaugWT design analysed in two dimensions. Figures 7–9 provide details for each case of the 86 runs: the area-weighted averaged throat velocity, maximum velocity, and throat velocity augmentation ratio. The values of the six parameters used in each of the 86 simulated runs in two-dimensional (2D) ANSYS Fluent are listed in Table 2. Figure 7 shows that the maximum velocity was recorded in case run 67, with a velocity of 5.48 m/s, while the lowest value for maximum velocity was recorded in case run 47 at 5.23 m/s. The average maximum velocity for all 86 runs in Figure 7 was 5.35 m/s. Correspondingly, the average maximum velocity augmentation ratio $\left(\frac{V_{th}}{V_{\infty}}\right)$ was determined as 2.68, given an inlet velocity of 2 m/s. This finding suggests that the inlet velocity increased approximately 2.68 times compared to the initial velocity. Notably, this value is slightly higher than the 2.22 times reported in a study conducted by [9].

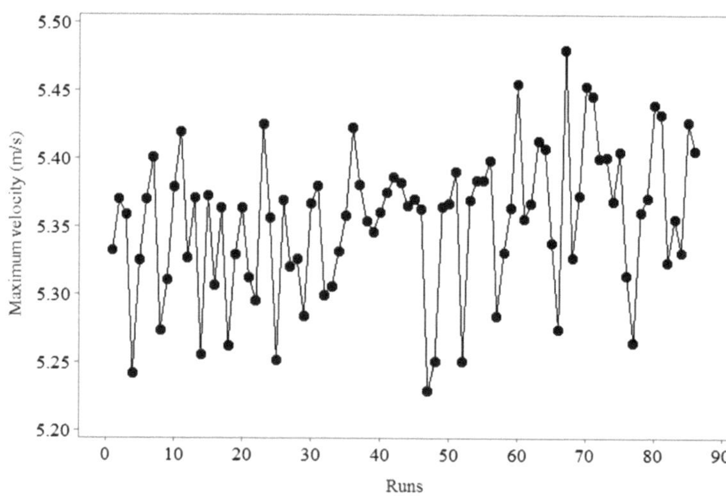

Figure 7. Maximum velocities for each case of the 86 runs.

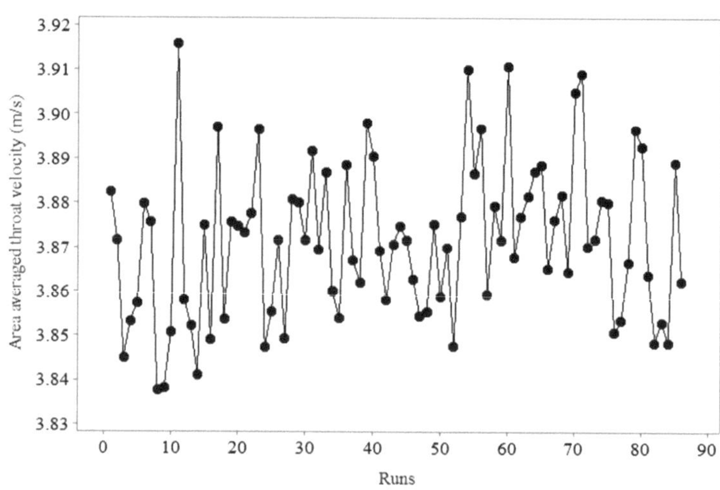

Figure 8. Area weighted averaged throat velocities for each case of the 86 runs.

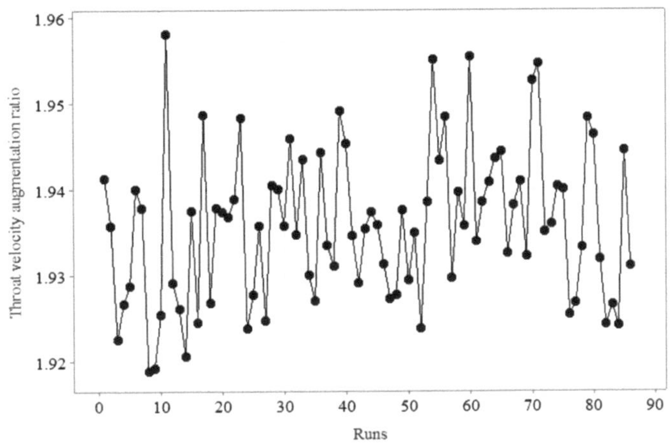

Figure 9. Throat velocity augmentation ratio for each of the 86 runs.

Figure 8 shows that the highest velocity, recorded in case run 11 at 3.92 m/s, corresponds to a throat velocity augmentation ratio of 1.96, as illustrated in Figure 9. The area-weighted averaged throat velocities in Figure 8 and their corresponding velocity augmentation ratios in Figure 9 were used to develop the model, as they represent a larger throat section area than the maximum velocity values depicted in Figure 7.

Validation of CFD Model

To validate the CFD model, we referenced [85] and adopted the same computational domain and baseline shroud geometry presented in Figure 10. The mesh structure employed for validation can be seen in Figure 11. The CFD model's validation involved comparing the graph of U/Uo versus x/D with the results of Abe and Ohya, as illustrated in Figure 12, where U, Uo, x, and D represent the free stream velocity, streamwise velocity, streamwise coordinate, and inlet diameter of the diffuser, respectively. Notably, the simulations indicated that the highest value of U/Uo was 1.29% lower than the results of Abe and Ohya. This small deviation is likely attributable to minor shroud geometry and mesh structure differences but remains within acceptable margins.

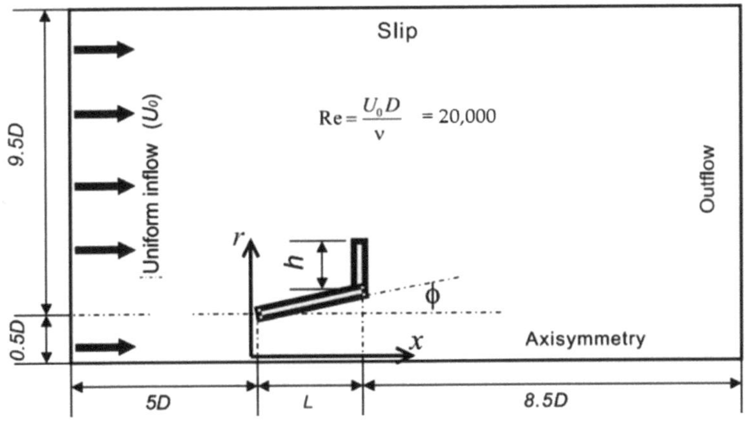

Figure 10. Reference study's computational domain dimensions [85].

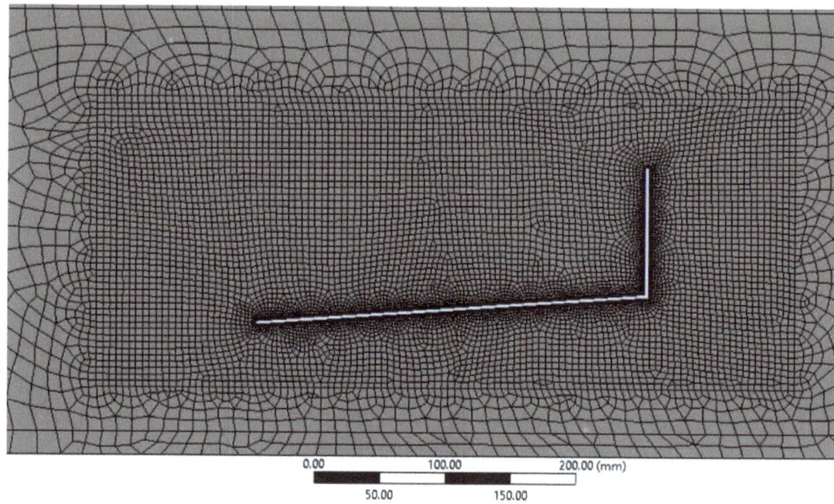

Figure 11. Mesh structure.

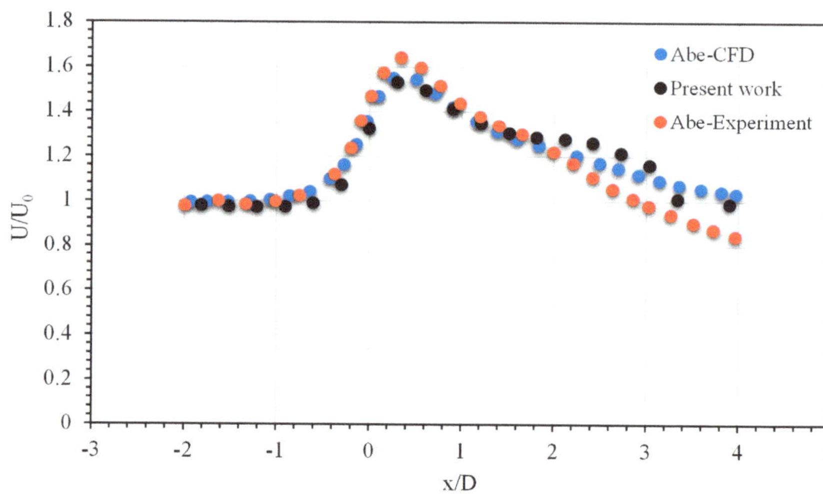

Figure 12. Velocity profile along the central axis.

3.2. ANOVA Analysis

In the present study, response surface methodology (RSM) utilising the central composite design method was employed to investigate the interaction of geometrical parameters and their effects on the velocity augmentation ratio. Design Expert 13 software was used to analyse empirically collected data and develop a second-order surface response quadratic model by estimating model component coefficients. The coefficients for the model components were determined through one-way ANOVA. Additionally, ANOVA was applied to evaluate the adequacy of results concerning input parameters, with a model deemed significant if it had a *p*-value below 0.05. The model's validity was assessed using the coefficient of regression (R^2), and its statistical significance was measured by the F-value.

Furthermore, the study examined pure errors and residuals at repeated points through the lack-of-fit test; 2D and 3D response surface plots were also generated. Additionally, a numerical optimisation method with the desirability function in Design Expert 13 software was carried out to determine optimal geometrical parameters and confirm the accuracy of the velocity augmentation ratio model in predicting the desired results. Finally, CFD analysis was performed with the obtained optimal set of geometrical parameters to validate the throat velocity predicted by the developed velocity augmentation ratio model.

3.3. Mathematical Model Fitting and Assessment

The RSM proposed a reduced quadratic model that relates the velocity augmentation ratio to the six independent geometrical parameters. The model summary statistics for the velocity augmentation ratio are given in Table 5. The reduced quadratic model was not aliased and was adequately significant to represent the correlation between the velocity augmentation ratio and its six independent geometrical parameters. It is on this basis that the model was selected for model fitting.

Table 5. Summary of the model.

Source	Std. Dev.	R^2	Adjusted R^2	Predicted R^2	PRESS	
Linear	0.0030	0.8905	0.8822	0.8642	0.0009	
2FI	0.0030	0.9128	0.8842	0.8684	0.0009	
Quadratic	0.0022	0.9581	0.9387	0.8743	0.0008	Suggested
Cubic	0.0021	0.9785	0.9430	−1.1482	0.0144	Aliased

The wind speed augmentation model $\left(\frac{V_{th}}{V_\infty}\right)$, in terms of coded factors, is given by Equation (12), and in terms of actual factors is given by Equation (13).

$$\begin{aligned}\frac{V_{th}}{V_\infty} = &\ 1.93782 + 0.001748310\theta_d + 0.0008569080\theta_c + 0.00306857L_c \\ &+0.00758128L_d + 0.00153922L_{th} + 0.000909229H_f \\ &+0.0006690890\theta_d L_c + 0.0007775L_c L_d + 0.000452187L_d L_{th} \\ &-0.001367210{\theta_d}^2 - 0.0004387250{\theta_c}^2 - 0.000500042{L_c}^2\end{aligned} \quad (12)$$

$$\begin{aligned}\frac{V_{th}}{V_\infty} = &-0.752974 + 0.355948\theta_d + 0.280702\theta_c - 0.00431128L_c \\ &-0.00217402L_d - 0.0133117L_{th} + 0.000363692H_f \\ &+0.000214108\theta_d L_c + 4.976 \times 10^{-6} L_c L_d + 1.447 \times 10^{-5} L_d L_{th} \\ &-0.021875\theta_d^2 - 0.00701959{\theta_c}^2 - 3.20027 \times 10^{-6} {L_c}^2\end{aligned} \quad (13)$$

where V_{th} is the throat velocity, V_∞ is the inlet upstream velocity, θ_d is the diffuser angle, θ_c is the concentrator angle, L_c is the length of the concentrator, L_d is the length of the diffuser, L_{th} is the length of the throat, and H_f is the flange height.

The velocity augmentation serves as the response, and θ_d, θ_c, L_c, L_d, L_{th}, and H_f, represent the coded terms for the examined parameters. The value and sign of each coefficient of the response parameters indicate the nature of the effects, with a negative sign suggesting antagonistic effects and a positive sign signifying synergistic effects. In accordance with the proposed correlation in Equation (12), the relative significance of the independent parameters is as follows: θ_d (diffuser angle) with a coefficient value of 0.00174831, θ_c (concentrator angle) with a coefficient value of 0.000856908, L_c (length of the concentrator) with a coefficient value of 0.00306857, L_d (length of the diffuser) with a coefficient value of 0.00758128, L_{th} (length of the throat) with a coefficient value of 0.00153922, and H_f (flange height) with a coefficient value of 0.000909229. This confirms that all of the parameters had a positive effect on the velocity augmentation. Additionally, the maximum increasing impact of the dependent factor was 0.0007775, related to the interaction between L_c and L_d parameters. The coded parameters in the quadratic model were beneficial for predicting the relative significance of the factors by comparing the coefficients of the factors [86]. By default, the factors' low and upper levels were coded as −1 and +1, respectively [66]. The coded equations were useful for understanding the relative effects of the parameters by comparing the coefficients of the factors [86]. In contrast, Equation (13) with actual factors specified in original units was used for predictions in the original factor units. Unlike the coded Equation (12), it was not suitable for determining the relative impact of factors, as coefficients were scaled for unit compatibility, and the intercept was not centred in the design space. The findings presented in Figure 13 reveal a consistent order of influence on velocity augmentation, as observed in the sequence: D > C > A > E > F > B, aligning with the patterns described in Equation (12). Notably, the length of the concentrator and the length of the diffuser emerged as the most influential factors, exerting the most significant impact on velocity augmentation.

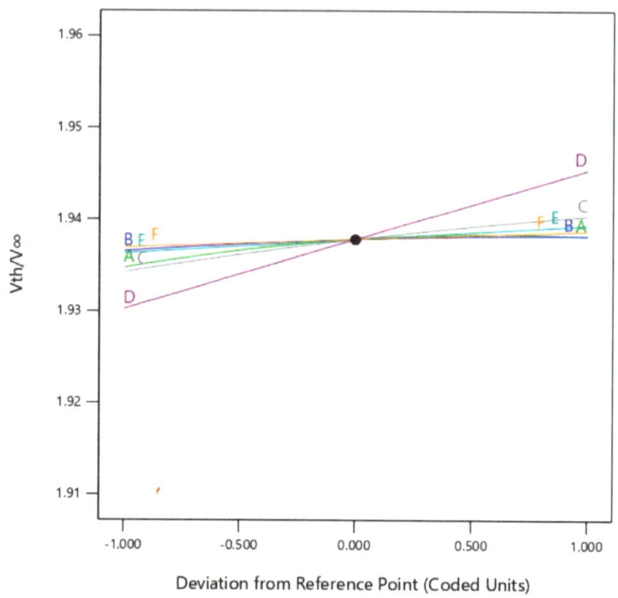

Figure 13. The perturbation plot for velocity augmentation within the design space, showing parameters such as (A) diffuser angle, (B) concentrator angle, (C) length of concentrator, (D) length of diffuser, (E) length of throat, and (F) flange height.

The analysis of model fitness included a lack-of-fit test and analysis of variance (ANOVA). The ANOVA results in Table 6 revealed a calculated p-value of <0.0001, signifying the statistical significance (p-value < 0.05) of the quadratic model with a low probability of error. The p-values less than 0.0500 in the ANOVA indicated the significance of model terms A, B, C, D, E, F, AC, CD, A^2, B^2, and C^2 in this study. For the velocity augmentation model, the maximum R^2 and adjusted R^2 were 0.9581 and 0.9387, respectively, with a standard deviation of 0.0022. The R^2 value of 0.9581 (Table 5) indicated that independent parameters accounted for 95.81% of the variation in the velocity augmentation ratio. A quadratic model with an R^2 value exceeding 0.9 has the strongest positive correlation [66,87]. Adequate precision, measured by a ratio value of 49.655, indicated a satisfactory signal-to-noise ratio, considering a ratio of adequate precision greater than four is acceptable [7,88]. Model significance and proper variable fitting were affirmed by the F-value of 106.67, with only a 0.01% probability that such a high F-value could arise due to noise. The non-significant lack of fit (p-value = 0.0866) can be used to navigate design spaces with a low probability of error [86]. Furthermore, the low coefficient of variation (C.V.% = 0.1149) and the good agreement between predicted and actual velocity augmentation values in Figure 14A suggest the experiment's reliability.

Table 6. Analysis of variance for velocity augmentation model.

Source	Sum of Squares	df	Mean Square	F-Value	*p*-Value	
Model	0.0063	12	0.0005	106.67	<0.0001	significant
$A\text{-}\theta d$	0.0003	1	0.0003	50.98	<0.0001	
$B\text{-}\theta c$	0.0001	1	0.0001	12.25	0.0008	
$C\text{-}Lc$	0.0008	1	0.0008	152.04	<0.0001	
$D\text{-}Ld$	0.0046	1	0.0046	929.89	<0.0001	
$E\text{-}Lth$	0.0002	1	0.0002	38.37	<0.0001	
$F\text{-}Hf$	0.0001	1	0.0001	13.39	0.0005	
AC	0.0000	1	0.0000	5.84	0.0182	
CD	0.0000	1	0.0000	7.82	0.0066	
DE	0.0000	1	0.0000	2.65	0.1082	
A^2	0.0003	1	0.0003	54.09	<0.0001	
B^2	0.0000	1	0.0000	5.57	0.0209	
C^2	0.0000	1	0.0000	5.85	0.0180	
Residual	0.0004	73	4.947×10^{-6}			
Lack of Fit	0.0003	64	5.320×10^{-6}	2.32	0.0866	not significant
Pure Error	0.0000	9	2.294×10^{-6}			
Cor Total	0.0067	85				

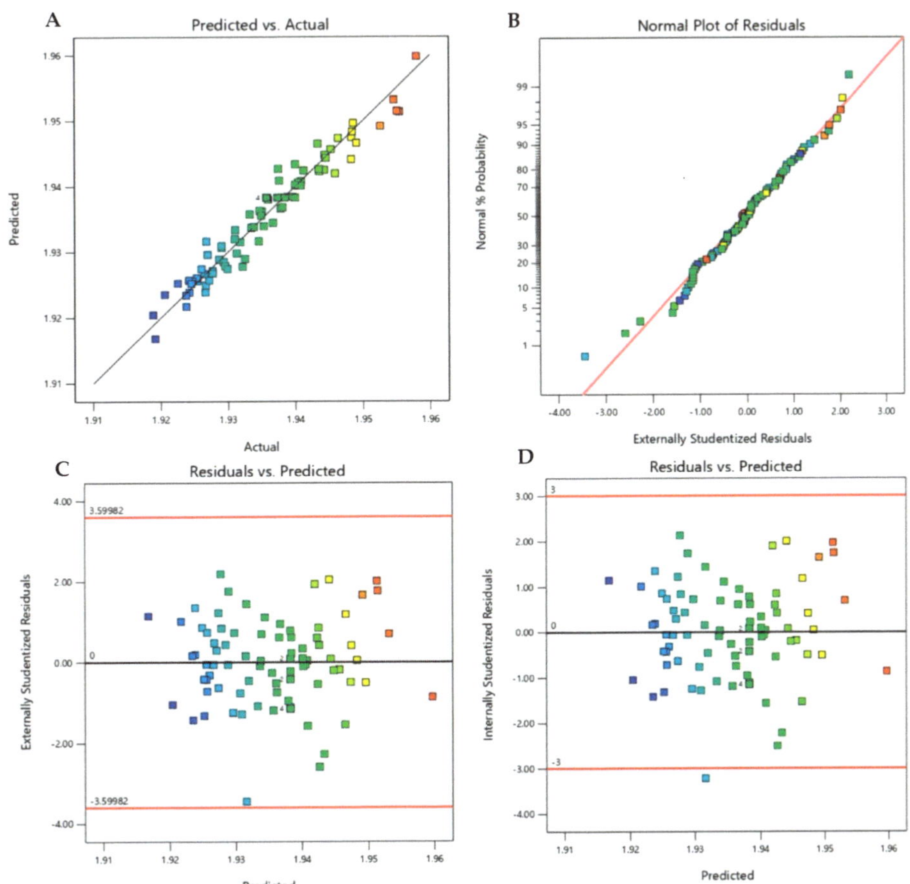

Figure 14. Diagnostic plots of the model: (**A**) predicted vs. actual, (**B**) normal % probability vs. externally studentised residuals, (**C**) externally studentised residuals vs. predicted, and (**D**) internally studentised residuals against predicted.

3.4. Regression Analysis

Figure 14A shows a plot of the actual value against the predicted value for velocity augmentation, in which the points were randomly placed on a straight line. These results confirmed that the predicted and actual values were in good agreement with the high acceptability of the models and, therefore, can be utilised to analyse and predict the velocity augmentation [7,86]. Figure 14B depicts the normal probability chart of the studentised residuals utilised to assess the adequacy and suitability of the velocity augmentation model. As shown in the graph, the points form a straight line, affirming that the errors were distributed normally with a mean of zero and a fixed value, thereby confirming the model's adequacy [37,62,65,86,89]. Similar to the findings of [7], the results show no evidence to dispute any independence violations or hypotheses of permanent variance. In Figure 14C, a random distribution of points is displayed, varying up and down the x-axis within the range of −3.59982 to +3.59982, enclosed by the red line, without exhibiting any noticeable trends. This observation serves as an assessment of the adequacy and reliability of the velocity augmentation model [65]. This scenario suggests that the velocity augmentation model is reasonably exempt from any violations of the independence or constant variance assumptions [86]. In Figure 14D, a chart displays internally studentised residuals plotted against the predicted velocity augmentation ratio. The residuals exhibit a random distribution within the range of +3.00 to −3.00, suggesting that the velocity augmentation model effectively captures the relationship between the independent geometrical parameters and the velocity augmentation ratio [59,90].

3.5. Interaction Effects of Geometrical Parameters on Velocity Augmentation

The surface plots (three-dimensional and contour) for the results achieved are shown in Figures 15–17, where the data demonstrate the combined effect of the geometrical parameters. The interaction between the two geometrical parameters (length of diffuser and length of concentrator) on the velocity augmentation ratio was significant compared to the other parameters in Figures 15 and 17. It must noted that the concentrator length and diffuser angle in Figure 15 interact in the same way that the diffuser length and concentrator length do in Figure 16. In Figure 15, the contour and response surface plots illustrate the velocity augmentation ratio as a function of concentrator length and diffuser angle. The results reveal that as the concentrator length increased with an increase in the diffuser angle, the velocity augmentation ratio also increased. However, once an optimal diffuser angle was reached, as depicted in Figure 15, the velocity augmentation ratio started to decrease. The interaction of these parameters was weak, as indicated by the ANOVA analysis. Consistent with these findings, previous studies such as [31,91] observed that an increase in the diffuser angle leads to an increase in both velocity inside the diffuser and flow separation. Subsequently, further increasing the diffuser angle starts to decrease the velocity inside the diffuser.

The curved slope of the response surface plot in Figure 16 indicates that the length of the diffuser had a greater influence on the velocity augmentation than the length of the concentrator and diffuser angle in Figure 15. This indicates its influential role in increasing the throat velocity in a CDaugWT design. As shown in Figure 16, a rise in the length of the diffuser with the increase in the length of the concentrator was responsible for the increase in the velocity augmentation ratio. However, it must be noted that once an optimal concentrator length was reached, any further increase in length started to impact the flow negatively, hence the decrease in the velocity augmentation ratio, as shown in the graph [32,92]. In addition, Ref. [37] asserted that an extended concentrator length intensifies flow separation in the diffuser zone with CDaugWT design, causing a decrease in the velocity augmentation ratio, as shown in Figure 16.

In Figure 17, the velocity augmentation ratio is depicted as a function of the length of the throat and length of the diffuser at constant values (diffuser angle = 9.75°, concentrator angle = 19.75°, length of concentrator = 362.5 mm, flange height = 97.5 mm). By keeping the other parameters constant at those given values, the maximum velocity augmentation ratio showed a positive correlation with the length of the diffuser. Still, it remained constant in relation to the length of the throat. This observation is not surprising, as this combination in the ANOVA analysis had a *p*-value of 0.1082, greater than 0.1, indicating that the model term was classified as insignificant. However, it was required to support the hierarchy in the model. On the other hand, it is interesting to note that the increase in velocity augmentation ratio with the increase in diffuser length was reported in previous publications [33,36,54,91,93,94].

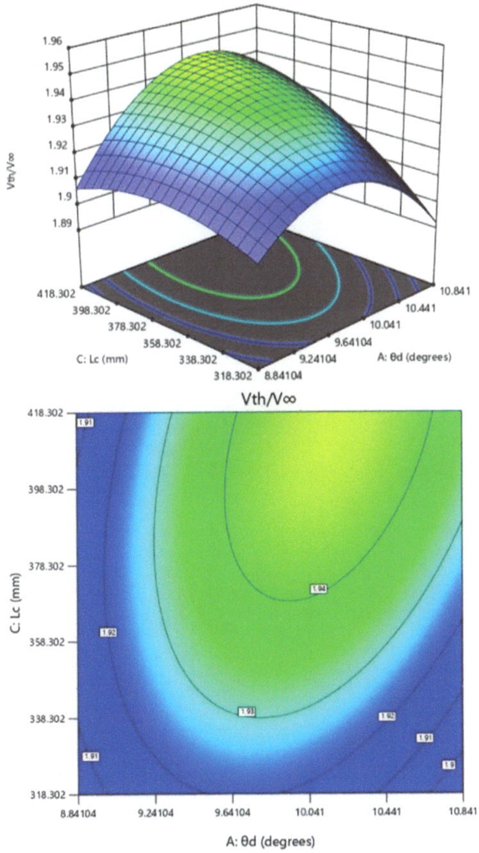

Figure 15. The contour and response surface plots of velocity augmentation ratio as a function of concentrator length and diffuser angle.

3.6. Optimisation of Design Parameters through the Desirability Function (SRO)

The optimisation process in this study employed the desirability function, with values ranging from 0 (undesirable) to 1 (desirable). Utilising a numerical optimisation method (SRO), the procedure identified points that maximise the desirability function [66]. Numerical optimisation allows for selecting desirable values for input parameters and responses. Various input optimisations, such as range, maximum, minimum, target, and none (for responses), can be chosen to establish an optimised output value under specific conditions [7,88]. In this study, specific ranged values were assigned to the input variables, while the objective was to achieve a maximum response. The optimisation ramp in Figure 18 and Table 7 displays the optimal values, with a maximum velocity augmentation ratio of 1.953 and a desirability function value of 0.9. The optimal projected geometrical parameters were as follows: (A) diffuser angle = 10°, (B) concentrator angle = 20°, (C) length of concentrator = 375 mm, (D) length of diffuser = 975 mm, (E) length of throat = 70 mm, and (F) flange height = 100 mm. The optimisation proved successful, given the proximity of the desirability value to 1 (0.9). A confirmatory CFD analysis yielded a throat velocity of 3.898 m/s with the optimal geometrical parameters values obtained from the desirability function (SRO) utilised. This result is slightly below the model-predicted throat velocity of 3.906 m/s. This demonstrates the suitability and accuracy of the developed model.

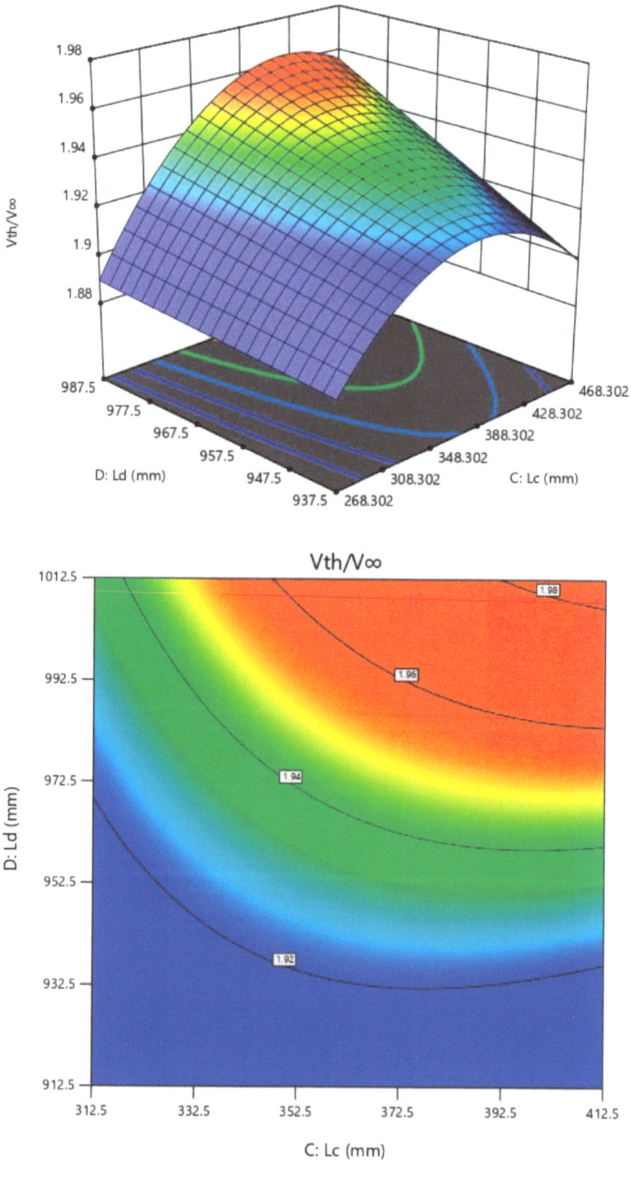

Figure 16. The contour and response surface plots of velocity augmentation as a function of diffuser length and concentrator length.

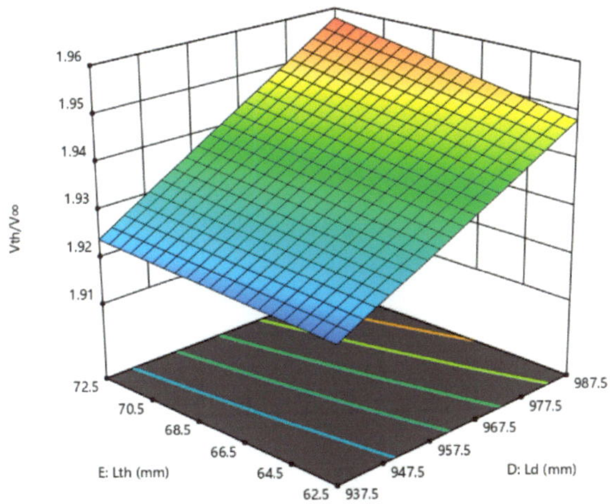

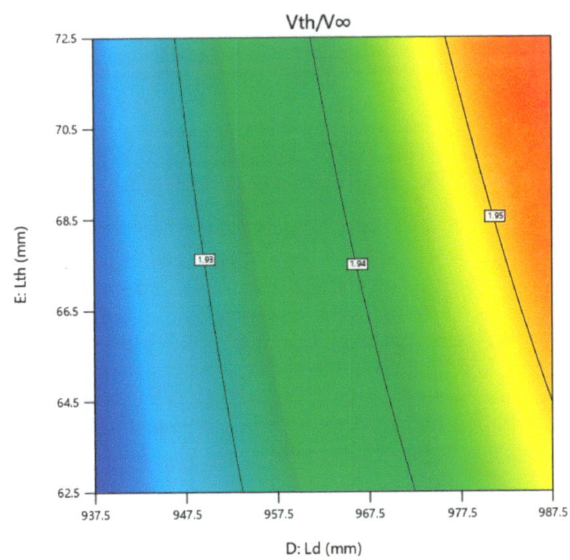

Figure 17. The contour and response surface plots of velocity augmentation as a function of the length of the throat and the diffuser length.

Table 7. Ideal ranges for response and designated parameters.

Parameter	Target	Minimum	Maximum	Optimised Value
Diffuser angle (θd)	In range	9.50	10.00	10°
Concentrator angle (θc)	In range	19.50	20.00	20°
Concentrator length (Lc)	In range	350.00	375.00	375
Diffuser length (Ld)	In range	950.00	975.00	975
Length of the throat (Lth)	In range	65.00	70.00	70
Flange height (Hf)	In range	95.00	100.00	100
Velocity augmentation ratio ($\frac{V_{th}}{V_\infty}$)	Maximum	1.91887	1.95795	1.95313
Desirability	-	0	1	0.9

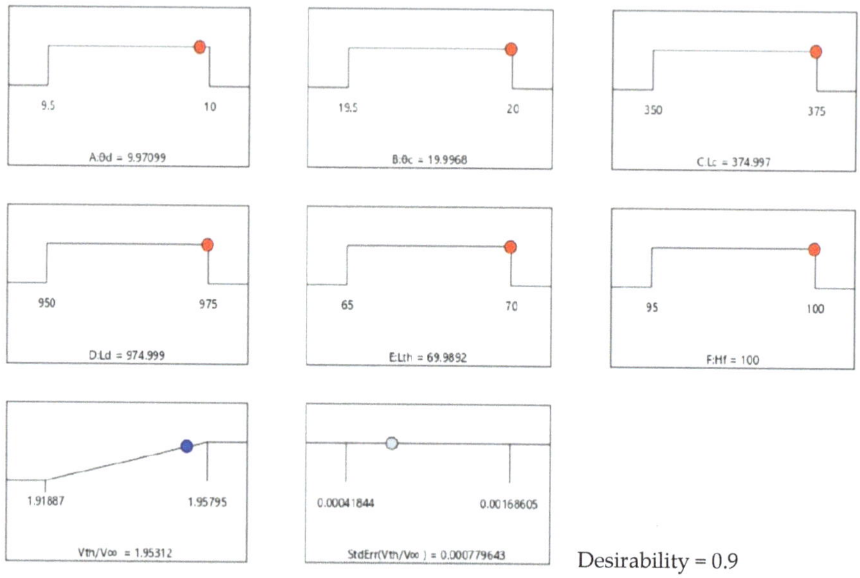

Figure 18. Desirability ramp for adjustment of settings for maximum available velocity augmentation ratio optimised values for (A) diffuser angle, (B) concentrator angle, (C) length of concentrator, (D) length of diffuser, (E) length of throat, and (F) flange height.

3.7. Validation of the Velocity Augmentation Model with CFD Results

Table 8 displays the optimal geometrical parameter values, specifically the throat velocity V_{th} values obtained from the surface response optimisation algorithms (SRO) alongside $CFDV_{th}$, which represents the throat velocity values from CFD analysis in ANSYS Fluent, utilising the optimal geometrical parameter values from the optimisation method (SRO). The results indicate a consistent alignment of the SRO outcomes with the values derived from CFD analysis. The percentage difference of 0.2% was obtained, highlighting the developed model's excellent accuracy and reliability in predicting the CDaugWT design's throat velocity.

Table 8. Validation of model-predicted values using CFD analysis.

Method	θ_d	θ_c	L_c	L_d	L_{th}	H_f	V_{th}	$CFDV_{th}$	% Difference
SRO	10	20	375	975	70	70	3.906	3.898	0.2

4. Conclusions

This study developed a velocity augmentation model using RSM with CcompD, illustrating the relationship between throat section velocity and six geometrical parameters of the CDaugWT design. The model accurately predicted the simulation data with an R^2 accuracy of 0.9581. Our primary conclusions are as follows:

(1) The non-aliased reduced quadratic model is significant in representing the relationship between the velocity augmentation ratio and its six independent geometrical parameters. The choice of this model for fitting was grounded in these considerations.

(2) The *p*-value for the velocity augmentation ratio model was below 0.05, indicating its significance. On the other hand, the model's F-value was 106.67, suggesting a less than 0.01% probability that noise could result in such a substantial F-value. Moreover, the low coefficient of variation (C.V.% = 0.1149) highlighted a favourable agreement between predicted and actual velocity augmentation values.

(3) The coded Equation (12) confirmed the relative importance of the independent parameters and that each parameter positively influenced velocity augmentation. Notably, the concentrator length and diffuser length emerged as the most impactful factors, exerting the greatest influence on velocity augmentation.
(4) The optimum geometrical parameters for the CDaugWT, resulting in a maximum velocity augmentation ratio of 1.953 and a desirability function value of 0.9 at the throat section, were determined as follows: diffuser angle of 10°, concentrator angle of 20°, concentrator length of 375 mm ($0.62R_{th}$), diffuser length of 975 mm ($1.61R_{th}$), throat length of 70 mm ($0.12R_{th}$), and flange height of 100 mm ($0.17R_{th}$) with R_{th} representing the throat radius.
(5) The results of CFD validation indicated that the average throat velocity obtained through RSM optimisation, 3.906 m/s, showed a percentage difference of 0.2% compared to the value obtained from the CFD simulation of 3.898 m/s. This underscores the excellent accuracy and reliability of the developed model in predicting the throat velocity of the CDaugWT design.

This study confirms the effectiveness of the combination of RSM and CFD methodologies in developing a mathematical model that relates the throat velocity with its geometrical parameters and the ability to optimise the CDaugWT to obtain optimal geometrical parameters. This study also highlights and magnifies the need to utilise RSM in developing mathematical models that consider the interaction effects of the parameters a deviation from one-factor analysis, which is not a holistic approach to the study of the impact of geometrical parameters of the CDaugWT design. In addition, similar to the point noted in [7], comparing RSM predictions with CFD data suggests that RSM effectively predicts shrouded wind turbine performance. Therefore, utilising RSM will significantly reduce both the time and cost compared to CFD computations.

Author Contributions: Conceptualisation, methodology, software, formal analysis, and writing—original draft preparation, N.S.; writing—review and editing, methodology, formal analysis, supervision, and funding acquisition, G.M. and P.M. All authors have read and agreed to the published version of the manuscript.

Funding: This research received no external funding. It received internal funding from Govan Mbeki Research and Development Centre (GMRDC).

Informed Consent Statement: Not applicable.

Data Availability Statement: The data presented in this study are available on request from the corresponding author.

Conflicts of Interest: The authors declare no conflict of interest.

References

1. Lesala, M.E.; Shambira, N.; Makaka, G.; Mukumba, P. The Energy Poverty Status of Off-Grid Rural Households: A Case of the Upper Blinkwater Community in the Eastern Cape Province, South Africa. *Energies* **2023**, *16*, 7772. [CrossRef]
2. Shonhiwa, C.; Makaka, G.; Mukumba, P.; Shambira, N. Investigation of Wind Power Potential in Mthatha, Eastern Cape Province, South Africa. *Appl. Sci.* **2023**, *13*, 12237. [CrossRef]
3. Shambira, N.; Makaka, G.; Mukumba, P.; Shonhiwa, C. Comparative Analysis of Numerical Methods for Assessing Wind Potential in Fort Beaufort, South Africa, using Two- Parameter Weibull Distribution Model. Weather stations. In Proceedings of the SAIP2023, the 67th Annual Conference of the South African Institute of Physics, Zululand, South Africa, 3–7 July 2023; Prinsloo, P.A., Ed.; South African Institute of Physics, University of Zululand: Zululand, South Africa, 2023; pp. 428–436.
4. Shambira, N.; Makaka, G.; Mukumba, P.; Lesala, M.; Roro, K.; Julies, J.; Tazvinga, H. Wind Resource Assessment in the Upper Blinkwater area in the Province of Eastern Cape, South Africa. *Int. J. Eng. Res. Technol.* **2020**, *9*, 387–402.
5. Bontempo, R.; Di Marzo, E.M.; Manna, M. Diffuser augmented wind turbines: A critical analysis of the design practice based on the ducting of an existing open rotor. *J. Wind Eng. Ind. Aerodyn.* **2023**, *238*, 105428. [CrossRef]
6. Ahmad, S.; Rafat Al, W.; El Haj, A.; Khalil, K.; Salim, B.N.M. Analysis of Accelerating Devices for Enclosure Wind Turbines. *Int. J. Astronaut. Aeronaut. Eng.* **2017**, *2*, 9. [CrossRef] [PubMed]

7. Taghinezhad, J.; Abdoli, S.; Silva, V.; Sheidaei, S.; Alimardani, R.; Mahmoodi, E. Computational fluid dynamic and response surface methodology coupling: A new method for optimization of the duct to be used in ducted wind turbines. *Heliyon* **2023**, *9*, e17057. [CrossRef]
8. Shambira, N.; Makaka, G.; Mukumba, P. Analytical models for concentrator and diffuser augmented wind turbines: A review. *Int. J. Smart Grid Clean Energy* **2021**, *10*, 123–132. [CrossRef]
9. Rahmatian, M.A.; Shahrbabaki, A.N.; Moeini, S.P. Single-objective optimization design of convergent-divergent ducts of ducted wind turbine using RSM and GA, to increase power coefficient of a small-scale horizontal axis wind turbine. *Energy* **2023**, *269*, 126822. [CrossRef]
10. Danish, S.N.; Almutairi, Z.; Alshareef, M. CFD studies of the effect of holes and angles of upstream duct of horizontal axis wind turbines. *AIP Adv.* **2023**, *13*, 035326. [CrossRef]
11. Hameed, H.S.A.; Hashem, I.; Nawar, M.A.A.; Attai, Y.A.; Mohamed, M.H. Shape optimization of a shrouded Archimedean-spiral type wind turbine for small-scale applications. *Energy* **2023**, *263*, 125809. [CrossRef]
12. Rezek, T.J.; Camacho, R.G.R.; Manzanares-filho, N. A novel methodology for the design of diffuser-augmented hydrokinetic rotors. *Renew. Energy* **2023**, *210*, 524–539. [CrossRef]
13. Quispe-abad, R.; Müller, N. Finding an absolute maximum theoretical power coefficient for ducted wind turbines. *J. Wind Eng. Ind. Aerodyn.* **2023**, *236*, 105335. [CrossRef]
14. Hesami, A.; Nikseresht, A.H. Towards development and optimization of the Savonius wind turbine incorporated with a wind-lens. *Energy* **2023**, *274*, 127263. [CrossRef]
15. Khamlaj, T.; Rumpfkeil, M. Analysis and optimization of ducted wind turbines. *Energy* **2018**, *162*, 1234–1252. [CrossRef]
16. Ohya, Y.; Karasudani, T.; Nagai, T.; Watanabe, K. Wind lens technology and its application to wind and water turbine and beyond. *Renew. Energy Environ. Sustain.* **2017**, *2*, 2. [CrossRef]
17. Richmond-Navarro, G.; Uchida, T.; Calderón-Muñoz, W.R. Shrouded wind turbine performance in yawed turbulent flow conditions. *Wind Eng.* **2022**, *46*, 518–528. [CrossRef]
18. Sagar, G.; Kalyankar, R.; Chikhale, P. Power generation by diffuser augmented wind turbine. *Int. J. Innov. Eng. Res. Technol.* **2021**, 1–5. Available online: https://repo.ijiert.org/index.php/ijiert/issue/view/41 (accessed on 26 December 2023).
19. Joshy, A.; Sreejith, B. Performance Analysis of Modified Nozzle-Diffuser Augmented Horizontal Axis Hydrokinetic Turbine. In Proceedings of the Yukthi 2021-The International Conference on Emerging Trends in Engineering, GEC Kozhikode, Kerala, India, 24–26 September 2021; pp. 1–9. [CrossRef]
20. Yeh, P.; Tsai, S.; Chen, W.; Wu, S.; Hsieh, M.; Chen, B. A Simple Nozzle-Diffuser Duct Used as a Kuroshio Energy Harvester. *Processes* **2021**, *9*, 1552. [CrossRef]
21. Rajendra Prasad, K.; Manoj Kumar, V.; Swaminathan, G.; Loganathan, G.B. Computational investigation and design optimization of a duct augmented wind turbine (DAWT). *Mater. Today Proc.* **2020**, *22*, 1186–1191. [CrossRef]
22. Sri, P.U.; Jeevesh, C. Performance analysis of a horizontal axis wind turbine. In Proceedings of the Conference on Emerging Trends in Engineering (ICETE) Emerging Trends in Smart Modelling Systems and Design, Hyderabad, India, 22–23 March 2019; Springer International Publishing: Berlin/Heidelberg, Germany, 2020; pp. 440–448.
23. Ye, J.; Cheng, Y.; Xie, J.; Huang, X.; Zhang, Y.; Hu, S.; Salem, S.; Wu, J. Effects of Divergent Angle on the Flow Behaviors in Low Speed Wind Accelerating Ducts. *Renew. Energy* **2020**, *152*, 1292–1301. [CrossRef]
24. Siavash, N.K.; Najafi, G.; Hashjin, T.T.; Ghobadian, B.; Mahmoodi, E. An innovative variable shroud for micro wind turbines. *Renew. Energy* **2020**, *145*, 1061–1072. [CrossRef]
25. Al-quraishi, B.A.J.; Zelawati, N.; Asmuin, B. Review on Diffuser Augmented Wind Turbine (DAWT). *Int. J. Integr. Eng.* **2019**, *11*, 178–206. [CrossRef]
26. Ohya, Y.; Karasudani, T.; Sakurai, A.; Abe, K.I.; Inoue, M. Development of a shrouded wind turbine with a flanged diffuser. *J. Wind Eng. Ind. Aerodyn.* **2008**, *96*, 524–539. [CrossRef]
27. Thangavelu, S.K.; Goh, C.Y.; Sia, C.V. Design and flow simulation of concentrator augmented wind turbine. *IOP Conf. Ser. Mater. Sci. Eng.* **2019**, *501*, 012041. [CrossRef]
28. Shikha, S.; Bhatti, T.S.; Kothari, D.P. Air concentrating nozzles: A promising option for wind turbines. *Int. J. Energy Technol. Policy* **2005**, *3*, 394–412. [CrossRef]
29. Orosa, J.A.; Garca-Bustelo, E.J.; Oliveira, A.C. An experimental test of low speed wind turbine concentrators. *Energy Sources Part A Recover. Util. Environ. Eff.* **2012**, *34*, 1222–1230. [CrossRef]
30. Abdelrazek, A.M.; Abdelrazik, A.M.; Kassab, S. Performance of a flanged diffuser augmented wind turbine structure under variable geometrical conditions. *Am. J. Eng. Res.* **2021**, *10*, 142–153.
31. Maulana, M.I.; Syuhada, A.; Nawawi, M. Analysis of diffuser augmented wind turbine (DAWT) with flange and curved interior using CFD. *AIP Conf. Proc.* **2018**, *1984*, 2–7. [CrossRef]
32. Ghorani, M.M.; Karimi, B.; Mirghavami, S.M.; Saboohi, Z. A numerical study on the feasibility of electricity production using an optimized wind delivery system (Invelox) integrated with a Horizontal axis wind turbine (HAWT). *Energy* **2023**, *268*, 126643. [CrossRef]
33. Elsayed, A.M. Design optimization of diffuser augmented wind turbine. *CFD Lett.* **2021**, *13*, 45–59. [CrossRef]
34. Takeyeldein, M.M.; Ishak, I.S.; Lazim, T.M. the Effect of the Number of the Blades on Diffuser Augmented Wind Turbine Performance. *J. Eng. Sci. Technol.* **2023**, *18*, 1019–1037.

35. Jauhar, T.A.; Hussain, M.I.; Kiren, T.; Arif, W.; Miran, S.; Lee, G.H. Effect of flanged diffuser divergence angle on wind turbine: A numerical investigation. *PLoS ONE* **2023**, *18*, e0287053. [CrossRef]
36. Maw, Y.Y.; Tun, M.T. Sensitivity analysis of angle, length and brim height of the diffuser for the small diffuser augmented wind turbine using the numerical investigation. *ASEAN Eng. J.* **2021**, *11*, 280–291. [CrossRef]
37. Hosseini, S.R.; Ganji, D.D. A novel design of nozzle-diffuser to enhance performance of INVELOX wind turbine. *Energy* **2020**, *198*, 117082. [CrossRef]
38. Koç, E.; Yavuz, T. Effect of flap on the wind turbine-concentrator combination. *Int. J. Renew. Energy Res.* **2019**, *9*, 551–560. [CrossRef]
39. Kaseb, Z.; Rahbar, M. Towards CFD-based optimization of urban wind conditions: Comparison of Genetic algorithm, Particle Swarm Optimization, and a hybrid algorithm. *Sustain. Cities Soc.* **2022**, *77*, 103565. [CrossRef]
40. Elsakka, M.M.; Ingham, D.B.; Ma, L.; Pourkashanian, M.; Moustafa, G.H.; Elhenawy, Y. Response Surface Optimisation of Vertical Axis Wind Turbine at low wind speeds. *Energy Rep.* **2022**, *8*, 10868–10880. [CrossRef]
41. Cheng, B.; Yao, Y. Design and optimization of a novel U-type vertical axis wind turbine with response surface and machine learning methodology. *Energy Convers. Manag.* **2022**, *273*, 116409. [CrossRef]
42. Ramayee, L.; Supradeepan, K. Influence of Axial Distance and Duct Angle in the Improvement of Power Generation in Duct Augmented Wind Turbines. *J. Energy Resour. Technol.* **2022**, *144*, 091302. [CrossRef]
43. Bouvant, M.; Betancour, J.; Velásquez, L.; Rubio-Clemente, A.; Chica, E. Design optimization of an Archimedes screw turbine for hydrokinetic applications using the response surface methodology. *Renew. Energy* **2021**, *172*, 941–954. [CrossRef]
44. Netto, D.C.; Camacho, R.R.; Filho, N.M. Surrogate-Based Design Optimization of a H-Darrieus Wind Turbine Comparing Classical Response Surface, Artificial Neural Networks, and Kriging. *J. Appl. Fluid Mech.* **2023**, *16*, 703–716.
45. Kaseb, Z.; Montazeri, H. Data-driven optimization of building-integrated ducted openings for wind energy harvesting: Sensitivity analysis of metamodels. *Energy* **2022**, *258*, 124814. [CrossRef]
46. Panda, D.; Satapathy, A.K.; Sarangi, S.K. Thermo-hydrodynamic analysis and optimal design of a GM cycle cryorefrigerator using response surface methodology and particle swarm optimization. *Sci. Technol. Built Environ.* **2019**, *25*, 1467–1481. [CrossRef]
47. Taghinezhad, J.; Alimardani, R.; Masdari, M.; Mahmoodi, E. Performance optimization of a dual-rotor ducted wind turbine by using response surface method. *Energy Convers. Manag. X* **2021**, *12*, 100120. [CrossRef]
48. Djimtoingar, S.S.; Derkyi, N.S.A.; Kuranchie, F.A.; Yankyera, J.K. A review of response surface methodology for biogas process optimization. *Cogent Eng.* **2022**, *9*, 2115283. [CrossRef]
49. Atalan, Y.A.; Tayanç, M.; Erkan, K.; Atalan, A. Development of nonlinear optimization models for wind power plants using box-Behnken design of experiment: A case study for Turkey. *Sustainability* **2020**, *12*, 6017. [CrossRef]
50. Aydar, A.Y. Utilization of Response Surface Methodology in Optimization of Extraction Extraction of Optimization of Plant Plant Materials. In *Statistical Approaches with Emphasis on Design of Experiments Applied to Chemical Processes*; InTech: London, UK, 2018; pp. 157–169.
51. Tang, J.; Avallone, F.; Bontempo, R.; Van Bussel, G.J.W.; Manna, M. Experimental investigation on the effect of the duct geometrical parameters on the performance of a ducted wind turbine. *J. Phys. Conf. Ser.* **2018**, *1037*, 022034. [CrossRef]
52. Aldhufairi, M.; Muda, M.K.H.; Mustapha, F.; Ahmad, K.A.; Yidris, N. Design of Wind Nozzle for Nozzle Augmented Wind Turbine. *J. Adv. Res. Fluid Mech. Therm. Sci.* **2022**, *95*, 36–43. [CrossRef]
53. Lipian, M.; Dobrev, I.; Massouh, F.; Jozwik, K. Small wind turbine augmentation: Numerical investigations of shrouded- and twin-rotor wind turbines. *Energy* **2020**, *201*, 117588. [CrossRef]
54. Ramayee, L.; Supradeepan, K.; Ravinder Reddy, P.; Karthik, V. Design of shorter duct for wind turbines to enhance power generation: A numerical study. *J. Braz. Soc. Mech. Sci. Eng.* **2022**, *44*, 160. [CrossRef]
55. Putra, F.Q.; Rifai, D.; Suryopratomo, K.; Budiarto, R. Multilevel Diffuser Augmented for Horizontal Axis Wind Turbine. *E3S Web Conf.* **2018**, *42*, 01001. [CrossRef]
56. Susandi, A.; Arifin, F.; Kusumanto, R. Simulation of Diffuser Parameters in the Performance of Horizontal Axis Wind Turbine using Computational Fluid Dynamics. *Eng. Technol.* **2021**, *63*, 7739–7749.
57. Beller, C. *Layout Design for a Venturi to Encase a Wind Turbine Integrated in a High Rise*; Risø National Laboratory for Sustainable Energy: Roskilde, Denmark, 2008; Volume 6, pp. 3527–3537.
58. Yadav, V.; Tiwari, P.K. A Review on CFD Analysis of Effects of Convergence and Divergence Angles on the Performance of a Nozzle. *Int. J. Sci. Res. Eng. Trends* **2020**, *6*, 33–35.
59. Behera, S.K.; Meena, H.; Chakraborty, S.; Meikap, B.C. Application of response surface methodology (RSM) for optimization of leaching parameters for ash reduction from low-grade coal. *Int. J. Min. Sci. Technol.* **2018**, *28*, 621–629. [CrossRef]
60. Sahu, S.; Behera, S. A review on modern control applications in wind energy conversion system. *Energy Environ.* **2022**, *33*, 223–262. [CrossRef]
61. Qader, B.S.; Supeni, E.E.; Ariffin, M.K.A.; Talib, A.R.A. RSM approach for modeling and optimization of designing parameters for inclined fins of solar air heater. *Renew. Energy* **2019**, *136*, 48–68. [CrossRef]
62. Aslani, A.; Masoumi, H.; Gilani, H.G.; Ghaemi, A. Improving adsorption performance of l - ascorbic acid from aqueous solution using magnetic rice husk as an adsorbent: Experimental and RSM modeling. *Sci. Rep.* **2023**, *13*, 10860. [CrossRef]
63. Khamlaj, T.A.; Rumpfkeil, M.P. Theoretical analysis of shrouded horizontal axis wind turbines. *Energies* **2017**, *10*, 38. [CrossRef]

64. Khalid, W.; Sherbaz, S.; Maqsood, A.; Hussain, Z. Design and optimization of a diffuser for a horizontal axis hydrokinetic turbine using computational fluid dynamics based surrogate modelling. *Mechanika* **2020**, *26*, 161–170. [CrossRef]
65. Pashaei, H.; Mashhadimoslem, H.; Ghaemi, A. Modeling and optimization of CO2 mass transfer flux into Pz-KOH-CO2 system using RSM and ANN. *Sci. Rep.* **2023**, *13*, 4011. [CrossRef]
66. Kumari, S.; Rajput, V.D.; Minkina, T.; Rajput, P.; Sharma, P.; Verma, A.K.; Agarwal, S.; Garg, M.C. Application of RSM for Bioremoval of Methylene Blue Dye from Industrial Wastewater onto Sustainable Walnut Shell (Juglans regia) Biomass. *Water* **2022**, *14*, 3651. [CrossRef]
67. Sheshadri, S.M.; Dighe, V.V. Ground effect for ducted wind turbines: A computational study. *J. Phys. Conf. Ser.* **2022**, *2265*, 042079. [CrossRef]
68. Refaie, A.G.; Hameed, H.S.A.; Nawar, M.A.A.; Attai, Y.A.; Mohamed, M.H. Comparative investigation of the aerodynamic performance for several Shrouded Archimedes Spiral Wind Turbines. *Energy* **2022**, *239*, 122295. [CrossRef]
69. Hashem, I.; Hafiz, A.A.; Mohamed, M.H. Characterization of aerodynamic performance of wind-lens turbine using high-fidelity CFD simulations. *Front. Energy* **2020**, *16*, 661–682. [CrossRef]
70. Taghinezhad, J.; Alimardani, R.; Masdari, M.; Mosazadeh, H. Parametric study and flow characteristics of a new duct for ducted wind turbines system using analytical hierarchy process: Numerical & experimental study. *Energy Syst.* **2022**, *3*, 1–30. [CrossRef]
71. Gaden, D.L.F.; Bibeau, E.L. A numerical investigation into the effect of diffusers on the performance of hydro kinetic turbines using a validated momentum source turbine model. *Renew. Energy* **2010**, *35*, 1152–1158. [CrossRef]
72. Ajayi, O.O.; Unser, L.; Ojo, J.O. Implicit rule for the application of the 2-parameters RANS turbulence models to solve flow problems around wind turbine rotor profiles. *Clean. Eng. Technol.* **2023**, *13*, 100609. [CrossRef]
73. Wilberforce, T.; Alaswad, A. Performance analysis of a vertical axis wind turbine using computational fluid dynamics. *Energy* **2023**, *263*, 125892. [CrossRef]
74. Yiyin, K.; AM, S.Y.; Muhammad, A.; Sri, S. Assessment of Flange Diffuser Structures to Improve the Power Generation of a Diffuser Augmented Wind Turbine. *Przegląd Elektrotechniczny* **2022**, *4*, 21–26. [CrossRef]
75. Auyanet, A.G.; Verdin, P.G. Numerical Study of the Effect of Flap Geometry in a Multi-Slot Ducted Wind Turbine. *Sustainability* **2022**, *14*, 12032. [CrossRef]
76. Sridhar, S.; Zuber, M.; Shenoy, B.S.; Kumar, A.; Ng, E.Y.K.; Radhakrishnan, J. Aerodynamic comparison of slotted and non-slotted diffuser casings for Diffuser Augmented Wind Turbines (DAWT). *Renew. Sustain. Energy Rev.* **2022**, *161*, 112316. [CrossRef]
77. Badawy, Y.E.M.; Nawar, M.A.A.; Attai, Y.A.; Mohamed, M.H. Co-enhancements of several design parameters of an archimedes spiral turbine for hydrokinetic energy conversion. *Energy* **2023**, *268*, 126715. [CrossRef]
78. Rahmatian, M.A.; Hashemi Tari, P.; Mojaddam, M.; Majidi, S. Numerical and experimental study of the ducted diffuser effect on improving the aerodynamic performance of a micro horizontal axis wind turbine. *Energy* **2022**, *245*, 123267. [CrossRef]
79. Al-Quraishi, B.A.J.; Asmuin, N.Z.; Nasir, N.F.; Latif, N.A.; Taweekun, J.; Mohd, S.; Mohammed, A.N.; Al-Wahid, W.A.A. CFD Investigation of Empty Flanged Diffuser Augmented Wind Turbine. *Int. J. Integr. Eng.* **2020**, *12*, 22–32.
80. Khalid, M.W.; Ahsan, M. Computational Fluid Dynamics Analysis of Compressible Flow Through a Converging-Diverging Nozzle using the k-ε Turbulence Model. *Eng. Technol. Appl. Sci. Res.* **2020**, *10*, 5180–5185. [CrossRef]
81. Ramayee, L.; Supradeepan, K. Numerical study on flow characteristics of shroud with and without flap for wind turbine applications. *Wind Eng.* **2023**, *47*, 546–563. [CrossRef]
82. Ramayee, L.; Supradeepan, K. Grid convergence study on flow past a circular cylinder for beginners. *AIP Conf. Proc.* **2021**, *2317*, 1. [CrossRef]
83. Calle, A.R.; Baca, G.A.; Gonzales, S. Optimization of the Eolic Cell to improve the wind velocity augmentation effect through the metamodel of optimal prognosis. *Energy Convers. Manag. X* **2022**, *16*, 100330. [CrossRef]
84. Görgülü, Y.F.; Özgür, M.A.; Ramazan, K.Ö.S.E. CFD analysis of a NACA 0009 aerofoil at a low reynolds number. *Politek. Derg.* **2021**, *24*, 1237–1242. [CrossRef]
85. Abe, K.I.; Ohya, Y. An investigation of flow fields around flanged diffusers using CFD. *J. Wind Eng. Ind. Aerodyn.* **2004**, *92*, 315–330. [CrossRef]
86. Azhar, F.N.A.; Taha, M.F.; Ghani, S.M.M.; Ruslan, M.S.H.; Yunus, N.M.M. Experimental and Mathematical Modelling of Factors Influencing Carbon Dioxide Absorption into the Aqueous Solution of Monoethanolamine and 1-butyl-3-methylimidazolium Dibutylphosphate Using Response Surface Methodology (RSM). *Molecules* **2022**, *27*, 1779. [CrossRef] [PubMed]
87. Chaib, M.; Slimane, A.; Slimane, S.A.; Ziadi, A.; Bouchouicha, B. Optimization of Ultimate Tensile Strength with DOE approach for application FSW process in the aluminum alloys. *Frat. ed Integrità Strutt.* **2021**, *57*, 169–181. [CrossRef]
88. Pashaei, H.; Ghaemi, A.; Nasiri, M.; Karami, B. Experimental Modeling and Optimization of CO2 Absorption into Piperazine Solutions Using RSM-CCD Methodology. *ACS Omega* **2020**, *5*, 8432–8448. [CrossRef]
89. Khoshraftar, Z.; Masoumi, H.; Ghaemi, A. Experimental, response surface methodology (RSM) and mass transfer modeling of heavy metals elimination using dolomite powder as an economical adsorbent. *Case Stud. Chem. Environ. Eng.* **2023**, *7*, 100329. [CrossRef]
90. Peng, H.; Guo, J.; Qiu, H.; Wang, C.; Zhang, C.; Hao, Z.; Rao, Y.; Gong, Y. Efficient Removal of Cr (VI) with Biochar and Optimized Parameters by Response Surface Methodology. *Processes* **2021**, *9*, 889. [CrossRef]
91. Thangavelu, S.K.; Wan, T.G.L.; Piraiarasi, C. Flow Simulations of Modified Diffuser Augmented Wind Turbine. *IOP Conf. Ser. Mater. Sci. Eng.* **2020**, *886*, 012023. [CrossRef]

92. Abdelrazek, A.M.; Abdelrazik, A.M.; Kassab, S.Z. Performance of a flanged diffuser augmented wind turbine structure with and without the presence of the turbine: 3D Study. In Proceedings of the 4th International E-Conference on Advances in Engineering, Technology and Management—ICETM, Changsha, China, 8–10 November 2024; pp. 115–125.
93. Arifin, F.; Kusumanto, R.D.; Bow, Y.; Rusdianasari; Taqwa, A.; Susandi, A.; Herlambang, Z.D.; Wang, M.W.; Sitompul, C.R. Study the Effect Diffuser Length and Degree to Horizontal Wind Turbine. In Proceedings of the 4th International Conference on Applied Science and Technology on Engineering Science (iCAST-ES 2021), Sion, Mumbai, 7 May 2021; Science and Technology Publications: Setúbal, Portugal, 2021; pp. 681–685.
94. Bekele, N.; Bogale, W. Parametric study of a diffuser for horizontal axis wind turbine power augmentation. *AIMS Energy* **2019**, 7, 841–856. [CrossRef]

sustainability

MDPI

Article

The Optimization of PEM Fuel-Cell Operating Parameters with the Design of a Multiport High-Gain DC–DC Converter for Hybrid Electric Vehicle Application

B. Karthikeyan [1], Palanisamy Ramasamy [2,*], M. Pandi Maharajan [3], N. Padmamalini [4], J. Sivakumar [5], Subhashree Choudhury [6] and George Fernandez Savari [7]

1 Department of EEE, K. Ramakrishnan College of Technology, Trichy 621112, India; karthikeyanb.eee@krct.ac.in
2 Department of Electrical and Electronics Engineering, SRM Institute of Science and Technology, Chennai 603203, India
3 Department of EEE, Nadar Saraswathi College of Engineering and Technology, Theni 625531, India; pandimaha.net@gmail.com
4 Department of Physics, St. Joseph's Institute of Technology, Chennai 600119, India; sirishaaa73@gmail.com
5 Department of Electronics and Communication Engineering, St. Joseph's College of Engineering, Chennai 600119, India; sivakumarj@stjosephs.ac.in
6 Department of EEE, Siksha 'O' Anusandhan (Deemed to be University), Bhubaneswar 751030, India; subhashreechoudhury@soa.ac.in
7 OES Technologies, 4056 Blakie Road, London, ON N6L 1P7, Canada; gsavari@oes-inc.com or george.electrix@gmail.com
* Correspondence: palanisr@srmist.edu.in or krspalani@gmail.com

Abstract: The fossil fuel crisis is a major concern across the globe, and fossil fuels are being exhausted day by day. It is essential to promptly change from fossil fuels to renewable energy resources for transportation applications as they make a major contribution to fossil fuel consumption. Among the available energy resources, a fuel cell is the most affordable for transportation applications because of such advantages as moderate operating temperature, high energy density, and scalable size. It is a challenging task to optimize PEMFC operating parameters for the enhancement of performance. This paper provides a detailed study on the optimization of PEMFC operating parameters using a multilayer feed-forward neural network, a genetic algorithm, and the design of a multiport high-gain DC–DC converter for hybrid electric vehicle application, which is capable of handling both a 6 kW PEMFC and an 80 AH 12 V heavy-duty battery. To trace the maximum power from the PEMFC, the most recent SFO-based MPPT control technique is implemented in this research work. Initially, a multilayer feed-forward neural network is trained using a back-propagation algorithm with experimental data. Then, the optimization phase is separately carried out in a neural-power software environment using a genetic algorithm (GA). The simulation study was carried out using the MATLAB/R2022a platform to verify the converter performance along with the SFO-based MPPT controller. To validate the real-time test bench results, a 0.2 kW prototype model was constructed in the laboratory, and the results were verified.

Keywords: PEMFC; hybrid electric vehicle; multiport; SFO; DC–DC converter

Citation: Karthikeyan, B.; Ramasamy, P.; Pandi Maharajan, M.; Padmamalini, N.; Sivakumar, J.; Choudhury, S.; Savari, G.F. The Optimization of PEM Fuel-Cell Operating Parameters with the Design of a Multiport High-Gain DC–DC Converter for Hybrid Electric Vehicle Application. *Sustainability* **2024**, *16*, 872. https://doi.org/10.3390/su16020872

Academic Editors: Ayman Al-Quraan and Ahmad M. A. Malkawi

Received: 4 December 2023
Revised: 6 January 2024
Accepted: 12 January 2024
Published: 19 January 2024

1. Introduction

1.1. Sustainable Development

As the fossil fuel crisis is a common issue across the globe, researchers have been harnessing renewable energy resources. There are two approaches to encouraging sustainable development: one is to optimally utilize existing conventional energy resources in such a way that energy efficiency improves and energy consumption reduces, and the other is to adopt new energy-conversion technologies and their development. Though there are plenty of renewable resources in use, fuel cells have attracted attention in the transportation field because of salient features like low operating temperature, high energy density, quick start-up, and scalable size. The chemical energy that exists inside a fuel cell leads to the production of electricity and hot-water by-products. A fuel cell can be classified into many types based on the electrolyte it uses.

In this research, a PEM fuel cell is used, in which the porous polymer membrane can transfer protons and resist the electron flow. Its salient features make it possible to use it for transportation applications. The construction of hybrid electric vehicles using fuel cells is a challenge for researchers because their output voltage depends on many operating parameters. Therefore, researchers are encouraged to use this technology to optimize the operating parameters and make it economically viable for end-users, promoting sustainable development.

1.2. Literature Review

The operating parameters of a 25 cm^2 proton-exchange membrane fuel cell were set to improve output power and efficiency. In this study, the researchers considered one design parameter and two operating parameters. The parameters considered were the landing-to-channel ratio, operating temperature, and hydrogen partial pressure. The results revealed that hydrogen partial pressure contributed more than the other two parameters [1]. The effect of different electrode materials used in a PEM fuel-cell electrode was investigated [2]. Three different materials were used in this study, namely aluminum, copper, and steel. In this study, a detailed three-dimensional PEM fuel cell was constructed and simulated using fluid dynamics in the ANSYS ANSYS FLUENT 18.0 software environment. Then, validation of the investigation was carried out by conducting an experimental verification study, and it was reported that the aluminum-based bipolar electrode performed better than the other materials used. A multidimensional optimization study on PEM fuel cells was carried out [3]. A set of parameters was considered for optimization that included efficiency, power density, and oxygen uniformity in the cathode material used. The optimization techniques used in this study were a computational fluid dynamics model, a surrogate model, and, finally, a multi-objective genetic algorithm. The optimized parameters performed better than the other set.

The importance of heating, cooling, and power systems was stressed [4]. The researchers used a modified mayfly optimization-based algorithm to optimize the design specification of a PEM fuel cell. The simulated results were compared against the conventional mayfly algorithm and showed that the proposed optimization-based algorithm could yield better results. There are two ways to improve a PEM fuel cell's performance: optimizing the operating parameters or optimizing the design specifications [5]. Therefore, it is necessary to optimize the cell dimensions to enhance performance. Bearing that in mind, these researchers optimized the bipolar plate dimension since it is directly related to the water management and thermal management of PEM fuel cells. The results of this study revealed that the square baffled channel could produce more power than bipolar plates of other shapes, as found through the application of an ANFIS-based model.

The optimal design of PEM fuel cells has been investigated [6]. The design variables included both design parameters and operating parameters. They employed a Box–Behnken model for numerical calculations. Then, a regression model constructed via the RSM method was analyzed using the NDRG-II algorithm (non-dominated ranking genetic algorithm), and the results revealed that the proposed optimization method offered better design parameters. A novel sunflower optimization technique to select the optimized operating parameters was implemented [7]. This method was purely based on a PEMFC circuit-based model, which reduced the SSE (sum of square error) of the actual value and the estimated value. The results obtained using this model were compared against seagull optimization, shuffled frog-leaping optimization, and multiverse optimization methods and showed that the proposed method yielded better results than the other three methods.

A suitable combination of parameters of nano-coolants to maximize the performance of PEM fuel cells, such as relative humidity, thermal conductivity, and empirical coefficients, has been investigated [8]. The optimal combination was identified using a genetic algorithm. The performance of the PEM fuel cells was improved when the nano-coolant was added with an optimized thermal conductivity value. An Elman Neural Network for selecting the suitable parameters of a PEM fuel cell has been proposed [9]. After the identification of suitable parameters for PEM fuel cells, the researchers used combined algorithms, including TLBO and DE, to optimize the parameter selection. After the implementation of this method, the fuel cell could operate efficiently on the maximum output voltage and output power.

PEM fuel-cell parameters have been analyzed using a novel optimization technique called the deer-hunting optimization algorithm [7]. The operating conditions of the PEM fuel cell were analyzed under different fuel pressure conditions. The proposed optimization yielded a very fast convergence with reduced weight function when applied to a convolution neural network. The proposed optimization performed better than other conventional methods used in the state of the art. The operating conditions of the proton-exchange membrane fuel cell were optimized using the Tuguchi method, in which many operating parameters, such as operating temperature, electrode inlets, and the fuel flow channel parameters, were considered [10].

The COSMOL multiphysics environment was used to optimize the operating parameters. A novel optimization technique known as the chaos-embedded PSO algorithm was proposed for the first time to determine the unknown parameters of PEMFC [11]. A new objective function was formulated and yielded a better convergence rate. It is worth mentioning that the proposed objective function yielded the minimum error in finding the global maximum value. The polarization curve of PEMFC in which maximum performance was obtained was important [12]. The researchers used the real-time operating data of 50 cm^2 using a data analytical model of one-dimensional size. Here, the results were validated with real-time data, and a good agreement was found between analytical data results and real-time polarization curve data. There were 2% and 3% error deviations in obtaining maximum and minimum power between simulation and real-time experimental results, respectively.

An accurate modeling of PEMFC performance analysis is inevitable [13]. Therefore, researchers proposed the whale optimization technique, which was aimed to increase the accuracy of the predicted model and reduce the error between simulated results and data obtained from the polarization curve. After simulation and real-time data validation, it was obvious that the proposed model could perform well. The performance of PEMFC was not only altered by the operating parameters but also influenced by the material used in the electrode [2]. From that perspective, researchers analyzed the performance of PEM fuel cells with electrodes made from Cu, Al, and stainless-steel material alloys. The simulation analysis was carried out using CFD-ANSYS software, and then the results were validated through experimental work. The polarization curve obtained from this work showed the effectiveness of the material used in the electrode.

The hybrid electric vehicle is usually fed from multiple sources like solar PV and fuel cells with a battery backup. There are multiple converters adopted in the vehicle to handle different sources, which leads to many difficulties like bulk in size, occupying more space, complex control circuits, etc. To overcome these practical difficulties, a multiport DC–DC converter, which handles two input sources effectively, has been proposed in the literature [14–19]. There was power-quality improvement for a grid-connected PEMFC in which modified pelican optimization was used to generate a switching pulse to the buck converter [20]. A multiport DC–DC converter has been shown to be capable of handling clean energy resources [21]. This multiport converter could integrate with three energy resources along with a battery, and the output was isolated from the source. A novel meta-heuristic technique, known as converged collective animal behavior optimization for

PEMFC operating parameters, has been proposed [22]. Pulse generation for high-voltage converters using SiC-MOSFET switches has been implemented [23].

1.3. Main Contribution to This Research

To trace the optimum power from a practical source, the converter is designed with an SFO-based MPPT controller. The output port of the DC–DC converter drives the BLDC motor through three phase inverters, as shown in Figure 1. A smooth mode transition is an inherent feature of the proposed DC–DC converter.

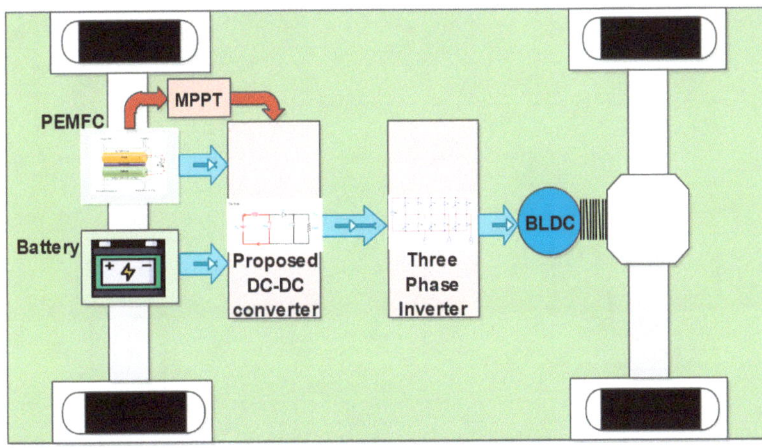

Figure 1. Architecture of a fuel-cell-energized hybrid electric vehicle.

1. The fuel cell is the most obvious choice of power source for sustainable transportation applications as it does not pollute the environment, offers high energy density, noise-free operations, etc. This research deals with the optimization of fuel-cell operating parameters, and based on that, a multiport high-gain DC–DC converter is designed for electric vehicle applications.
2. To contribute to sustainable development, we worked on the PEM fuel-cell field because it is challenging to obtain constant output voltage at all times. It is necessary to optimize the operating parameters. In this context, a deep-learning neural network has been developed and trained to optimize the significant parameters.
3. Regulating the DC power when it is used in electric vehicle applications is another challenge. A multiport high-gain DC–DC converter is designed along with an SFO-MPPT controller, and the performance of the power converter was experimentally validated. We hope these two objectives and their outcomes will be helpful for future researchers to work on the PEM fuel-cell field, therefore contributing to sustainable development in the transportation field.

2. Modeling of Input Sources

2.1. Physical Model of PEMFC

The electrochemistry modeling of fuel cells is more effective and advantageous than the thermodynamics model. Though the thermodynamics model has been used by researchers for theoretical performance analysis, the rate at which reactant reacts to produce electricity cannot be analyzed. Moreover, the electrochemistry model can be used to predict the loss that occurs during the chemical reaction. With that aspect in mind, the following section elucidates the electrochemical modeling of PEM fuel cells. Figure 2a shows the construction of PEMFC, and Figure 2b shows the equivalent circuit model of PEMFC.

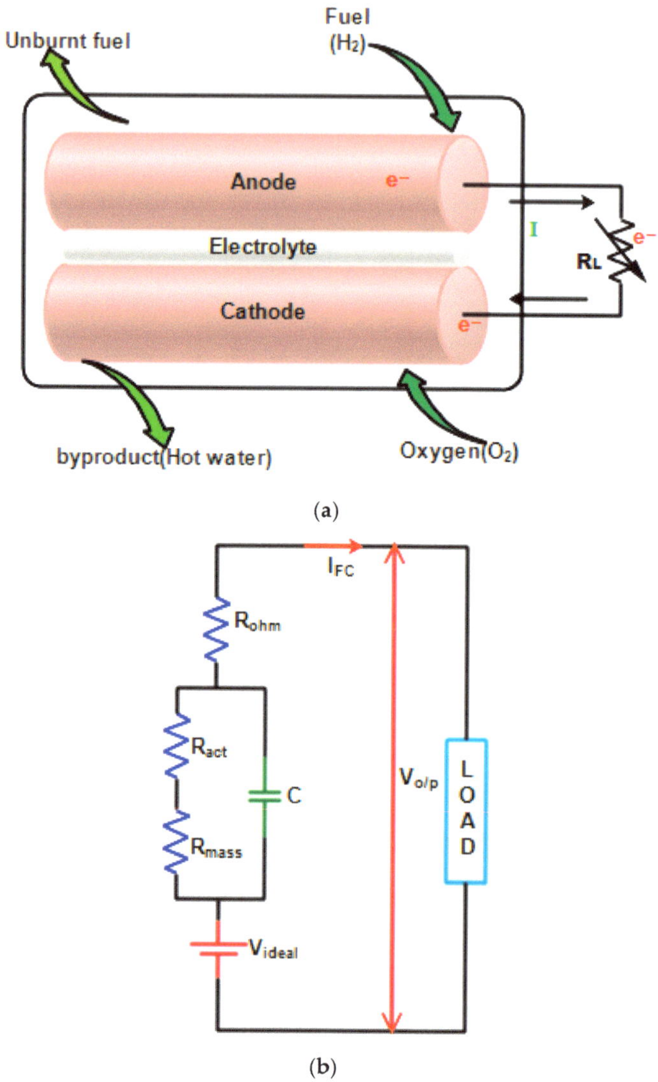

Figure 2. (**a**) The physical model of PEMFC; (**b**) Equivalent circuit of the PEMFC.

From Figure 2b, the operating voltage of the PEM fuel cell can be written as

$$V_{o/p} = V_{ideal} - V_{act(a+c)} - V_{mass(a+c)} - V_{ohm} \quad (1)$$

$$V_{ideal} = -(\frac{G}{2F} + \frac{RT}{2F}\ln\frac{P_{H_2O}}{P_{O_2}\sqrt{P_{H_2}}}) \quad (2)$$

a—anode, c—cathode, G—Gibbs constant, F—Faraday's constant, R—gas constant, T—operating temperature in K, P_{H_2}—hydrogen pressure, P_{O2}—oxygen pressure, P_{H2O}—water pressure.

$$V_{act(a+c)} = \frac{RT}{\alpha F}\ln\left(\frac{i_d}{i_r}\right) \quad (3)$$

α—coefficient of charge transfer, i_d—current density and i_r—current density corresponding to reaction exchange, i_L—limiting current density

$$V_{mass(a+c)} = \frac{RT}{\alpha F}(\frac{\alpha+1}{\alpha})\ln(1-\frac{i_L}{i_d}) \tag{4}$$

$$V_{ohm} = iR_{(a+c)} \tag{5}$$

$R_{(a+c)}$—resistance of anode and cathode. From the modeling of the PEM fuel cell, it is obvious that the operating parameters like temperature, fuel supply pressure, air supply pressure, fuel flow rate, and airflow rate decide the output voltage of the PEM fuel cell at any cost.

2.2. Physical Model of the Battery

The physical model of a lead-acid battery is depicted in Figure 3a. To understand the electrical behavior of any battery, it is essential to obtain a mathematical model of the battery. Three models of the batteries have been given in the literature [24,25], namely the equivalent circuit model, electrochemical model, and data-driven model. The equivalent circuit model (ECM) is considered to be an obvious choice because it gives good accuracy without complexity. It consists of purely electrical components like controlled voltage sources, resistors, and capacitors. Second-order ECM is a well-proven model for accuracy without complexity. It consists of one controlled voltage source representing the State of Charge (SoC) and series resistance connected with two parallel RC branches, as depicted in Figure 3b. The charging and discharging behavior of the battery can be understood by a mathematical model of the battery, as given below.

The mathematical modeling of the battery is derived as follows:

The terminal voltage of the battery is expressed as

$$V_b = V_{oc} - I_b R_o - V_{(RC)1} - V_{(RC)2} \tag{6}$$

The current flowing through the battery is written as

$$I_b = \frac{V_{(RC)1}}{R_1} + C_1\dot{V}_1 = \frac{V_{(RC)2}}{R_2} + C_2\dot{V}_2 \tag{7}$$

The voltage across each pair of RC parallels is expressed as

$$V_{(RC)1} = \left(\frac{Q}{C_1} + I_b R_1\right)e^{\left(-\frac{t}{R_1 C_1}\right)} - I_b R_o \tag{8}$$

$$V_{(RC)2} = \left(\frac{Q}{C_2} + I_b R_2\right)e^{\left(-\frac{t}{R_2 C_2}\right)} - I_b R_o \tag{9}$$

For the RC parallel branch, the voltage and current are related as follows:

$$\dot{V}_1 = -\frac{V_{(RC)1}}{R_1 C_1} + \frac{I_b}{C_1} \tag{10}$$

$$\dot{V}_2 = -\frac{V_{(RC)2}}{R_2 C_2} + \frac{I_b}{C_2} \tag{11}$$

Here, V_b—voltage across the battery, I_b—current through the battery, $V_{(RC)1}$, $V_{(RC)2}$—voltage across the parallel branch RC, Q—usable capacity of the battery, V_{oc}—voltage across the open circuit in the battery.

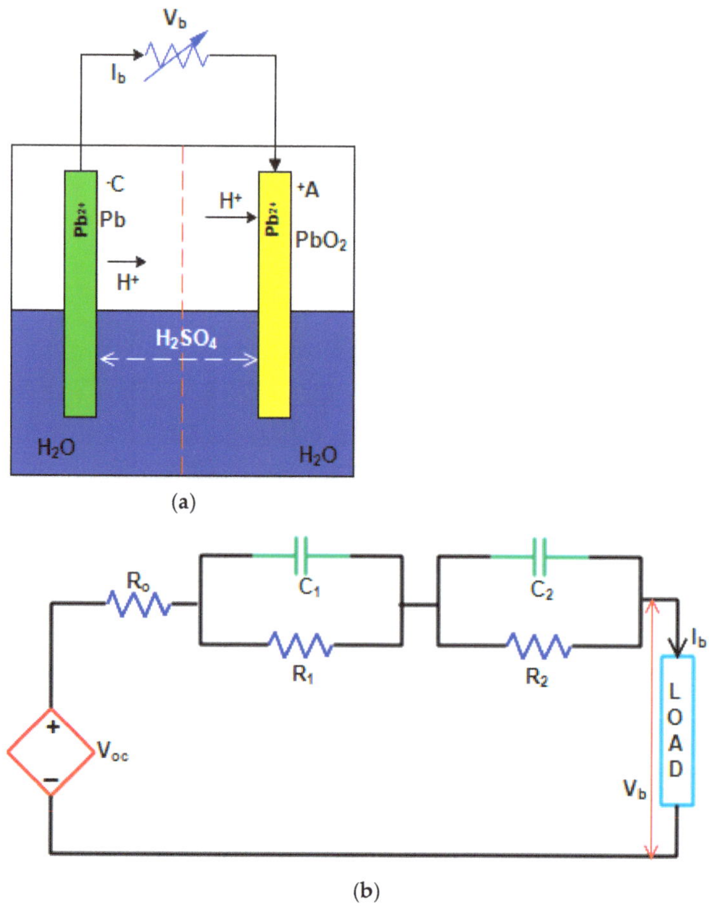

Figure 3. (**a**) Physical model of a lead-acid battery (**b**). Equivalent circuit of a lead-acid battery.

3. Optimization of PEMFC Operating Parameters

The modeling of PEMFC using design and operating parameters has been reviewed, and it is concluded that numerous parameters influence the output voltage of the system [26], but their contribution towards deciding the output voltage must be analyzed using machine learning for better understanding. In this experimental work, there are five parameters, namely system temperature, fuel supply pressure, air supply pressure, fuel flow rate, and airflow rate, which vary from minimum value to maximum value as given in Table 1, and the corresponding output voltages and output powers are noted. To analyze the most significant factor, a certain number of parameter experiments must be conducted. Therefore, here, three levels are considered for five factors, which results in 243 data points obtained from experiments. The interaction of other parameters with the most significant factors vs. output parameters is shown in Figure 4. The importance of the parameters is analyzed using statistical analysis, and its results are depicted in Figure 5. It is found that the operating temperature is the most significant factor, whereas the airflow rate is the least significant factor among the five parameters.

System parameters vs. Output voltage System parameters vs. Output power

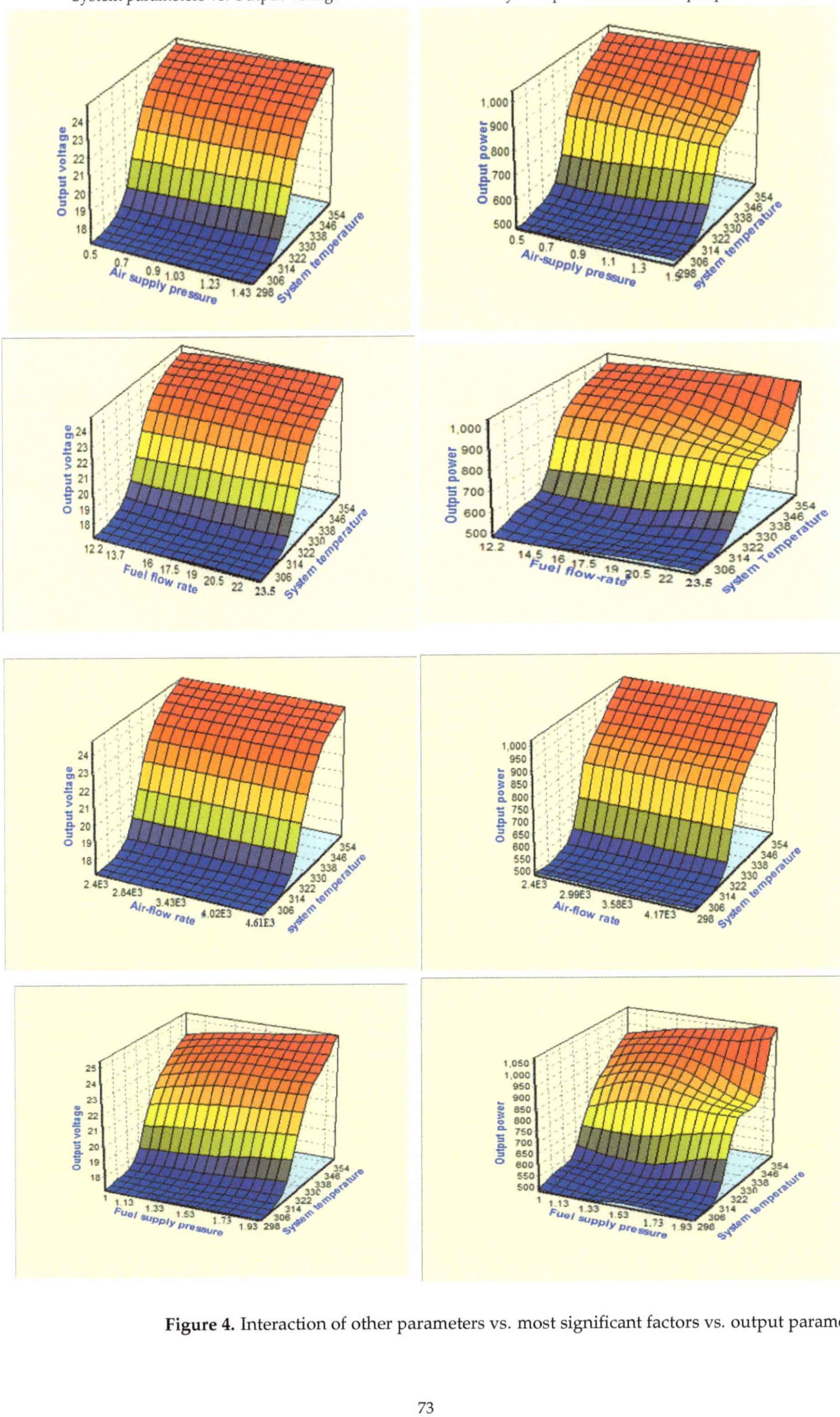

Figure 4. Interaction of other parameters vs. most significant factors vs. output parameters.

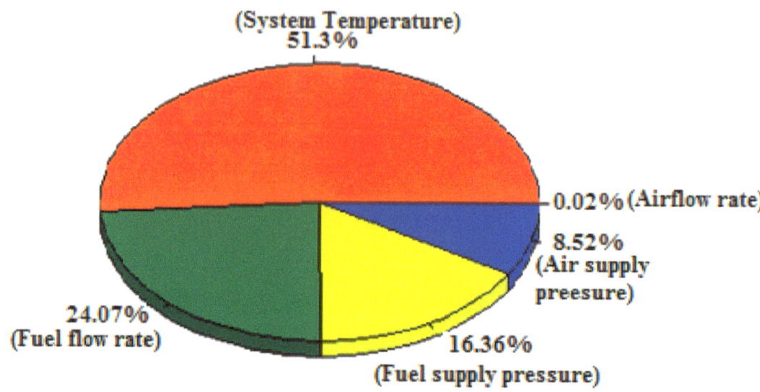

Figure 5. The contribution breakdown of operating parameters.

Table 1. The range of PEMFC operating parameters.

System Parameter	Max. Value	Min. Value
System temperature	358 K	298 K
Fuel supply pressure	2 bar	1 bar
Air supply pressure	1.5 bar	0.5 bar
Fuel flow rate	23.46 lpm	12.2 lpm
Airflow rate	4615 lpm	2400 lpm
Output voltage	25.63 V	17.06 V
Output current	1094.91 W	485.01 W

3.1. Multilayer Feed-Forward Neural Network

To optimize the input parameters of a 6 kW PEMFC, a simple multilayer feed-forward neural network is taken in which one input layer with five nodes, two hidden layers with five nodes each, and one output layer with two nodes are set in the proposed network as shown in Figure 6. The relationship between input and output of the hidden layer is expressed as

$$h_{in_j} = b_{oj}^{hid} + \sum_{i=1}^{n} x_i v_{i_i} \tag{12}$$

The net output of the hidden layer is obtained by applying the sigmoid activation function, and the input and output of the layer is expressed as

$$O_{in_k} = b_{ok}^{Out} + \sum_{j=1}^{n} h_j w_{j_k} \tag{13}$$

The net output is obtained from the output layer by applying a linear activation function. The network is trained using an incremental back-propagation algorithm with 80% input data. During each iteration, the bias and weight are updated in the hidden and output layers using the following equation.

$$v_{ij(next)} = v_{ij(pre)} + \Delta v_{ij} \tag{14}$$

$$w_{jk(next)} = w_{jk(pre)} + \Delta w_{jk} \tag{15}$$

$$b_{oj(next)}^{hidd} = b_{oj(pre)}^{hidd} + \Delta b_{oj}^{hidd} \tag{16}$$

$$b_{ok(next)}^{Out} = b_{ok(pre)}^{Out} + \Delta b_{ok}^{Out} \tag{17}$$

h_{inj} and O_{ink} are the inputs of the hidden and output layers. b_{oj} and b_{ok} are the bias terms in the hidden and output layer. V_{ij} and w_{jk} are the connection weights of the hidden and output layers.

A linear activation function is used in the output layer to obtain the net output. The learning rate of the network is 0.8, with 0.8 as momentum, and it is trained until the root mean square error (RMSE) reaches 0.01. It takes 13 min with 10,320 iterations for learning. After successful completion of learning, it is tested with the remaining 20% of input data. The results have been stored for optimization.

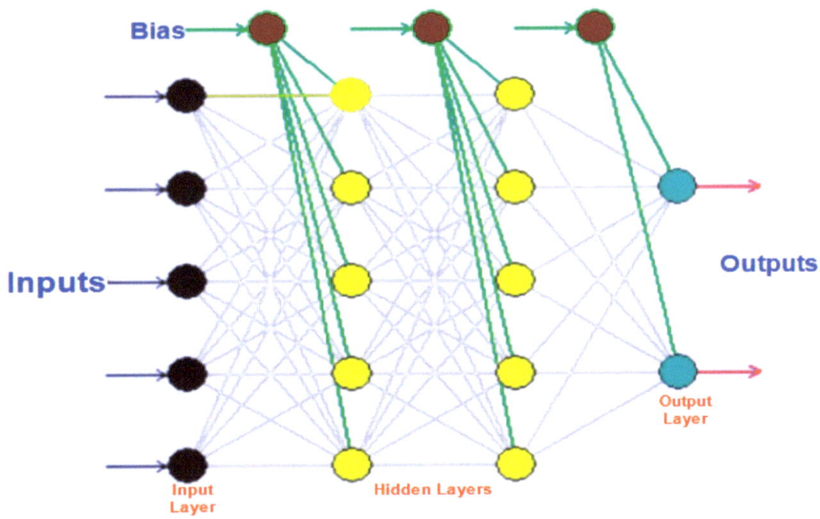

Figure 6. Multilayer feed-forward neural network.

3.2. Optimization of PEMFC Operating Parameters Using the Genetic Algorithm

The genetic algorithm (GA) is a meta-heuristic optimization technique that imitates the natural process of Darwin's principle of evolution. The conventional optimization process suffered from local optimum and failed to provide accurate results, whereas GA is a global search optimization algorithm and is implemented in the following sequence of processes: selection, cross-over, and mutation. The first step in GA is to generate parental chromosomes with population size P. Each chromosome is represented as

$$P = \{X_1, X_2, X_3 \ldots X_n\} \tag{18}$$

Here, n is the number of individual chromosomes in the population size.

Each chromosome in the population is an N-dimensional vector and can be represented as

$$a_m < x_m < b_m \tag{19}$$

Here, a_m and b_m are the lower and upper limits of variables to be optimized, and x_m is the total number taken into account for optimization. The parent generation chromosomes are evaluated at the end of each generation with its fitness function $f(X_n)$. After that, the cross-over process begins. Two new offspring, X_p and X_q, are produced from two random parents, X_r and X_s, and the process of cross-mating is limited by the cross-mate rate r.

$$x_{p,i} = (1 - r)x_{ri} + rx_{si} \tag{20}$$

$$x_{q,i} = (1 - r)x_{si} + rx_{ri} \tag{21}$$

The population size doubles at the end of the cross-over process. Then, chromosomes with high fitness values are selected among the doubled population. The mating process begins with reliable genetic diversity with a probability of 0.1. This process is repeated with newly generated chromosomes as parental chromosomes until a satisfied fitness value is attained or the required number of iterations is reached. The optimization phase is separately carried out in a neural-power software environment. The genetic algorithm is used with population size = 30, cross-over rate = 0.8, and mutation rate = 0.1 to optimize the results obtained from a multilayer feed-forward neural network. It takes less than a minute with 20,501 iterations to yield the optimized result. The maximum output voltage and power are obtained when the input parameters are set as follows: system temperature = 328 K, fuel supply pressure = 1.9998894 bar, air supply pressure = 1.4995502 bar, fuel flow rate = 19.490219 lpm, and airflow rate = 4613.4466 lpm.

4. The Multiport High-Gain DC–DC Converter

The proposed DC–DC converter consists of two input ports and one output port, as depicted in Figure 7. One input port is fed with a polymer-exchange membrane fuel cell, and the second input port is connected to a battery. There are three active switches used in the proposed converter, namely M1, M2, and AS, in which main switch M1 is operated when the fuel cell is sufficient to supply the load, and M2 is operated when the fuel cell is drained off and shut down for fuel filling. An auxiliary switch AS is operated when the fuel cell and battery are together enough to meet the demand. The load RL is supplied from the output capacitor Co. The mode of operation of the proposed converter is detailed as follows:

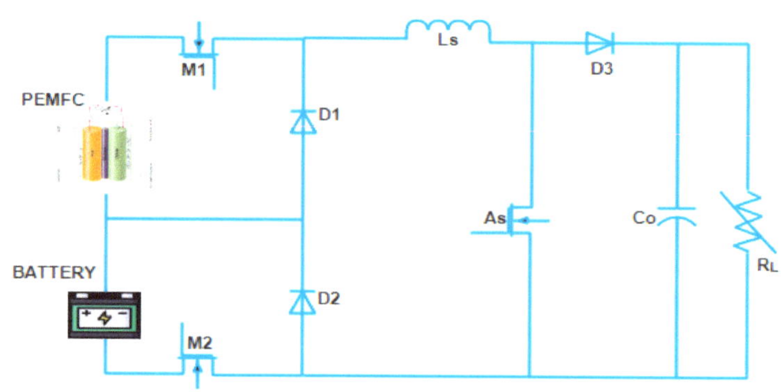

Figure 7. The multiport high-gain DC–DC converter.

During Mode I operation, the fuel cell alone is sufficient to supply load, the switch M1 is turned on, and the diodes D2 and D3 cause the source inductance Ls to become charged, as shown in Figure 8. The load RL is supplied from the output capacitor Co. The steady-state equations of this mode are given below.

$$V_{FC} = L\frac{di_s}{dt} + V_{Co} \tag{22}$$

When main switch M1 is in the OFF condition, VFC becomes zero, and the inductor Ls supplies load, as shown in the key waveforms in Figure 11a.

$$V_{Co} = -L\frac{di_s}{dt} \tag{23}$$

During Mode II operation, hydrogen fuel is completely drained off, and it looks for a new hydrogen cylinder. In this case, the battery comes into action to drive the load. The active switch M2 is turned on during Mode II, as depicted in Figure 9a. When the main switch M2 is in the OFF condition, both diodes D1 and D2 come into conduction, as depicted in Figure 9b. When the main switch M2 is in the OFF condition, the source inductance supplies the load. The key waveforms are shown in Figure 11b.

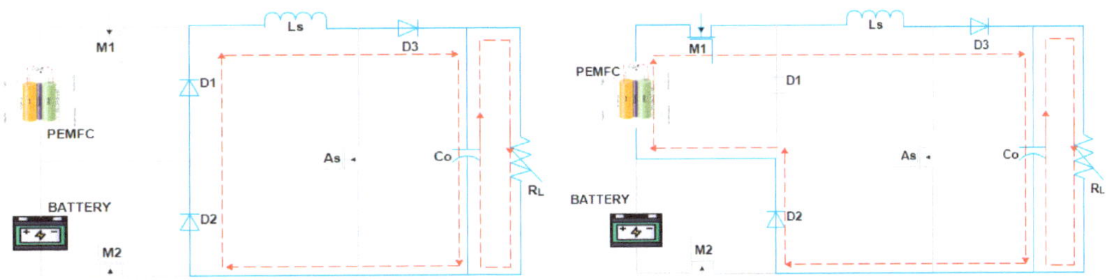

(a) (b)

Figure 8. Mode I operation (**a**) M1 is turned ON; (**b**) Both M1 and M2 are turned OFF.

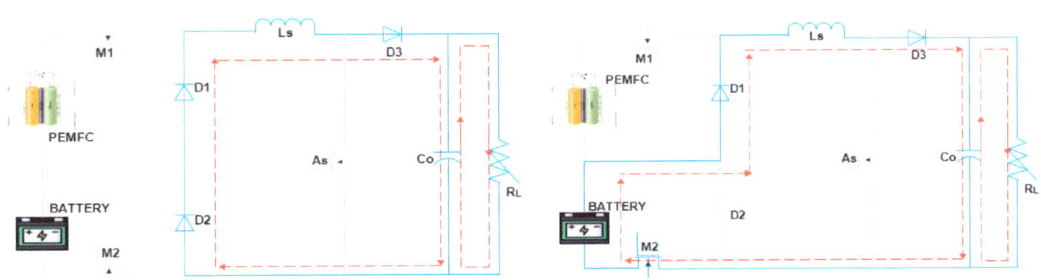

(a) (b)

Figure 9. Mode II operation (**a**) M2 is turned ON; (**b**) Both M1 and M2 are turned OFF.

$$V_{batt} = L\frac{di_s}{dt} + V_{Co} \tag{24}$$

In this particular mode, both the battery and fuel cell are operated simultaneously since they are supposed to meet the demand together. In this case, all three switches, M1, M2, and AS, come into action, during which the source inductance absorbs energy from both the battery and fuel cell, as shown in Figure 10. When all three switches are in the OFF condition, diodes D1 and D2 come into action. The key waveforms are shown in Figure 11c.

$$V_{batt} + V_{FC} = L\frac{di_s}{dt} + V_{Co} \tag{25}$$

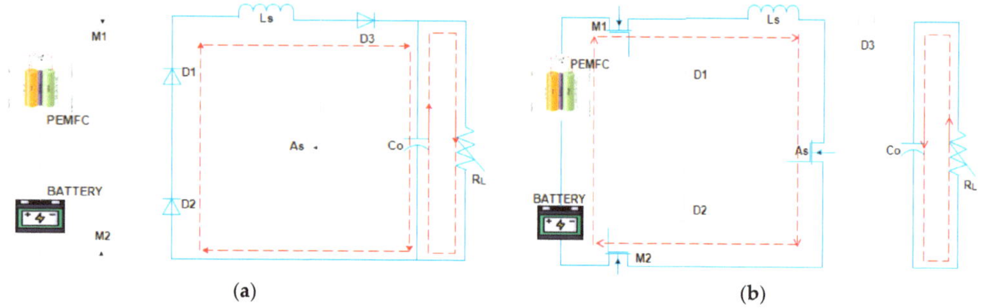

Figure 10. Mode III operation (**a**) M1 and M2 are turned ON; (**b**) Both M1 and M2 are turned OFF.

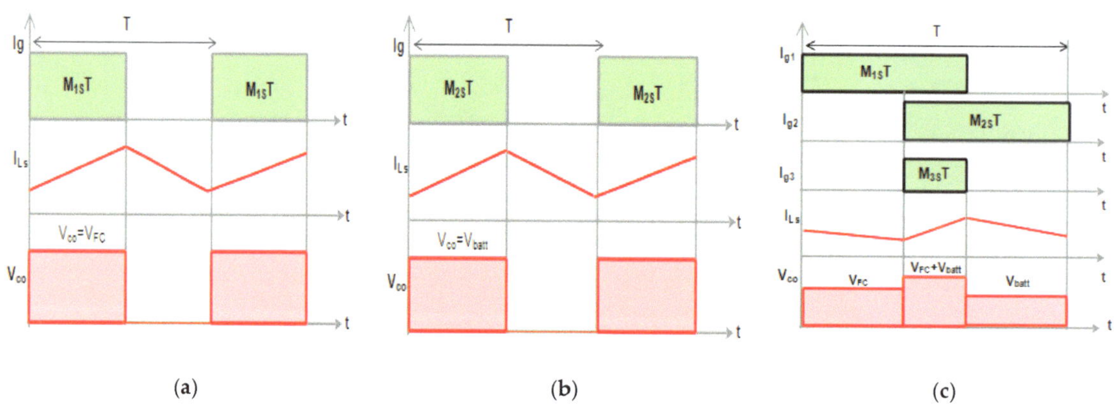

Figure 11. (**a**) Key waveform of Mode I operation; (**b**) Key waveform of Mode II operation; (**c**) Key waveform of Mode III operation.

5. Voltage Conversion Ratio

It is assumed that the inductance is sufficient to supply load continuously, and all semiconductor devices used in the converter are ideal. According to the voltage–time balance equation, the voltage applied across inductance over the period of time in Mode I can be expressed as

$$V_{out}(1-(M_{S1}+M_{S2})) = (V_{FC}-V_{out})M_{S1}+(V_{batt}-V_{out})M_{S2} \quad (26)$$

On simplifying the expression (26), the output voltage can be expressed as

$$V_{out} = V_{FC}M_{S1}+V_{batt}M_{S2} \quad (27)$$

Similarly, the voltage–time balance equation for the source inductance Ls for Mode II can be written as

$$(V_{FC}+V_{batt})M_{S3} = (-V_{FC}+V_{out})M_{S1}+(-V_{batt}+V_{out})M_{S2} \quad (28)$$

On simplifying the expression (29), the output voltage can be expressed as

$$V_{out}(2-(M_{S1}+M_{S2})) = V_{FC}M_{S1}+V_{batt}M_{S2} \quad (29)$$

$$V_{out} = \frac{V_{FC}M_{S1} + V_{batt}M_{S2}}{(2 - (M_{S1} + M_{S2}))} \tag{30}$$

Assuming V_{FC} and V_{batt} are equal for the ideal case ($V_{FC} = V_{batt} = V_{in}$), the voltage conversion ratio can be generalized as

$$G = \frac{V_{out}}{V_{in}} = \frac{M_{S1} + M_{S2}}{(2 - (M_{S1} + M_{S2}))} \tag{31}$$

6. SFO Algorithm

The sunflower optimization (SFO) algorithm has recently been introduced into the engineering era to solve non-linear problems. It imitates the orientation process of sunflowers towards the Sun to obtain maximum solar irradiance for the whole day. It is a well-known global search optimization technique rather than a local search optimization one. Sunflower fertilization happens during the orientation of the sunflower towards the Sun to produce the next generation of sunflowers. For simplicity, it is assumed that one pollen comes from the sunflower during the fertilization process. The sunflower inverse square law relates solar irradiance intensity and power from the Sun, as denoted below.

$$I_{SR} = \frac{P_{sun}}{4\pi r^2} \tag{32}$$

where I_{SR}—represents intensity of solar radiation, P_{sun}— represents Sun power, and r—distance between Sun and sunflower. The alignment of the sunflower towards the Sun is based on the following expression.

$$D_j = \frac{S^{best} - S_j}{\left\|S^{best} - S_j\right\|} \tag{33}$$

where S^{best}—represents the best solution, and S_j—jth iteration solution. The movement of the Sun in each step of the iteration is denoted as follows.

$$d_j = \delta \times P_j(\|S_j + S_{j-1}\|) \times \|S_j + S_{j-1}\| \tag{34}$$

where δ—the inertial displacement, and $P_j(\|S_j + S_{j-1}\|)$—the probability of pollination. To limit the control parameter within the control range, the following mathematical expression is followed.

$$d_{maxlim} = \frac{\|S_{max} - S_{min}\|}{2j} \tag{35}$$

where S_{max} and S_{min} are the maximum and minimum values of the solution, and j is the population size. The minimization of the cost function is updated for each iteration by following the below expression.

$$S_{j+1} = S_j + (d_j * D_j) \tag{36}$$

where S_{j+1} is the new sunflower orientation, S_j is the current position of the sunflower.

7. Simulation Results

The proposed converter is constructed using the following components listed in Table 2. Both the simulation and experimental setup are constructed using the same components as given below. To analyze the results, an extensive simulation analysis has been carried out using the MATLAB/R2002a platform. In this section, a detailed analysis of the mode of operation and mode transition is clearly elucidated.

Table 2. The proposed converter components.

S. No	Components Used	Symbol	Rating
1.	Power switch	M_1, M_2	5 A
2.	Auxiliary switch	A_s	2 A
3.	Diodes	D_1, D_2, D_3	5 A
4.	Source inductor	L_s	300 μH
5.	Output capacitor	C_o	230 μF (electrolytic)
6.	Load resistance	R_L	100 Ω

7.1. PEMFC Is Meeting the Demand Alone

A 6 kW 45 V PEM fuel cell is taken for this study. This stack contains 65 cells connected in series with nominal operating points of 133 A nominal current and 45 V open-circuit voltage, respectively. The nominal stack efficiency of this converter is around 46%. The nominal composition of hydrogen, oxygen, and water present in the air is taken in a ratio of 99:21:1. During Mode I, the fuel cell is able to meet the demand alone. As the fuel cell is fully supplied from hydrogen fuel, it can produce 45 V in the ideal case, and the same voltage is stepped up to 220 V by the proposed converter during the optimum power extraction conditions with the help of the SFO-based control technique. The duty cycle is maintained as 0.5, and the corresponding inductor current waveform is depicted in Figure 12.

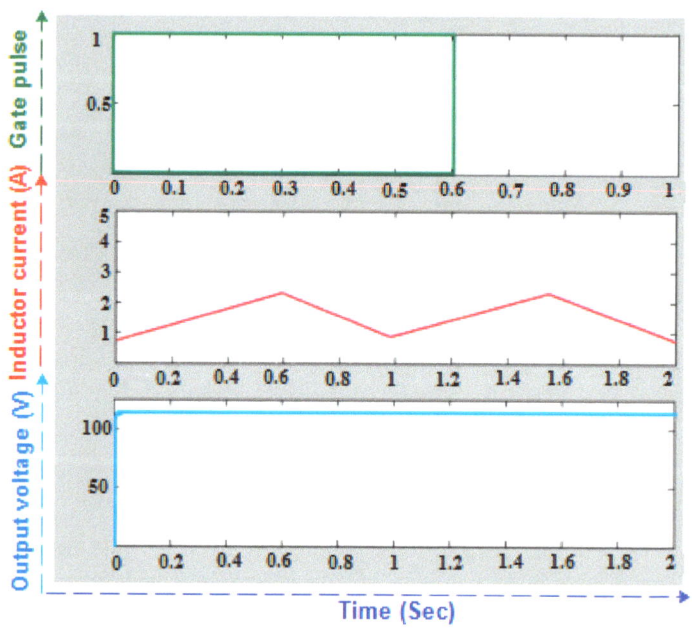

Figure 12. Mode I waveforms.

7.2. Battery Is Meeting the Demand Alone

In this mode, the battery takes over demand when the fuel for PEMFC is completely exhausted. A 24 V 80 AH heavy-duty lead-acid battery is taken as the source. The charging voltage of each cell is 2.2 V/cell under high temperature, and it is 245 V/cell under low temperature. Since heat dissipation is a major concern for hybrid electric vehicle applications, it is calculated as 0.45 W/cell for 80 AH. For a long battery life-span, the ripple content must be less than 5 A. The duration of this mode must be minimal since the battery rating is low for the sake of advantages to the hybrid vehicle. The duty cycle is maintained at 0.6, which can give the output voltage of 120 V, while the input voltage supplied from the battery is 24 V. The corresponding inductor current waveform is shown in Figure 13.

7.3. Both Sources Are Meeting Demand Together

This mode comes into action when the PEMFC's fuel is partially exhausted, it is in need of fuel filling, and the lead-acid battery is partially full enough to supply the demand. During this Mode III, both the switches, M1 and M2, are turned ON for the duty cycle of 0.5 and 0.6, respectively, which in turn conduct the auxiliary switch AS for the duty cycle of 0.1, as shown in the figure. Both input voltages (45 V, 24 V) are stepped up to 345 V, as depicted in the figure. The duration of this mode is greater than Mode II and less than Mode I. The inductor current waveform is presented in Figure 14.

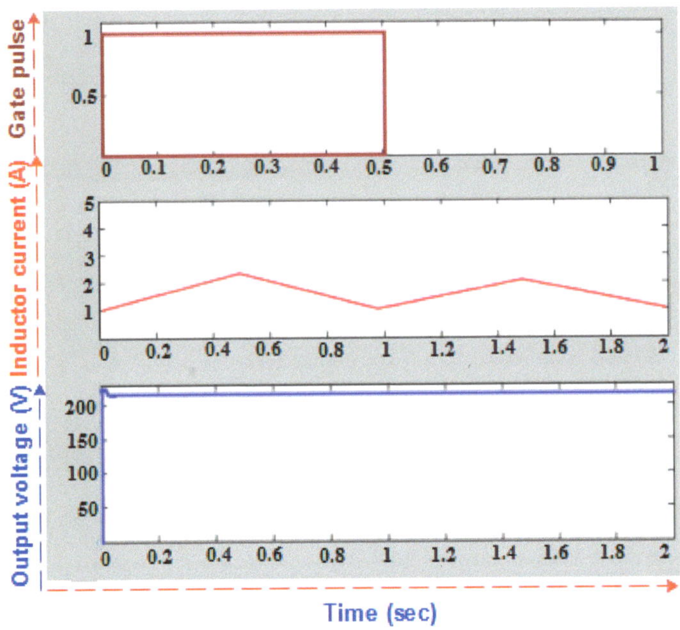

Figure 13. Mode II waveforms.

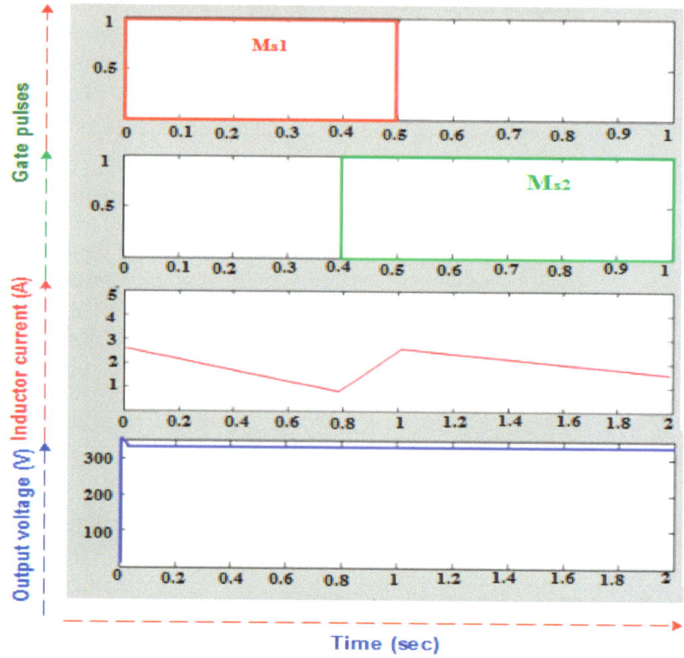

Figure 14. Mode III waveforms.

7.4. Performance of SFO Controller

To verify the effectiveness of the SFO control technique in tracing maximum power from PEMFC, the operating temperature is taken as a variable parameter, as discussed in Section 3. The variation in operating temperature is as follows: 54.84 °C for the first 0.3 s, 24.85 °C for the next 0.3 s, and 84.85 °C for the next 0.3 s. During this variation in temperature, the SFO controller is supposed to generate dynamic pulses to the switches to extract high power from PEMFC. The power-tracing capability is checked against the ANN controller and variable step-size FLC-based controller, as shown in Figure 15. The proposed SFO-based controller can yield better power-tracing capability compared with the other two controllers in terms of maximum power traced and transient parameters.

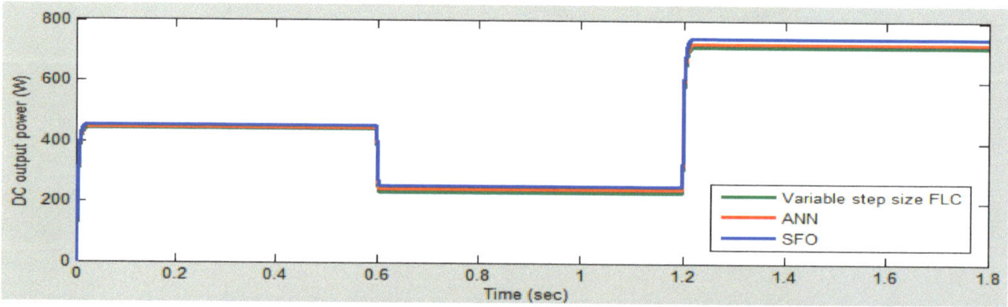

Figure 15. The performance comparison of the MPPT control technique.

8. Experimental Results

To validate the simulation results on a real-time test bench, a 0.2 kW hardware prototype was constructed in the laboratory using the hardware components listed in Table 2.

The entire experimental arrangement is shown in Figure 16. A 2 kW PEMFC and 80 AH 24 V Exide-made heavy-duty lead-acid battery are used as the sources for a multiport high-gain DC–DC converter.

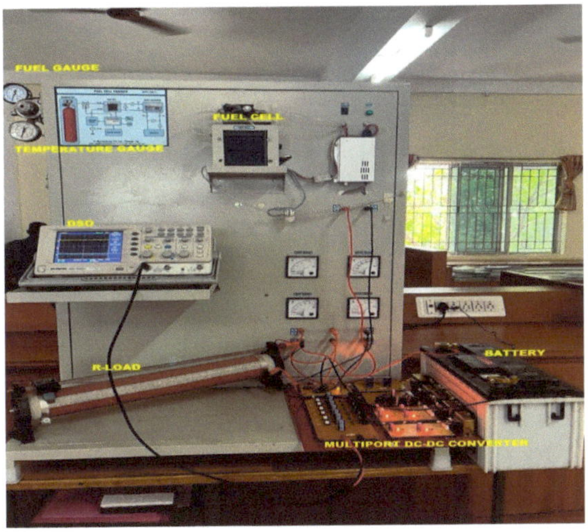

Figure 16. Experimental setup.

During Mode I, the fuel-cell voltage is stepped up to 215.6 V, and its corresponding inductor waveform is shown in Figure 17a. During Mode II, the battery voltage is stepped up to 115.6 V, and its corresponding inductor current waveform is depicted in Figure 17b. When both these sources supply demand, the corresponding output voltage is stepped up to 339.8 V, and its inductor current waveform is given in Figure 17c. The output power is measured across the load resistance 1 kΩ by adjusting from its minimum value to maximum value using the Tektronix SMU 2450 power-quality analyzer. It should be noted that the maximum power conversion efficiency is 95.12% for a 105 W load, as given in Figure 18. The various losses contributing to overall loss are measured using the same analyzer and shown in Figure 19. The performance comparison of the proposed converter is made against the state of the art, as listed in Table 3. It is observed that the proposed converter can handle more than one source with a minimum number of circuit components and yield a maximum conversion efficiency.

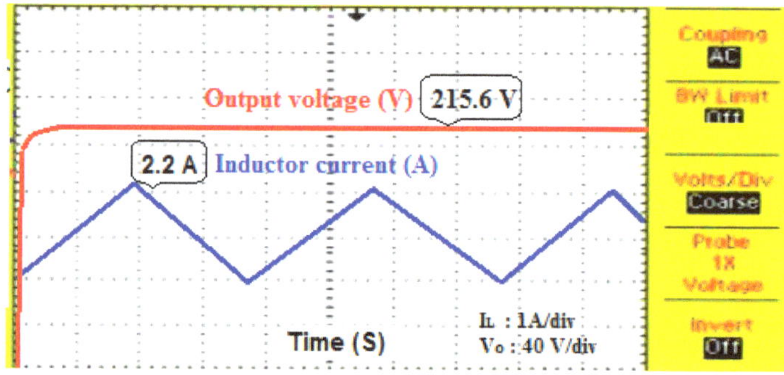

(a)

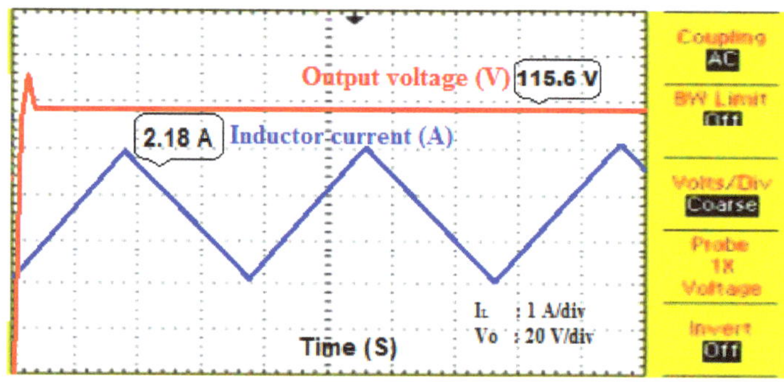

(b)

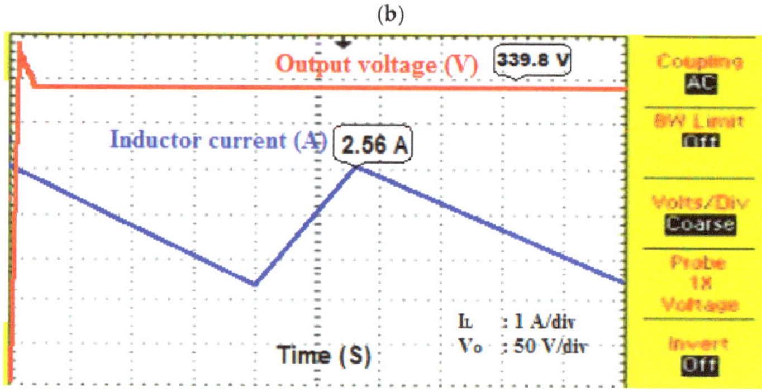

(c)

Figure 17. (**a**) Mode I inductor current and output voltage; (**b**) Mode II inductor current and output voltage; (**c**) Mode III inductor current and output voltage.

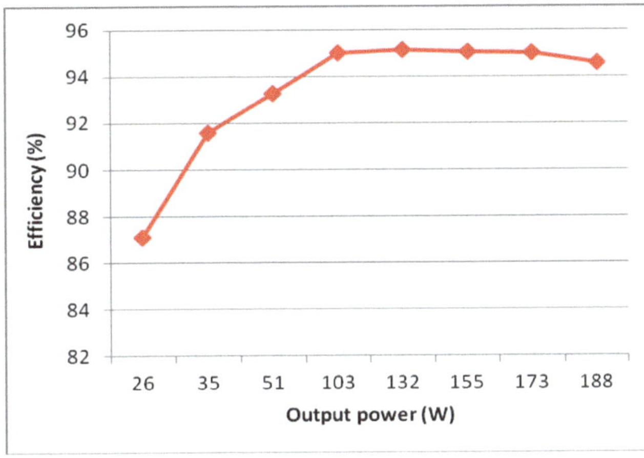

Figure 18. Conversion efficiency.

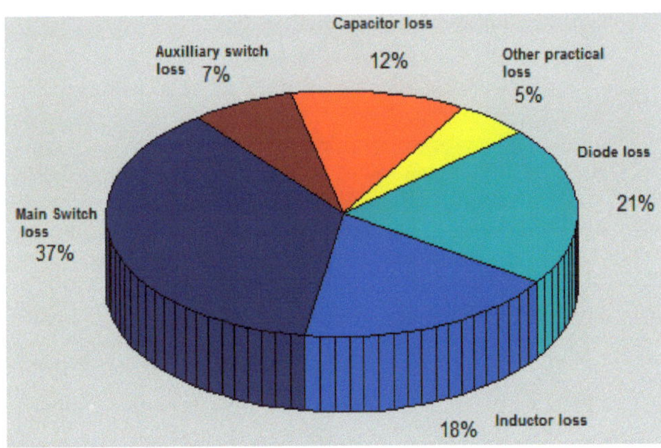

Figure 19. Converter loss breakdown.

Table 3. The performance comparison.

S. No	Reference No	Circuit Components				Peak Power Conversion η (%)	No of Ports
		No of Switches (M)	No of Diodes (D)	No of Capacitors (C)	No of Inductors (L)		
1	[14]	2	2	3	2	98.41	4
2	[15]	5	9	7	9	98.12	2
3	[16]	3	4	2	1	94.80	3
4	[17]	2	2	3	2	95.13	2
5	[18]	2	8	3	5	Not reported	2
6	[19]	4	3	5	5	98.1	3
7	Proposed converter	3	3	1	1	95.12	3

9. Conclusions

In this research, the optimization of 6 kW PEMFC operating parameters was carried out using a neural-power software environment. The operating parameters varied, and the corresponding output voltage and output power were measured experimentally. This dataset was used for training a multilayer feed-forward neural network, and then the results obtained after testing were optimized using a genetic algorithm. It was worth mentioning that the operating temperature was the most significant parameter in deciding the performance of PEMFC. This inference led to the design of the MPPT controller, which extracts optimum power from PEMFC under variable temperatures. A multiport high-gain DC–DC converter was designed to handle two sources, namely 6 kW PEMFC and an 80 AH 24 V heavy-duty battery with a smooth transition. A meta-heuristic algorithm called an SFO-based controller was designed to obtain optimum output power from PEMFC under dynamic operating temperature. The performance of the DC–DC converter, along with the SFO-based MPPT control technique, was tested using MATLAB/R2022a software environment. To validate the real-time test bench results, a 0.2 kW DC–DC converter was designed in the laboratory, and the results were compared against the literature. Future researchers may use advanced artificial intelligence (AI) techniques to optimize the operating parameters and design a Hardware In Loop (HIL)-based operating condition management system for hybrid electric vehicle applications.

Highlights of this research work:

- The PEMFC operating parameters were optimized, and it was experimentally proven that the operating temperature is the most significant factor.
- A multiport high-gain DC–DC converter can effectively handle PEMFC and batteries with inherent smooth mode transition.
- The SFO-based MPPT control technique outperforms ANN and variable fuzzy-based controllers under dynamic operating temperature conditions.
- The maximum conversion efficiency of a multiport high-gain DC–DC converter is 95.12%, with fewer components used in the converter.

Author Contributions: Conceptualization, Methodology, Original draft preparation; B.K.; Investigation, M.P.M.; Supervision, S.C.; Software, N.P.; Visualization, J.S.; Project Administration, G.F.S.; Writing—review and editing, formal analysis, P.R. All authors have read and agreed to the published version of the manuscript.

Funding: This research received no external funding.

Institutional Review Board Statement: Not applicable.

Informed Consent Statement: Not applicable.

Data Availability Statement: The data presented in this study are available on request from the corresponding author. The data are not publicly available due to further research being carried out by the same team.

Conflicts of Interest: Author George Fernandez Savari was employed by the company OES Technologies. The remaining authors declare that the research was conducted in the absence of any commercial or financial relationships that could be construed as a potential conflict of interest.

References

1. Pavanan, V.; Varadharajan, L. Optimization of various parameters for the performance enhancement of PEM Fuel Cell. *Indian J. Sci. Technol.* **2018**, *11*, 1–7. [CrossRef]
2. Wilberforce, T.; Ijaodola, O.; Emmanuel, O.; Thompson, J.; Olabi, A.G.; Abdelkareem, M.A.; Sayed, E.T.; Elsaid, K.; Maghrabie, H.M. Optimization of Fuel Cell Performance Using Computational Fluid Dynamics. *Membranes* **2021**, *11*, 146. [CrossRef] [PubMed]
3. Li, H.; Xu, B.; Lu, G.; Du, C.; Huang, N. Multi-objective optimization of PEM fuel cell by coupled significant variables recognition, surrogate models and a multi-objective genetic algorithm. *Energy Convers. Manag.* **2021**, *236*, 114063. [CrossRef]
4. Wei, D.; Ji, J.; Fang, J.; Yousefi, N. Evaluation and optimization of PEM Fuel Cell-based CCHP system based on Modified Mayfly Optimization Algorithm. *Energy Rep.* **2021**, *7*, 7663–7674. [CrossRef]

5. Wilberforce, T.; Olabi, A.; Monopoli, D.; Dassisti, M.; Sayed, E.T.; Abdelkareem, M.A. Design optimization of proton exchange membrane fuel cell bipolar plate. *Energy Convers. Manag.* **2023**, 277. [CrossRef]
6. Chen, Z.; Zuo, W.; Zhou, K.; Li, Q.; Huang, Y.; Jiaqiang, E. Multi-objective optimization of proton exchange membrane fuel cells by RSM and NSGA-II. *Energy Convers. Manag.* **2023**, *277*, 116691. [CrossRef]
7. Yuan, Z.; Wang, W.; Wang, H.; Razmjooy, N. A new technique for optimal estimation of the circuit-based PEMFCs using developed Sunflower Optimization Algorithm. *Energy Rep.* **2020**, *6*, 662–671. [CrossRef]
8. Kumar, P.; Kannaiah, S.K.; Choudhury, S.R.; Rajasekar, N. Genetic Algorithm-based Modeling of PEM Fuel Cells Suitable for Integration in DC Microgrids. *Electr. Power Compon. Syst.* **2017**, *45*, 1152–1160. [CrossRef]
9. Guo, C.; Lu, J.; Tian, Z.; Guo, W.; Darvishan, A. Optimization of critical parameters of PEM fuel cell using TLBO-DE based on Elman neural network. *Energy Convers. Manag.* **2019**, *183*, 149–158. [CrossRef]
10. Karthikeyan, P.; Mahadevan, K. Investigation on the effects of SiC particle addition in the weld zone during friction stir welding of Al 6351 alloy. *Int. J. Adv. Manuf. Technol.* **2015**, *80*, 1919–1926. [CrossRef]
11. Özdemir, M.T. Optimal parameter estimation of polymer electrolyte membrane fuel cells model with chaos embedded particle swarm optimization. *Int. J. Hydrogen Energy* **2021**, *46*, 16465–16480. [CrossRef]
12. Salva, J.A.; Iranzo, A.; Rosa, F.; Tapia, E.; Lopez, E.; Isorna, F. Optimization of a PEM fuel cell operating conditions: Obtaining the maximum performance polarization curve. *Int. J. Hydrogen Energy* **2016**, *41*, 19713–19723. [CrossRef]
13. Danoune, M.; Djafour, A.; Wang, Y.; Gougui, A. The Whale Optimization Algorithm for efficient PEM fuel cells modeling. *Int. J. Hydrogen Energy* **2021**, *46*, 37599–37611. [CrossRef]
14. Suresh, K.; Bharatiraja, C.; Chellammal, N.; Tariq, M.; Chakrabortty, R.K.; Ryan, M.J.; Alamri, B. A Multifunctional Non-Isolated Dual Input-Dual Output Converter for Electric Vehicle Applications. *IEEE Access* **2021**, *9*, 64445–64460. [CrossRef]
15. Lee, W.-S.; Kim, J.-H.; Lee, J.-Y.; Lee, I.-O. Design of an Isolated DC/DC Topology with High Efficiency of Over 97% for EV Fast Chargers. *IEEE Trans. Veh. Technol.* **2019**, *68*, 11725–11737. [CrossRef]
16. Mamouri, L.; Mesbahi, T.; Bartholomeus, P.; Paul, T. Design of a DC/DC Power Converter for Li-ion Battery/Supercapacitor Hybrid Energy Storage System in Electric Vehicles. In Proceedings of the 2020 IEEE Vehicle Power and Propulsion Conference (VPPC), Gijon, Spain, 18 November–16 December 2020; pp. 1–5.
17. Hu, R.; Zeng, J.; Liu, J.; Yang, J. Double-input DC-DC converter for applications with wide-input-voltage-ranges. *J. Power Electron.* **2018**, *18*, 1619–1626.
18. Karthikeyan, B.; Sundararaju, K.; Palanisamy, R. A variable step size fuzzy logic controller based maximum power point tracking controller for proton exchange membrane fuel cell powered resonant pulse width modulation high step up converter with multicarrier sinusoidal pulse width modulation inverter fed induction motor. *Int. Trans. Electr. Energy Syst.* **2021**, *31*, e13093. [CrossRef]
19. Aljarajreh, H.; Lu, D.D.C.; Siwakoti, Y.P.; Tse, C.K.; See, K.W. Synthesis and Analysis of Three-Port DC/DC Converters with Two Bidirectional Ports Based on Power Flow Graph Technique. *Energies* **2021**, *14*, 5751. [CrossRef]
20. Guo, X.; Ghadimi, N. Optimal Design of the Proton-Exchange Membrane Fuel Cell Connected to the Network Utilizing an Improved Version of the Metaheuristic Algorithm. *Sustainability* **2023**, *15*, 13877. [CrossRef]
21. Ibrahim, N.F.; Ardjoun, S.A.E.M.; Alharbi, M.; Alkuhayli, A.; Abuagreb, M.; Khaled, U.; Mahmoud, M.M. Multiport Converter Utility Interface with a High-Frequency Link for Interfacing Clean Energy Sources (PV\Wind\Fuel Cell) and Battery to the Power System: Application of the HHA Algorithm. *Sustainability* **2023**, *15*, 13716. [CrossRef]
22. Mahdinia, S.; Rezaie, M.; Elveny, M.; Ghadimi, N.; Razmjooy, N. Optimization of PEMFC Model Parameters Using Meta-Heuristics. *Sustainability* **2021**, *13*, 12771. [CrossRef]
23. Tehrani, K.; Weber, M.; Rasoanarivo, I. Design Of High Voltage Pulse Generator With Back To Back Multilevel Boost Buck Converters Using Sic-Mosfet Switches. In Proceedings of the 2020 IEEE 15th International Conference of System of Systems Engineering (SoSE), Budapest, Hungary, 2–4 June 2020; Institute of Electrical and Electronics Engineers (IEEE): New York, NY, USA, 2020; pp. 507–512.
24. Maheshwari, A.; Nageswari, S. Real-time state of charge estimation for electric vehicle power batteries using optimized filter. *Energy* **2022**, *254*, 124328. [CrossRef]
25. Adaikkappan, M.; Sathiyamoorthy, N. Modeling, state of charge estimation, and charging of lithium-ion battery in electric vehicle: A review. *Int. J. Energy Res.* **2022**, *46*, 2141–2165. [CrossRef]
26. Ding, R.; Zhang, S.; Chen, Y.; Rui, Z.; Hua, K.; Wu, Y.; Li, X.; Duan, X.; Wang, X.; Li, J.; et al. Application of Machine Learning in Optimizing Proton Exchange Membrane Fuel Cells: A Review. *Energy AI* **2022**, *9*, 100170. [CrossRef]

sustainability

MDPI

Article

Optimizing a Single-Slope Solar Still for Fresh-Water Production in the Deserts of Arid Regions: An Experimental and Numerical Approach

Ibrahim M. Al-Helal [1,*], Abdullah Alsadon [2], Samy Marey [3], Abdullah Ibrahim [2] and Mohamed R. Shady [1]

[1] Department of Agricultural Engineering, College of Food and Agriculture Sciences, King Saud University, P.O. Box 2460, Riyadh 11451, Saudi Arabia
[2] Department of Plant Production, College of Food and Agriculture Sciences, King Saud University, P.O. Box 2460, Riyadh 11451, Saudi Arabia
[3] Sciences, Technology and Innovation Unit, King Saud University, P.O. Box 2454, Riyadh 11451, Saudi Arabia
* Correspondence: imhelal@ksu.edu.sa; Tel.: +966-504420213

Abstract: Solar desalination is a promising sustainable solution to overcome the scarcity of fresh water in the deserts of arid regions. The productivity of a solar still depends mainly on its design parameters and the meteorological conditions of its location (longitude and latitude angles). Therefore, this study aimed to optimize the main design parameters of a single-slope solar still for freshwater production in the arid climate of the central region of Saudi Arabia (24°4′ N, 32.89° E). Experiments were conducted on four identical solar stills, with the same basin surface area and air gap distances (*d*) of 14, 16, 18, and 20 cm, respectively. The stills operated using three basin water depths (*h*) of 0.5, 1, and 1.5 cm on clear sunny days. The performance and productivity of the four stills were evaluated. The results showed that reducing the air gap distance (*d*) and water depth (*h*) significantly enhanced the distillate freshwater yield, and the optimum ratio of the length/width is 2 and of the back/front wall height is 3.65. Specifically, at a low water depth (*h*) of 0.5 cm, the daily distillate yield of the solar still increased by about 11% when the air gap distance (*d*) decreased from 20 to 14 cm. For the lowest air gap distance (*d*) of 14 cm, the distillate yield increased by about 23% when *h* decreased from 1.5 to 0.5 cm. Using the measured parameters, several numerical correlations have been developed to estimate the desalination rate (m_c) as a function of the solar irradiance (I_s) and ambient temperature (T_{am}). The developed correlations can be used successfully to estimate the values of m_c instead of the prohibitive experimental measurements. The stills showed excellent performance in the arid climate and reduced water salinity from 31,250 to 495 ppm. This should encourage decision-makers to expand investment in solar desalination to sustainably develop the deserts of arid regions.

Keywords: solar desalination; still; water depth; gap distance; arid climate; single slope

check for updates

Citation: Al-Helal, I.M.; Alsadon, A.; Marey, S.; Ibrahim, A.; Shady, M.R. Optimizing a Single-Slope Solar Still for Fresh-Water Production in the Deserts of Arid Regions: An Experimental and Numerical Approach. *Sustainability* **2024**, *16*, 800. https://doi.org/10.3390/su16020800

Academic Editors: Ayman Al-Quraan and Ahmad M. A. Malkawi

Received: 16 December 2023
Revised: 7 January 2024
Accepted: 10 January 2024
Published: 17 January 2024

1. Introduction

Freshwater is an essential requirement of life and sustainable development of all sectors in the deserts of the Arabian Peninsula. Freshwater scarcity is the main challenge in the deserts, despite the huge amount of brackish water resources. In the deserts of arid regions, freshwater scarcities are due to the high salinity of water resources and the rapid increase in population and agriculture extension needs. Recently, about 40% of the world's population has been located in remote deserts or islands, which, in most cases, do not have access to clean freshwater [1]. The process of obtaining freshwater by using the available clean, eco-friendly, and sustainable energy (e.g., solar energy) is a promising solution for freshwater scarcity (known as solar–thermal desalination). Moreover, the desalination of saline water by using solar stills is economically profitable because of the low cost and simple design and construction of the stills [2]. Passive solar stills are more convenient and profitable because they depend only on solar energy for heating [3]. The yield of the low-cost passive solar still depends on several parameters, such as the climatic conditions (solar radiation intensity, wind speed, ambient temperature, etc.), structural design (slope angle of the cover, side walls heights, relative dimensions, absorber specifications, etc.), and operation conditions (water depth and flowrate, surface area of evaporation, etc.) [4,5].

Several studies have tested the effect of operation and design parameters on the performance and production rate of solar stills. The most common parameters in the previous studies are the air gap distance (the average distance between the basin water surface and cover) and the basin-water depth. Regarding the air gap distance, solar stills with various air gaps were examined under the climatic conditions of Ankara (40° N, 33° E), Turkey [6]. Their result showed that the distillate yield increased by 11% when the air gap was reduced from 13 to 8 cm. They [6] reported that under the climatic conditions of Ankara, Turkey, the air gap distance should be 8 cm or lower for desired freshwater production. The effect of the air gap distance on the yield of a solar still was studied under the climate of Tamil Nadu, India [7]. They used a cover slope angle of 10° which is approximately equal to the test site latitude (9.9° N). The results showed that productivity increased by 1.84 times when the air gap distance decreased from 0.45 m to 0.15 m [7]. The effect of two air gap distances (6.6 cm and 26.6 cm) with a water depth of 1 cm was examined at a location of 13° N and 80° E [8,9]. Their results showed that the productivity of the distillate freshwater increased when the average air gap distance of the solar still was decreased. This was due to the high rate of the convective heat transfer between the condensing cover and absorber. For the same absorber dimensions, the effect of changing the elevations of the north and south walls on the productivity of a single-slope solar still have been examined [10]. The results showed a significant increase in the yield of the solar still when the elevation of the north wall increased [10].

A survey in the literature revealed that similar solar stills, having a typical air gap distance, installed in different locations (latitude and longitude) produced different desalination rates. However, in a specific location, such as the central region of the Kingdom of Saudi Arabia (KSA), and under arid climatic conditions, such information is still unclear; therefore, in situ experiments are necessary to show the effect of air gap distance (d) on the performance of a solar still.

On the other hand, the effect of basin water depth (h) on the performance of solar stills has been evaluated in different locations worldwide and under various climatic conditions. A study in Bagdad (33.3° N, 44° E), Iraq [11] found that increasing the basin water depth, from 1 to 10 cm, reduced the still productivity by about 48%, even though increasing the water depth (h) increases the stored heat energy in the basin water during the daytime and makes the production continue even at night. However, increasing the water depth (h) from 1 cm to 7.5 cm decreased the productivity by about 77% [12]. A study in Tamil Nadu (9°11′ N, 77°52′ E), India examined the effect of increasing h from 1 cm to 5 cm on the productivity of both single- and double-basin solar stills [13]. They reported that the maximum productivity was at the lowest water depth of 1 cm, at which, the double-basin produced 17.38% more than the single-basin stills [13]. The effect of h as 2, 4, 8, and 16 cm on the solar still productivity in summer and in winter were examined under the climatic conditions of Shiraz (29.6° N, 52.5° E), Iran [14]. They reported that the productivity decreased as h increased; however, the nocturnal yield increased. The performance of passive and active solar stills at various h of 0.5, 1.0, and 1.5 cm were tested in the New Delhi (28.3° N, 77.2° E) climate [15]. They reported that increasing h decreased the temperature and internal energy of basin water, and thus the productivity of the solar still decreased. Different water depths, h (2, 3, 4, 5, and 6 cm) for passive and active solar stills were evaluated under the climatic conditions of Andhra Pradesh (16.51° N, 80.52° E), India [16]. They reported that h of 4 cm is optimum for the specific climate of the experimental site, and the active solar still enhanced the yield by about 57.55% compared with the passive solar still.

Based on the previous studies, large numbers of optimum air gap distances (d) and basin water depths (h) were recorded based on the locations of the experimental sites (latitude and longitude) and the meteorological conditions of these sites (solar radiation intensity, sunshine duration, air temperature, wind speeds, etc.). In addition, the available previous studies conducted under climatic conditions are completely different from the arid climate of the central region of KSA. In arid climates, intensive solar radiation flux, high air temperature, low relative humidity, and long periods of sunshine are common in most of the months of the year [17]. However, a similar study used to determine the optimum air gap distance (d) and basin water depth (h) for single-slope solar stills under harsh, arid climatic conditions of the KSA is still missing and urgently required.

Accordingly, this study aimed to determine the optimum design parameters of the simplest type and low-cost solar still (i.e., single-slope). These parameters are the air gap distance (d), basin water depth (h), and back/front wall height ratio; these would be optimized for the harsh arid climatic conditions of the Arabian Peninsula. Even though wind stream may have a positive effect on cooling the glass cover and enhancing the still productivity, the effect of wind was excluded because of the very low wind speed over the stills during the days of the experiments. For this purpose, four identical insulated wooden frame single-slope solar stills covered with glass sheets were constructed with different air gap heights (d) of 14, 16, 18, and 20 cm. The stills would be operated, in parallel, for several clear sunny days using various water depths (h) of 0.5, 1.0, and 1.5 cm, one for each day. The experiments would be conducted during the daytime; however, the nighttime period is beyond the scope of this study.

2. Materials and Methods

2.1. Experimental Setup and Measuring Procedure

The experiments were conducted on clear sunny days under solar and thermal irradiance on the roof of the building of the Agricultural Research and Experiment Station at King Saud University (Riyadh, Saudi Arabia, 46°47′ E, longitude and 24°4′ N, latitude). The measurements were carried out in October 2023 from 7:0 to 19:0 for 12 h every day. During the experiments, the wind speed was very low; in addition, the roof of the building was surrounded by an external wall of 2 m height, which was relatively far from the solar stills; therefore, the effect of wind over the still cover was neglected. Four wooden frame single-slope solar stills (designated as A, B, C, and D), with identical cover slope angles of 24.4° (the latitude angle of Riyadh), and different air gap distances (d) of 14, 16, 18, and 20 cm for stills A, B, C, and D, respectively, were constructed. Wood bars of 5 cm × 5 cm cross section for each were used for constructing the frames. The front wall height (H_o) of the lowest solar still (still A) was decreased as much as possible to allow the cover, trough, basin, and base (h_o) to be installed properly. The air gap distance (d) was estimated according to Figure 1 as $d = (H + H_o)/2 - (h + h_o)$. A schematic diagram showing the layout dimensions (length, L = 100 cm, and width, W = 50 cm) of the solar still used in the study is illustrated in Figure 1. A black-painted galvanized iron sheet, in the form of a tray, was constructed to be used as an absorber basin; the galvanized iron sheet was also used to perform the trough (Figure 1). The basin surface area (A_s) is 4050 cm^2, with a dimension of 90 cm length, and 45 cm width (Figure 1); the aspect ratio (length/width) of 2 was recommended by [18] as an optimal ratio to efficiently collect solar radiation in the solar still. A condensing glass cover of 5 mm thickness was fixed and sealed on the top of each solar still cavity, each having a surface area of about 5400 cm^2. To minimize the heat losses, the outer surfaces of the frame (side walls and base) were insulated using glass wool of 5 cm thickness (thermal conductivity, k = 0.03 W m^{-2} °C^{-1}). The covers of the solar stills faced to the south direction to collect most of the incident solar radiation during the day. For the stills A, B, C, and D, the layout dimensions are summarized in Table 1. A schematic diagram of the experimental setup, including the four solar stills, is illustrated in Figure 2. Before starting the actual measurements, the experiment was conducted for several days to fix any problems, such as leakage, measuring errors, etc., and to reach steady state operation conditions. During the 9 days of measurements (10–18 October), the water depth (h) in the basin was kept constant for each three consecutive days of measurements; it was 0.5, 1.0, and 1.5 cm. This was achieved by continuously supplying feed water, equal to the amount of distillate-collected water, to the solar still basin. The working fluid was Red Sea water taken from Duba Coastal, Tabuk, KSA, with a salinity of 31,250 ppm.

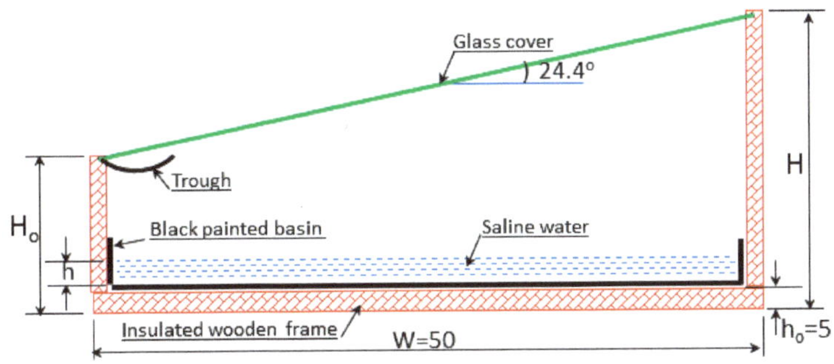

Figure 1. Schematic diagram and layout dimensions in cm, not to scale, for the solar still used in the study.

Table 1. Layout dimensions of the four solar stills used in the study.

Symbol	Definition	Still A	Still B	Still C	Still D	Units
L	(length)	100	100	100	100	cm
W	(width)	50	50	50	50	cm
H	(back wall height)	30.63	32.63	34.63	36.63	cm
H_o	(front wall height)	8.38	10.38	12.38	14.38	cm
h_o	(basin depth)	5	5	5	5	cm
α	(latitude angle)	24.4°	24.4°	24.4°	24.4°	degree
d	(air gap distance)	14	16	18	20	cm

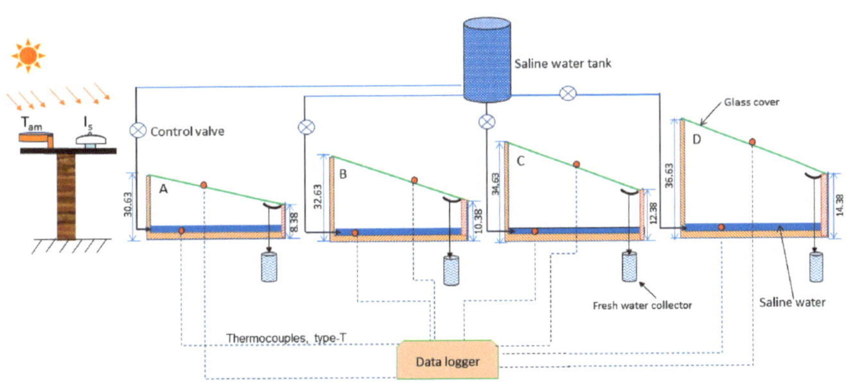

Figure 2. Schematic diagram of the experimental setup including the four solar stills (A, B, C, and D) used in the study. Dimensions in cm, not to scale.

The ambient air temperature (T_{am}) was measured using an aspirated psychrometer. The temperatures of the outer surface of the glass cover (T_g) and basin water (T_w) were measured using type-T copper constantan thermocouples of 0.3 mm in diameter, (Reotemp Instruments Co., San Diego, CA, USA). The thermocouple sensors used to measure T_g were covered with strips of aluminum foil to eliminate the effect of radiation on the thermocouple reading. The global solar radiation flux (I_s) was measured beside the stills, at a height of the covers level, using a CMP3 pyranometer (Kipp and Zonen, Sterling, VA, USA). The thermocouple sensors and pyranometer used in the experiment were calibrated by the supplier before use to eliminate any expected errors. All the required parameters were measured at every 10 min interval, averaged at every hour, and saved in a data logger (CR3000 Micrologger®, Campbell Scientific Inc., Logan, UT, USA). However, the distillate yield was collected manually and measured every hour. During the experiment, the four solar stills (A, B, C, and D) were operated in parallel, with the same water depth (h) in the four stills, for three consecutive days for each water depth, and the average value of each parameter was obtained. The average air gap distance (d) changes according to the water depth (h); the values of d according to h and the days of measurements are illustrated in Table 2.

Table 2. Values of d and h and the days of measurements for the solar stills.

Date of Experiments	Height (cm)	Still A	Still B	Still C	Still D
(10,11,12) October 2023	d	14	16	18	20
	h	0.5	0.5	0.5	0.5
(13,14,15) October 2023	d	13.5	15.5	17.5	19.5
	h	1	1	1	1
(16,17,18) October 2023	d	13	15	17	19
	h	1.5	1.5	1.5	1.5

2.2. Theoretical Approach

To evaluate the solar still, a mechanism to describe the evaporation process (heat and water vapor exchanges) is required. The key factors in this process are the evaporative and convective heat transfer coefficients (h_e and h_c). Once these coefficients were determined, the production rate (yield) of a solar still can be estimated theoretically without the need for measurements. Several analytical expressions (models) have been reported in the literature for these coefficients; we have selected three of them to calculate h_e and h_c (i.e., Dunkle [19], Kumar and Tiwari [20], and Zheng et al. [21]). For simplicity, among these models, several assumptions have been considered such as (i) the heat loss from inside to outside the solar still was neglected, (ii) there is no heat generation in the solar still, (iii) the basin-water depth is fixed as having a uniform temperature of T_w, (iv) all thermophysical properties of the humid air in the still are calculated at the mean temperature of the basin water surface (T_w) and condensation surface (T_g) of the solar still, and (v) the evaporation rate (m_{ev}) is equal to the condensation rate (m_c) which is completely collected to be as the production rate of the solar still.

The maximum evaporation or desalination rate of a solar still, (kg s^{-1}) can be calculated as follows:

$$m_{ev} = m_c = \frac{q_e}{h_{fg}} \tag{1}$$

where h_{fg} is the latent heat of vaporization of water (J kg^{-1}), and q_e is the evaporative heat transfer rate (W) and is given by the following:

$$q_e = h_e(T_w - T_g) \tag{2}$$

where h_e is the evaporative heat transfer coefficient (W m^{-2} °C^{-1}), T_w is the basin water temperature (°C), and T_g is the inner surface temperature of the glass cover (°C). An expression to estimate h_e is given by [19] as follows:

$$h_e = 0.016273 \times h_c \times \frac{P_w - P_g}{T_w - T_g} \tag{3}$$

where h_c is the convective heat transfer coefficient (W m^{-2} °C^{-1}), P_w (N m^{-2}) is the partial pressure of the humid air at the basin water temperature (T_w), and P_g (N m^{-2}) is the partial pressure of the humid air at the glass cover temperature (T_g).

In Equation (3), the convective heat transfer coefficient (h_c) is estimated using a semi-empirical expression proposed by the following numerical models:

(i) Dunkle model [19]

$$h_c = 0.884(\Delta T)^{1/3} \tag{4}$$

where ΔT (°C) is the temperature difference between the basin water and glass cover surface and is given by [19] as follows:

$$\Delta T = \left[T_w - T_g + \frac{(P_w - P_g)(T_w + 273)}{268.9 \times 10^3 - P_w}\right] \tag{5}$$

The partial pressures of water vapor in Equation (5) are given by [19] as follows:

$$P_w = exp\left(25.317 - \frac{5144}{T_w + 273}\right) \tag{6}$$

$$P_g = exp\left(25.317 - \frac{5144}{T_g + 273}\right) \tag{7}$$

(ii) Kumar and Tiwari model [20]

They considered the characteristic length of N_u as the average air gap distance (d) to express h_c in the following form:

$$N_u = \frac{h_c * d}{K} = c(Gr * \mathrm{Pr})^n \tag{8}$$

where Gr, Pr, d, and k are the Grashof number, Prandtl number, average air gap distance, and thermal conductivity of the humid air, respectively. The numerical constants c and n are dependent on the boundary conditions and the flow regime. Dunkle [19] has suggested that for humid air enclosed in a horizontal space, the numerical constants are $c = 0.075$ and $n = 0.333$ for ($3.22 \times 10^5 < Gr < 1 \times 10^7$). Hollands et al. [22] suggested that for $Ra > 5.5 \times 10^6$, the corresponding values are $c = 0.055$ and $n = 0.\ 333$. G_r and P_r dimensionless numbers are defined as follows:

$$G_r = \frac{g\ \beta\ \Delta T\ d^3}{\mu^2/\rho^2},\quad P_r = \frac{C_p\ \mu}{K} \tag{9}$$

where g is the gravitational acceleration (m s^{-2}), ρ is the humid air density (kg m^{-3}), β is the thermal expansion factor (=$1/T$, T is the humid air temperature in degree K), C_p is the specific heat of the humid air (J kg^{-1}.°C^{-1}), and μ is the dynamic viscosity of the humid air in the still (N s m^{-2}).

(iii) Zheng et al. model [21]

$$h_c = 0.2 \times (R_a)^{0.26}\left(\frac{K}{d}\right), \tag{10}$$

$$R_a = \frac{\rho\beta g d^3}{\mu\alpha}\left[T_w - T_g + \frac{P_w - P_g}{\frac{M_a P_t}{M_a - M_w} - P_w} \times (T_w + 273)\right] \tag{11}$$

where α is the thermal diffusivity (m^2 s^{-1}), M_a is the molecular weight of dry air (kg mol^{-1}), M_w is the molecular weight of water vapor (kg mol^{-1}), and P_t is the total pressure of the humid air in the solar still (N m^{-2}). For conventional passive solar still, the overall daily efficiency (η_d) is given by [23] as follows:

$$\eta_d = \frac{\int_{t1}^{t2} m_c(t) h_{fg} dt}{\int_{t1}^{t2} I_s(t)\, dt} \quad (12)$$

where t_1 and t_2 are the sunrise and sunset time; m_c(t) is the rate of distilled water at a specific time *t* (kg $s^{-1}m^{-2}$), and I_s(t) is the global solar radiation flux over the solar still (W m^{-2}). The physical properties of the humid air in the solar still, such as specific heat (C_p), density (ρ), thermal conductivity (*k*), viscosity (*μ*), latent heat of vaporization (h_{fg}), and thermal expansion factor (*β*), were estimated as a function of the glass and basin water mean temperature, T_m [$T_m = (T_g + T_w)/2$], using well-known correlations reported in [24].

3. Results and Discussions

3.1. Evaluation of the Solar Stills

(i) *Effect of water depth (h) and air gap distance (d)*

In order to optimize the most important parameters of a single-slope solar still (basin water depth, *h* and air gap distance, *d*), four identical solar stills (A, B, C, and D) having different *d* values (14, 16, 18, and 20 cm) were operated, in parallel, for nine days using different *h* values (0.5, 1, 1.5 cm); measurements were conducted for three consecutive days using constant water depth, and the measured parameters were averaged as one day for each water depth. The daily production for each solar still (in kg per m^2 of basin water surface) was collected for the three cases of *h* (0.5, 1, 1.5) and is illustrated in Figure 3. Based on the resulting yields in Figure 3, still A (*d* = 14 cm) showed the highest production rate (5.4 kg m^{-2}) at a water depth of 0.5 cm compared to the other three stills (B, C, and D). The proper design/construction considerations (e.g., well insulation, high transparency of the glass cover, clear sunny weather) that were considered have enhanced the desalination yields of the solar stills to reach 5.4 kg m^{-2} (for still A) and 4.7 kg/m^2 (for still D) at a constant water depth (*h*) of 0.5 cm.

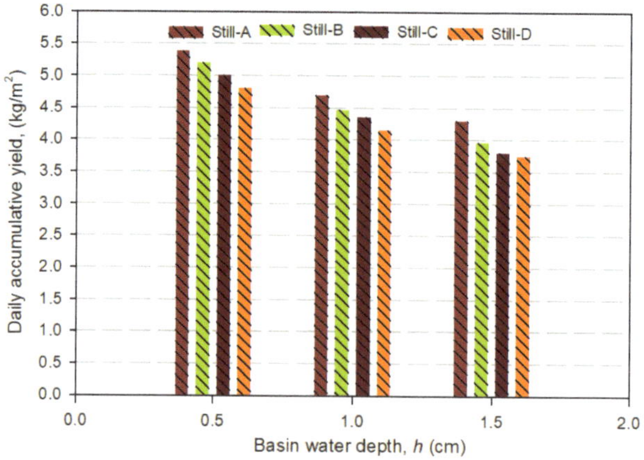

Figure 3. Daily accumulative yields (kg/m^2) at different basin water depths for the four solar stills tested in the study (10–18 October 2023).

For more clarification on the effects of *d* and *h* on the still performance, the daily yields of the four solar stills (A, B, C, and D) were estimated (in kg per m^2 of basin surface area) and are plotted in Figure 4a against the basin water depth (*h*). Under the same conditions of solar irradiance and ambient temperature, still A (having the lowest air gap distance of 14 cm) showed the highest daily yield compared to the other stills having higher air gap distances. In each solar still, the daily yield decreases as the water depth increases (Figure 4a). This is because for the same solar irradiance, increasing water depth would increase the amount of water in the basin and reduce the water temperature; thus, the distillate freshwater yield would decrease.

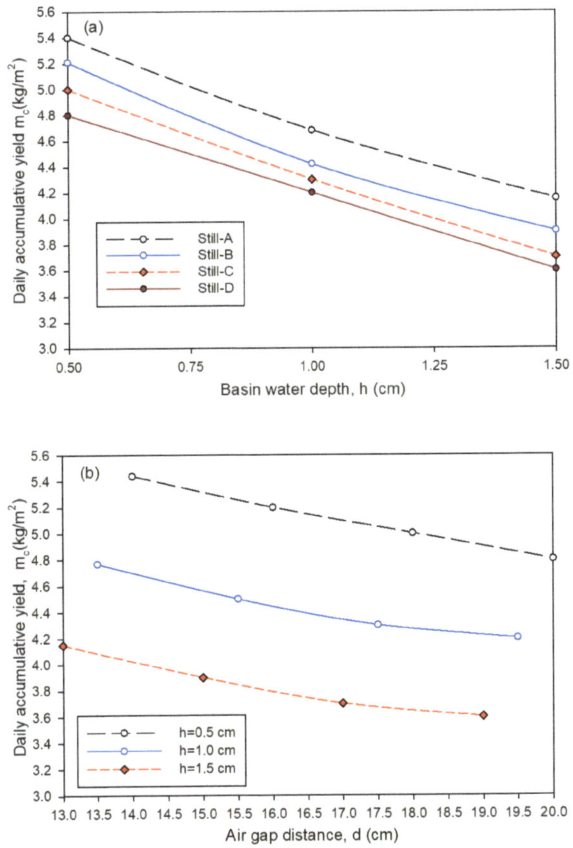

Figure 4. The daily accumulative distillate yields as affected by (**a**) the water depth (h), and (**b**) air gap distance (d) for the tested solar stills A, B, C, and D (10–18 October 2023).

On the other hand, increasing the basin water depth would attenuate the transmitted solar irradiance to reach and be absorbed by the absorber surfaces (i.e., the black-painted inner surface of the basin). For any water depth (h), the reduction in the air gap distance (d) improves the distillate yield, as shown in Figure 4b. The decrease in the air gap height would eliminate the shading effect of the side walls on the basin surface, allowing more solar irradiance to be absorbed by the basin water, and the evaporation as well as the desalination rate increases. Based on Figure 4a,b, the daily distillate yield of the solar still at a water depth (h) of 0.5 cm increased by about 11% when the air gap distance decreased from 20 to 14 cm. For an air gap distance (d) of 10 cm, the distillate yield increased by about 23% when the water depth decreased from 1.5 to 0.5 cm. This indicated that, during the daytime, the effect of basin water depth, h, on the still productivity, m_c, is much higher than the effect of the air gap distance, d.

The basin water depth (h) is an operator choice, whereas the air gap distance (d) is a designer choice; it depends on the still base dimensions (L, W, H, and H_o); d should be reduced as much as possible to enhance the still performance. To calculate the optimum value of d, designers should go through the following steps: (i) selecting L and W ($L/W = 2$, recommended) based on the requirements of customers (i.e., the amount of the freshwater), (ii) selecting the lowest height of the front wall (H_o) and the basin depth (h_o) based on the lower limit of design considerations, (iii) determining the cover slope angle (α), i.e., equal to the latitude angle of the location, and (iv) determining the back wall height (H) as $H = H_o + W \times \tan(\alpha)$. Then, the optimal air gap distance (d) can be calculated. In addition, an optimum value of the back-to-front wall height (H/H_o) of 3.65 should be taken into account. A proper design of a solar still would produce at least 5 kg m^{-2} per day, and according to the World Health Organization [25], a solar still having a basin surface area of 10–20 m^2 is required to meet the needs of one person per day. Scaling up of the proposed design is possible; the only limitation is the breakability of the glass cover. To avoid this, rigid supports (e.g., rigid plastic or stainless-steel bars) should be used below the glass cover to carry the weight of the cover; for a glass cover of 1 m width, one bar for every 1–1.5 m interval of glass span length is suggested based on our observations on the present experiments.

(ii) *The still daily overall efficiency, η_d*

The daily overall efficiency (η_d) of the four solar stills was calculated at the three water depths of 0.5, 1, and 1.5 cm using Equation (12), and the resulting values are depicted in Figure 5. The results showed that for a water depth of 0.5–1 cm, the solar still-A showed a higher daily overall efficiency (η_d = 46–53%) compared to the other stills tested. This result emphasizes that under the arid climatic conditions of Riyadh area, a water depth (h) of 0.5–1 cm and an air gap distance (d) of 14 cm or lower are the desired dimensions to design/construct a single-slope solar still; this can produce about 5 kg/m^2 of freshwater per day. On the other hand, values of η_d are quite low for the four solar stills (A, B, C, and D); this is because Equation (12) is the measure of how much the incident solar radiation over the still cover is absorbed by the basin water and converted to water vapor and then condensed on the cover surface. This hypothesis combined the cover transmittance with all the thermal losses from the still to the surroundings and considered the solar still as a solar collector. However, if Equation (12) was divided by the cover transmittance, then the resulting efficiency should be defined as the daily conversion efficiency of a solar still. In the worst case, by considering the cover transmittance of 85%, the resulting daily conversion efficiency (η_{co}) would increase by 17.6% for the four solar stills in Figure 5.

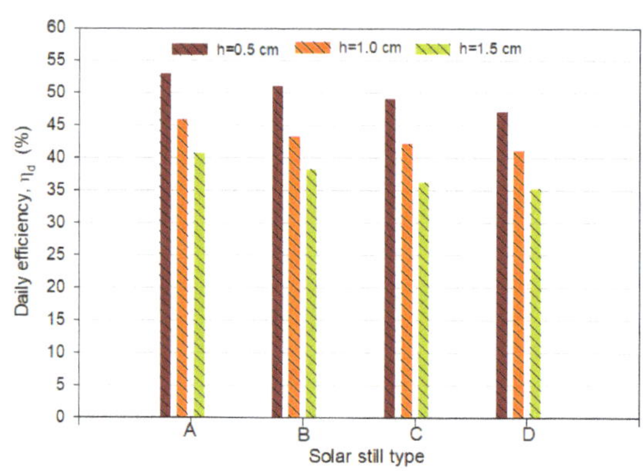

Figure 5. Daily overall efficiency of the tested solar stills (A, B, C, and D), estimated at the three water depths (h) of 0.5, 1, and 1.5 cm.

It is well known that the evaporation, as well as the desalination rate, of a solar still depends mainly on the intensity of solar irradiance and the design configuration of the still. In order to predict the desalination rate simply without the need for experiments, which is mainly induced by solar irradiance, the hourly distillate yields of the tested solar stills (A, B, C, and D having air gap distances of 14, 16, 18, and 20 cm, respectively) were obtained for the basin water depths of 0.5, 1, and 1.5 cm. All the yield data were gathered and plotted against the corresponding values of the hourly average solar radiation flux in Figure 6a (for the data before noon, from 7:0–12:0) and in Figure 6b (for the data after noon, from 13:0–19:0). Applying the regression analysis to the data in Figure 6a,b, the hourly distillate yields (m_c in kg m^{-2}) could be correlated before noon (R^2 = 0.92, standard error of estimate, SEE = 0.057, and $p < 0.0001$) and after noon (R^2 = 0.91, SEE = 0.072, and $p < 0.0001$) as a function of solar radiation flux (I_s in W m^{-2}) in the following form:

Before noon

$$m_c = -0.397 + 0.003(I_s) - 5.99 \times 10^{-6}(I_s)^2 + 4.4 \times 10^{-9}(I_s)^3, R^2 = 0.92 \quad (13)$$

After noon

$$m_c = 0.05 + 0.0009(I_s) - 7.1 \times 10^{-8}(I_s)^2, \quad R^2 = 0.91 \quad (14)$$

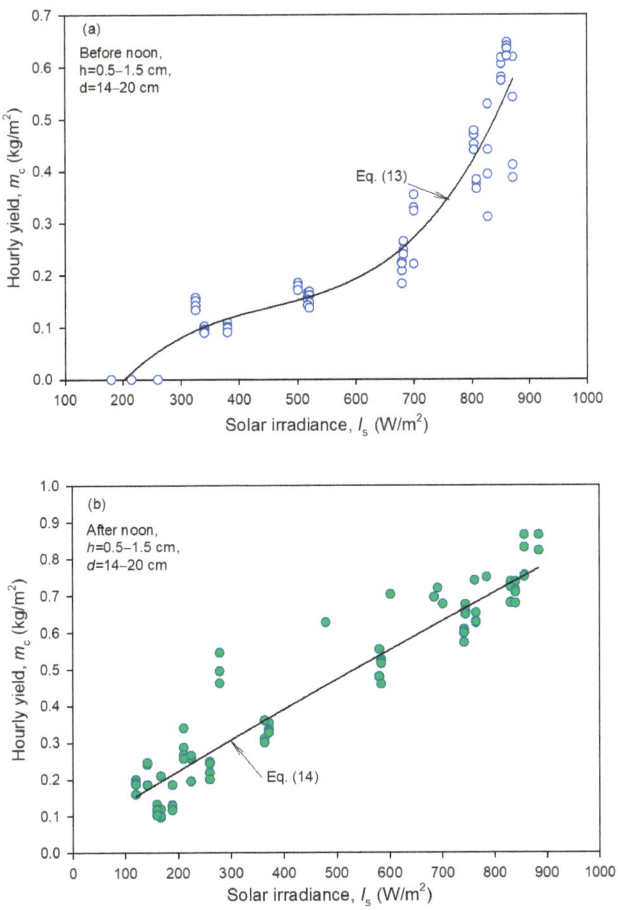

Figure 6. The hourly yields of the four solar stills (A, B, C, and D) as affected by solar irradiance intensity, estimated before noon (**a**), and after noon (**b**) for the three water depths (*h*) of 0.5, 1.0, and 1.5 cm (10–18 October 2023).

Equations (13) and (14) are valid for basin water depth of 0.5–1.5 cm and air gap distance of 14–20 cm and can be used, as a useful and easy tool, to estimate the expected freshwater yield, with a maximum possible error $\leq 8\%$ (before and after noon) in the arid area of the Arabian Peninsula as a function of the solar irradiance intensity (W m^{-2}), which is easily can be obtained from any meteorological station in this area. Based on such quick estimation, designers can perform feasibility studies and put the outline dimensions of a solar desalination project with a proper expectation of the outcomes.

(iii) *Diurnal variation of the desalination rate, m_c*

The results in the previous figures (Figures 3–5) revealed that operating the solar still-A, having the lowest air gap distance (d) of 14 cm, at the lowest water depth (h) of 0.5 cm is more efficient, giving higher freshwater production than the other solar stills and water depths. To show the daily behavior of the still A, the hourly yield of the still at three different water depths is illustrated in Figure 7a. The hourly distillate yield of the solar still A at h = 0.5 cm was the highest over all the solar stills and depths throughout the experiment, (Figure 7a). This is due to the highest solar energy absorbed (because d is the lowest), as well as the highest temperature of the basin water, T_w (because h is the lowest). Specifically, at d = 14 cm (still-A), when the basin water depth (h) decreased from 1.5 to 0.5 cm, the cumulative distillate yield increased by about 26% (Figure 7b).

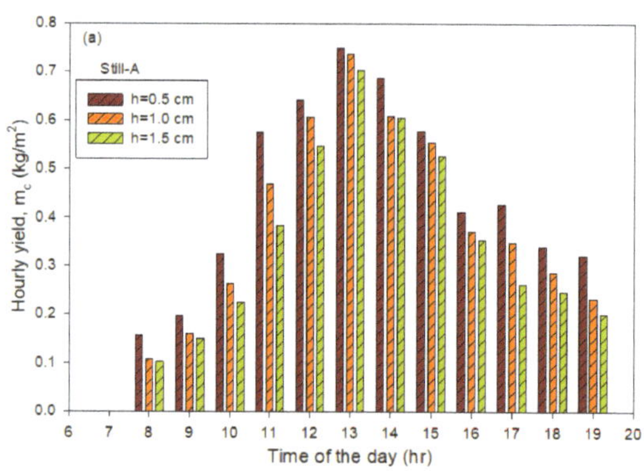

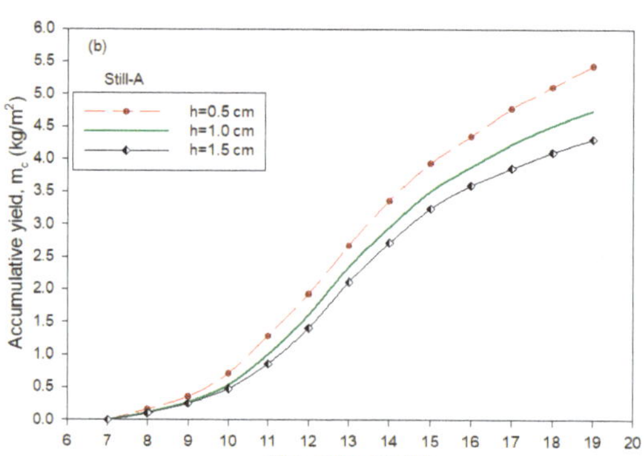

Figure 7. Diurnal variation of the hourly yield (**a**), and the accumulative yield (**b**) for the solar still A at different water depths (h), during the period of the experiment (10–18 October 2023) in Riyadh climate.

3.2. *Analytical Expressions to Estimate the Desalination Rate, m_c*

The desalination rate (m_c) can be predicted by conducting experiments on a real solar still, or by using analytical expressions (i.e., theoretical models, Section 2.2). Numerical expressions, as in Equations (13) and (14), provide a quick, less expensive, and more fixable and repeatable way compared with the expensive and time-consuming experimental prediction. Several analytical expressions have been developed in the past to predict m_c, and the input parameters to these expressions are the glass cover and basin water temperatures (T_g and T_w). In addition to Equations (13) and (14), another attempt was made to estimate m_c by using Equations (1)–(11). Therefore, the three selected models used to estimate the values of m_c without the need for measurements need evaluation to examine their validity for arid climates. For this reason, during all the experiments, the measured values of T_w (i.e., mainly depends on the solar irradiance, I_s, because of the frame insulation) were plotted against the corresponding values of I_s in Figure 8a for the before noon (7:0–12:0) and in Figure 8b for the after noon times (13:0–19:0). Applying the regression analysis to the data in Figure 8a,b, two expressions could be obtained for the before noon time (Equation (15), $R^2 = 0.97$, SEE = 3.68, and $p < 0.0001$), and after noon time (Equation (16), $R^2 = 0.91$, SEE = 4.68, and $p < 0.0001$) as follows:

$$T_w = 20.45 + 0.0353(I_s) + 3.2 \times 10^{-5}(I_s)^2, \quad R^2 = 0.97 \quad \text{(before noon)} \tag{15}$$

$$T_w = 20.87 + 0.129(I_s) - 7.72 \times 10^{-5}(I_s)^2, \quad R^2 = 0.91 \quad \text{(after noon)} \tag{16}$$

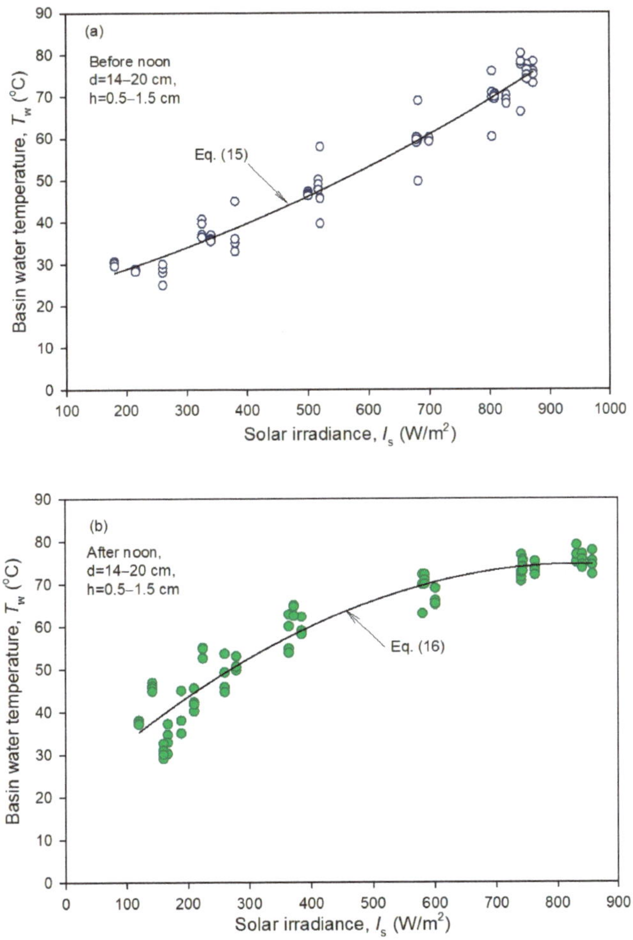

Figure 8. Basin water temperature (T_w) as affected by solar radiation flux (I_s) in the before noon (**a**) and after noon (**b**) times for the tested solar stills A, B, C, and D (10–18 October 2023).

Unlike the basin water temperature (T_w), the glass cover temperature (T_g) depends on the solar radiation flux as well as the ambient air temperature (T_{am}) because the glass cover is exposed directly to the ambient air and exchanges energy with it. In a similar manner to Equations (15) and (16), the measured data of T_g, T_{am}, and I_s for the stills A, B, C, and D were gathered for the three water depths, h (0.5, 1, and 1.5 cm), and then the values of T_g vs. I_s and T_{am} were plotted, as three dimensions, in Figure 9a for the data before noon and in Figure 9b for those after noon.

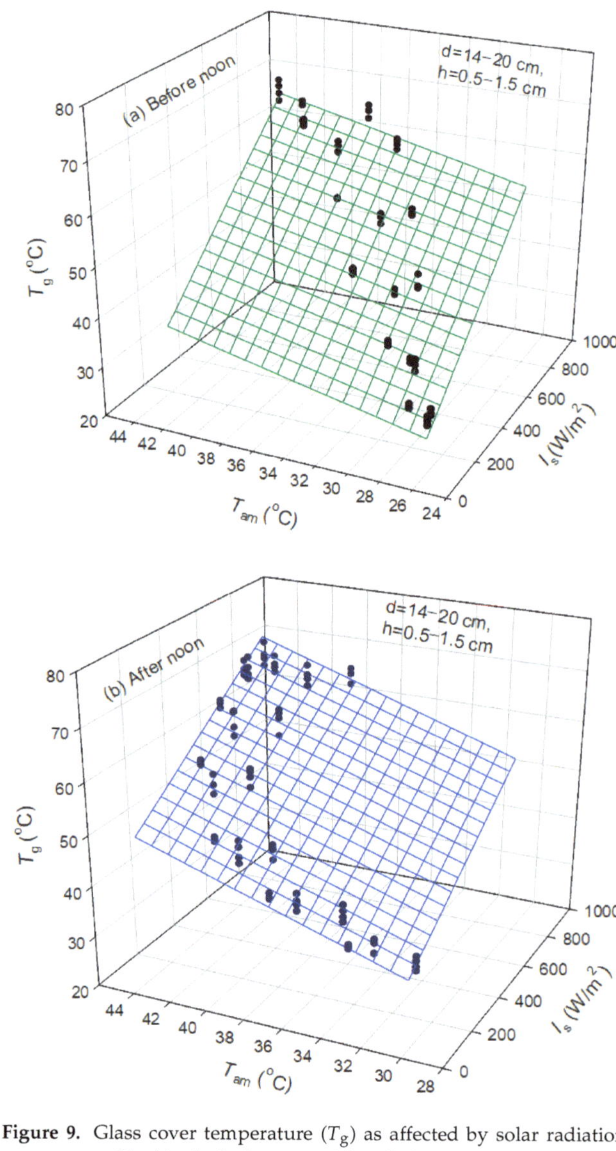

Figure 9. Glass cover temperature (T_g) as affected by solar radiation flux (I_s) and ambient air temperature (T_{am}) in the before noon (**a**) and after noon (**b**) times for the tested solar stills A, B, C, and D (10–18 October 2023).

Applying regression analysis to the data in Figure 9a,b, two numerical expressions could be obtained for the before noon time (Equation (17), R^2 = 0.97, SEE = 2.26, and

$p < 0.0001$) and the after noon time (Equation (18), $R^2 = 0.97$, SEE = 2.23, and $p < 0.0001$) in the following form:

$$T_g = -2.86 + 0.0457 I_s + 0.0668 T_{am}, \quad R^2 = 0.97 \text{ (before noon)} \tag{17}$$

$$T_g = -9.47 + 0.032 I_s + 1.185 T_{am}, \quad R^2 = 0.97 \text{ (afternoon)} \tag{18}$$

Most of the numerical models in the literature used to estimate the desalination rate (m_c) mainly used T_w and T_g as input parameters to the simulation. In order to make these models applicable, T_w and T_g (in °C) were correlated in Equations (15)–(18) as a function of solar irradiance (I_s, in W/m^2) and ambient air temperature (T_{am}, in °C). Then, an empirical model can be used easily to determine m_c, once the meteorological conditions are predetermined.

To validate the selected numerical models (Equations (1)–(11)), used for estimating the desalination rate (m_c), the values of T_w and T_g were calculated (Equations (15)–(18)), and then these values were substituted into Equations (1)–(11). The resulting values of m_c are depicted in Figure 10 as accumulative yields with time. In addition, the measured values of m_c (under the same conditions of h = 0.5 cm and d = 14 cm) are also depicted in Figure 10 for comparison. Overestimation errors are observed in Figure 10 on the predicted values of m_c, resulting from the three models, especially in the after noon time. This is mainly attributed to the assumptions allocated with these models; for example, the models have assumed that the evaporation rate is equal to the collected condensation rate ($m_c = m_{ev}$), and the condensed droplets that fall into the basin water were neglected. In addition, Kumar and Tiwari's model [20] and Zheng et al.'s model [21] are more convenient than Dunkle's model [19] to predict the values of m_c theoretically. Eventually, once the ambient air temperature (T_{am}) and solar radiation flux (I_s) were measured, the desalination rate (m_c) in kg per m^2 of basin surface can be calculated.

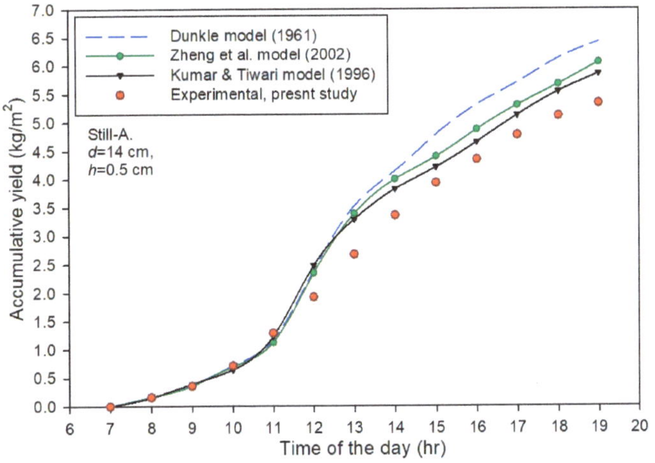

Figure 10. Time evolution of the measured accumulative distillate yield compared with those calculated using three models [19–21] for the solar still A (h = 0.5 cm, and d = 14 cm).

3.3. Distilled Water Analysis

The saline water used in the experiments was taken from the Red Sea (Duba Coastal, Tabuk, KSA). Two samples of seawater and the output distilled water were tested in the central laboratory, King Saud University. The total dissolved solids (TDS) in the seawater and distilled water were about 31,250 and 495 ppm, respectively. This is very acceptable because the maximum acceptable solids in the drinking water, according to the STD reference [25], is about 1200 ppm. Other chemical parameters were measured for the two samples and the results are illustrated in Table 3, and showed promising results, emphasizing that the distilled water can be used as drinkable water according to the STD [25]. However, iron (Fe) in the distilled water was higher than in the seawater; this may be attributed to the corrosion that occurred in the galvanized iron trough during the experiments. Therefore, anti-corrosive materials are highly recommended to make the trough. The distilled water can be remixed with brackish water (adjusting ppm of the mixture up to 1000) to be used for irrigation and greenhouse evaporative cooling in summer. This can solve the main challenge facing greenhouse growers in these regions, especially in summer, which is blocking the wet pad because of using brackish water for cooling the pad in the wet-pad fans system [26]. Solving this issue would significantly enhance the cooling performance of wet-pad fan systems, increase the lifetime of the pads, and reduce the fixed and operating costs of the greenhouses.

Table 3. Chemical analysis of distilled water and seawater.

Composition, ppm	Sample-1 (Seawater)	Sample-2 (Distilled Water)	STD [25]
Carbonates (CO_3)	560.504	zero	—
Chloride (Cl^-)	21,442	71	500
Phosphates (PO_4)	54.299	Not detected	0.3
Calcium (Ca)	537.072	16.032	200
Magnesium (Mg)	233.38	4.86	150
Iron (Fe)	0.0567	0.1008	1
Manganese (Mn)	Not detected	Not detected	—
Sodium (Na)	132.0029	22.6480	200
Potassium (K)	109.1886	2.8609	—
pH	6.18	6.23	6.5–9.2
Total dissolved solids (TDS)	31,250	495.36	1200

4. Conclusions and Recommendations

This study evaluated four solar stills to examine the effect of air gap distance (d) and basin water depth (h) on the performance and freshwater productivity (m_c) of these stills in the arid climate of Riyadh area. Based on the measured parameters, different analytical expressions have been developed to estimate the desalination rate (m_c) as a function of the meteorological parameters (i.e., solar radiation flux, I_s, and ambient air temperature, T_{am}). According to the obtained results, the main conclusion could be summarized as follows:

The solar radiation flux is the main power that induces the evaporation process and the solar still freshwater productivity; therefore, selecting the appropriate location to install solar stills is important.

Both the basin water depth and the air gap distance, between the water surface and glass cover, have a considerable effect on the distillate freshwater yield; the highest collected distilled yield could be obtained at the lowest air gap distance and lowest water depth due to the rapid evaporation and condensation rate as affected by the reduction in the still side walls. Low side walls reduce shading on the basin water and consequently increase the water temperature. Therefore, transparent side walls are recommended, in future work, to reduce the shading effects on the basin water, and enhance productivity. Moreover, the effect of wind speed over the glass cover on enhancing the still performance and productivity should be evaluated.

Based on our observation during the experiments, the trough, used to collect the condensed water, should be made from anti-corrosive materials to avoid corrosion. Hydrophilic paint is also recommended to be applied carefully on the inner surface of the glass cover, without jeopardizing the cover transmittance, to achieve film condensation and eliminate dropwise condensation on the cover.

Scaling up for the proposed design of solar still is possible to meet the requirements of any activity. Due to the breakability of glass cover, rigid supports (e.g., plastic or stainless-steel bars) are recommended below the glass cover to carry its load.

The values of the glass cover temperature (T_g) and basin water temperature (T_w) could be correlated as functions of the solar intensity (I_s), and ambient air temperature (T_{am}) with high R^2 values of 0.97. Thus, values of T_w and T_g can be used as input parameters for analytical models to finally estimate the desalination rate (m_c) theoretically instead of using expensive measurements. Other two simple correlations could be obtained to estimate m_c as a function of I_s (before and afternoon) with a maximum error of <8%; this will be useful for the feasibility studies of solar desalination projects in arid areas.

A high desalination rate per unit area of basin surface was achieved under the highly intensive solar radiation in the central region of Saudi Arabia; therefore, this climate is promising for solar energy applications, such as desalination and power generation.

The solar stills reduced the TDS from 31,250 ppm (sea water) to 495.36 ppm (drinkable water); the distilled water can be remixed with sea water up to ~1000 ppm to be used successfully for evaporative cooling and irrigation in the greenhouses.

More research is required for the night period and how to enhance the desalination rate in the absence of solar radiation by implementing phase change materials (PCM) and storage materials with high thermal capacities on the basin.

Author Contributions: Conceptualization, I.M.A.-H., S.M. and A.A.; methodology, S.M. and A.I.; software, M.R.S.; validation, A.I. and S.M.; formal analysis, I.M.A.-H. and S.M.; investigation, A.A.; resources, M.R.S.; data curation, A.I.; writing—original draft preparation, S.M. and A.I.; writing—review and editing, I.M.A.-H. and A.A.; visualization, M.R.S.; supervision, A.A.; project administration, S.M.; funding acquisition, I.M.A.-H. All authors have read and agreed to the published version of the manuscript.

Funding: The authors extend their appreciation to the Deputyship for Research and Innovation, "Ministry of Education" in Saudi Arabia for funding this research (IFKSUOR3-580-1).

Data Availability Statement: Data is contained within the article. The data that support the findings of this study are available upon request from the corresponding author.

Conflicts of Interest: The authors declare no conflicts of interest.

Nomenclature

A_s	Basin surface area of the solar still (m^2)
Cp	Specific heat capacity (J kg^{-1} $°C^{-1}$)
d	Air gap distance (m)
g	Gravitational acceleration (m s^{-2})
H	Back wall height (m)
H_o	Front wall height (m)
H	Basin water depth (m)
h_c	Convective heat transfer coefficient between water and glass cover (Wm^{-2} $°C^{-1}$)
h_{ev}	Evaporative heat transfer coefficient (Wm^{-2} $°C^{-1}$)
h_{fg}	Latent heat of vaporization of water (J kg^{-1})
h_o	Basin thickness (m)
I_s	Solar radiation intensity (Wm^{-2})
K	Thermal conductivity (Wm^{-1} $°C^{-1}$)
M_a	Molecular weight of dry air (kg mol^{-1})
M_w	Molecular weight of water vapor (kg mol^{-1})
m_c	Desalination rate (kg s^{-1})
m_{ev}	Evaporation rate (kg s^{-1})
P_T	Total pressure of the mixture of air and water vapor in still (Pa)

P_w	Partial pressure of the vapor at water temperature (Pa)
P_g	Partial pressure of the vapor at glass temperature (Pa)
SEE	Standard error of estimate
T_{am}	Ambient air temperature (°C)
T_g	Glass cover temperature (°C)
T_w	Basin water temperature (°C)
Greek symbols	
α	Slope angle of the glass cover (degree)
β	Thermal expansion factor (1/k)
η_d	Daily overall efficiency
ΔT	Temperature difference between water and glass (°C)
μ	Dynamic viscosity (Ns m^{-2})

References

1. Charlotte, B.; Marcel, W.; Michael, P. (Eds.) Roadmap for the Development of Desalination Powered by Renewable Energy, Fraunhofer Verlag. 2010. Available online: https://www.prodes-project.org/fileadmin/Files/ProDes_Road_map_on_line_version.pdf (accessed on 15 December 2023).
2. Abdallah, S.; Badran, O.; Abu-Khader, M.M. Performance evaluation of a modified design of a single slope solar still. *Desalination* **2008**, *219*, 222–230. [CrossRef]
3. Sampathkumar, K.; Arjunan, T.V.; Pitchandi, P.; Senthilkumar, P. Active solar distillation: A detailed review. *Renew. Sustain. Energy Rev.* **2010**, *14*, 1503–1526. [CrossRef]
4. Muftah, A.F.; Alghoul, M.A.; Fudholi, A.; Abdul-Majeed, M.M.; Sopian, K. Factors affecting basin type solar still productivity: A detailed review. *Renew. Sustain. Energy Rev.* **2014**, *32*, 430–447. [CrossRef]
5. Sharshir, S.W.; Yang, N.; Peng, G.; Kabeel, A. Factors affecting solar stills productivity and improvement techniques: A detailed review. *Appl. Thermal Eng.* **2016**, *100*, 267–284. [CrossRef]
6. Ghoneyem, A.; Ileri, A. Software to analyze solar stills and an experimental study on the effects of the cover. *Desalination* **1997**, *114*, 37–44. [CrossRef]
7. Rajaseenivasan, T.; Prakash, R.; Vijayakumar, K.; Srithar, K. Mathematical and experimental investigation on the influence of basin height variation and stirring of water by solar PV panels in solar still. *Desalination* **2017**, *415*, 67–75. [CrossRef]
8. Jamil, B.; Akhtar, N. Effect of specific height on the performance of a single slope solar still: An experimental study. *Desalination* **2017**, *414*, 73–88. [CrossRef]
9. Jamil, B.; Akhtar, N. Energy, Transportation and Global Warming. *Energy Transp. Glob. Warm.* **2016**, *5*, 161–163. [CrossRef]
10. Rajesh Tripathi, G.N.T. Performance evaluation of a solar still by using the concept of solar fractionation. *Desalination* **2004**, *169*, 69–80. [CrossRef]
11. Khalifa, A.J.N.; Hamood, A.M. On the verification of the effect of water depth on the performance of basin type solar stills. *Sol. Energy* **2009**, *83*, 1312–1321. [CrossRef]
12. Tiwari, A.K.; Tiwari, G.N. Effect of water depths on heat and mass transfer in a passive solar still in summer climatic condition. *Desalination* **2006**, *195*, 78–94. [CrossRef]
13. Elango, T.; Kalidasa, K.M. The effect of the water depth on the productivity for single and double basin double slope glass solar stills. *Desalination* **2015**, *359*, 82–91. [CrossRef]
14. Feilizadeh, M.; Karimi Estahbanati, M.R.; Ahsan, A.; Jafarpur, K.; Mersaghian, A. Effects of water and basin depths in single basin solar stills: An experimental and theoretical study. *Energy Convers. Manag.* **2016**, *122*, 174–181. [CrossRef]
15. Tripathi, R.; Tiwari, G.N. Thermal modeling of passive and active solar stills for different depths of water by using the concept of solar fraction. *Sol. Energy* **2006**, *80*, 956–967. [CrossRef]
16. Lalitha, R.N.; Ramachandra Raju, V. Experimental study on performance of passive and active solar stills in Indian coastal climatic condition. *Front. Energy* **2020**, *14*, 105–113. [CrossRef]
17. Abdel-Ghany, A.M.; Al-Helal, I.M.; Shady, M.R. Evaluation of human thermal comfort and heat stress in an outdoor urban setting in summer under arid climatic conditions. *Environ. Protec. Eng.* **2014**, *40*, 139–150. [CrossRef]
18. El-Swify, M.E.; Metias, M.Z. Performance of double exposure solar still. *Renew. Energy* **2002**, *26*, 531–547. [CrossRef]
19. Dunkle, R.V. Solar water distillation: The roof type still and a multiple effect diffusion still. In Proceedings of the International Heat Transfer Conference, Boulder, CO, USA, 28 August–1 September 1961.
20. Kumar, S.; Tiwari, G.N. Estimation of convective heat transfer in solar distillation systems. *Sol. Energy* **1996**, *57*, 459–464. [CrossRef]
21. Zheng, H.; Zhang, X.; Zhang, J.; Wu, Y. A group of improved heat and mass transfer correlations in solar stills. *Energy Convers. Manag.* **2002**, *43*, 2469–2478. [CrossRef]
22. Hollands, K.G.T.; Raithby, G.D.; Konicek, L. Correlation equations for free convection heat transfer in horizontal layers of air and water. *Int. J. Heat Mass Transf.* **1975**, *18*, 879–884. [CrossRef]
23. Duffie, J.A.; Beckman, W.A. *Solar Engineering of Thermal Processes*; John Wiley& Sons, Inc.: New York, NY, USA, 1991.

24. Dwivedi, V.K.; Tiwari, G.N. Comparison of internal heat transfer coefficients in passive solar stills by different thermal models: An experimental validation. *Desalination* **2009**, *246*, 304–318. [CrossRef]
25. World Health Organization. A Compendium of Drinking-Water Quality Standards in the Eastern Mediterranean Region. 2006. Available online: https://pesquisa.bvsalud.org/portal/resource/pt/who-116514 (accessed on 15 December 2023).
26. Al-Helal, I.M.; Abdel-Ghany, A.M. Energy partition and conversion of solar and thermal radiation into sensible and latent heat in a greenhouse under arid conditions. *Energy Build.* **2011**, *43*, 1740–1747. [CrossRef]

sustainability

MDPI

Article

Grid-Forming Inverter Control for Power Sharing in Microgrids Based on *P*/*f* and *Q*/*V* Droop Characteristics

Qusay Salem [1], Rafat Aljarrah [1], Mazaher Karimi [2,*] and Ayman Al-Quraan [3]

[1] Department of Electrical Engineering, Princess Sumaya University for Technology, Amman P.O. Box 1438, Jordan; q.salem@psut.edu.jo (Q.S.); r.aljarrah@psut.edu.jo (R.A.)
[2] School of Technology and Innovations, Electrical Engineering, University of Vaasa, 65200 Vaasa, Finland
[3] Department of Electrical Power Engineering, Yarmouk University, Irbid P.O. Box 566, Jordan; aymanqran@yu.edu.jo
* Correspondence: mazaher.karimi@uwasa.fi

Abstract: Grid-forming inverters are anticipated to be integrated more into future smart microgrids commencing the function of traditional power generators. The grid-forming inverter can generate a reference frequency and voltage itself without assistance from the main grid. This paper comprehensively investigates grid-forming inverter modelling and control methodology. A decentralized method employing an active power versus frequency $P-f$ droop and a reactive power versus voltage $Q-V$ droop is exploited to drive the operation of the grid-forming inverter. This decentralized method ensures balancing the supply and demand beside the power-sharing task between two or more inverters. The performance of the grid-forming inverter is examined by monitoring the frequency and RMS voltage of the inverter bus for three different periods of a varying PQ load. In addition, the performance of the resultant droop is compared with the assumed droop to validate the effectiveness of the proposed method. Finally, two grid-forming inverters equipped with the same droop characteristics are connected to a single load to observe the power-sharing concept among them. All simulations are implemented and executed using Matlab/Simulink version R2014b.

Keywords: grid-forming inverter; decentralized control; $P-f$ droop control; $Q-V$ droop control; power sharing

Citation: Salem, Q.; Aljarrah, R.; Karimi, M.; Al-Quraan, A. Grid-Forming Inverter Control for Power Sharing in Microgrids Based on *P*/*f* and *Q*/*V* Droop Characteristics. *Sustainability* **2023**, *15*, 11712. https://doi.org/10.3390/su151511712

Academic Editor: Luis Hernández-Callejo

Received: 10 June 2023
Revised: 11 July 2023
Accepted: 26 July 2023
Published: 28 July 2023

1. Introduction

Synchronous generators are usually dominant for traditional AC power networks, where the essential control targets of frequency and voltage regulation are satisfied through governor and exciter control, respectively. Furthermore, during load variations and short circuits, the frequency is kept within the permissible limits due to the higher inertia of the prime mover and rotor. This property of a synchronous generator makes the main utility power grid more stable. The deeper penetration of renewable energy sources (RES) into AC power networks entails interfacing the source with an inverter-based power electronic interface, which has low inertia. The most commonly used inverters are grid-following inverters, which are utilized to generate power, both active and reactive, into the main grid. This type of inverter requires a phase-locked loop (PLL) to coordinate with the main grid voltage and frequency, which are taken as references. The grid-forming inverters can extract their voltage and frequency alone without the need for the main grid voltage and frequency, contrary to grid-following inverters. This feature is beneficial for the distributed generator (DG) or microgrids operating in an autonomous mode. For an islanded microgrid system, some inverters can operate in grid-following mode while one or some others take the responsibility of grid-forming mode to create a neighborhood electricity grid [1–4].

Grid-forming inverters were primarily evolved to be applied in a microgrid on islands. However, their operation can be modified when integrating them in large-scale power networks, like wind and solar farms, placed in far-off places. This is because of the high-line impedance needed, which forms a weak grid side. Thus, voltage control at the common coupling point becomes a crucial problem, which can be solved using grid-forming inverter integration to strengthen the weak grid. Grid-forming inverters are a trending technology where the control methodologies, modelling techniques, and practical applications are not fully addressed. To remedy the shortage in the above, many academic and non-academic original and review papers have investigated the widespread aspects of grid-forming inverters [5–9].

Grid-forming inverters have to share the overall power generation according to their capacities. This is usually achieved through a supervisory control where a central controller sends control commands to the inverters through a communication medium. However, this strategy is less reliable, as the supervisory control can be threatened by cyberattacks or due to communication delays. Alternatively, the decentralized methods do not need a communication medium. Thus, cyberattacks and communication delays are not considered as a risk. That is why droop control as a decentralized method are preferred in the control of grid-forming inverters [10–12].

A hybrid grid-forming/-following inverter, which describes the inverter control system's dynamic behavior, was proposed in [13]. A method for damping the postfault oscillations observed in inverters that both form and follow the grid and based on droop control was investigated in [14] for changes in grid strength. The authors in [15] proposed a novel method to decouple the real and reactive power in grid-forming inverters based on droop control or any other control structure. A method of aggregation based on coherency for analyzing massive power networks involving grid-forming inverters with droop control was studied in [16] using hardware in the loop testbed. In [17], a creative decentralized control method for multiparallel grid-forming distributed generators in a microgrid on an island was proposed, where the system frequency was independent of the load conditions. In reference [18], the authors analyzed the effect that hybrid voltage-and-power-based synchronization control has on the transient stability of grid-forming inverters. The duality of grid-forming and -following inverters was analyzed in [19] through theoretical analysis and simulation results for single-, two-, and multi-inverter systems. In [20], a battery-supported inverter with an improved droop control was thought to function in parallel with a photovoltaic (PV)-based grid-forming inverter with modified virtual synchronous machine control under non-ideal grid voltage conditions and in the isolated mode of operation.

Some of the research articles discussed before used the droop control only as a tool for the grid-forming inverter to satisfy other objectives, like seamless transition or damping oscillations. Others proposed new control methods or strategies for the grid-forming inverter without employing droop control. However, to the best of the authors' knowledge, the applicability of droop-based control in the grid-forming inverter has not been addressed fully and comprehensively in the up-to-date literature. Moreover, the validity of the performance of such decentralized control has not been analyzed in depth, neither for the case of off-grid systems nor of a varying PQ load demand. Therefore, the innovation of this paper is represented by providing a comprehensive validation of the droop-based grid-forming inverter as an off-grid system with a varying PQ load demand to address such a literature gap. Among the existing control strategies of the grid-forming inverter in the literature, droop control as a decentralized control is investigated in depth and is utilized to drive the proposed grid-forming inverter operation, where $P-f$ and $Q-V$ droops are adjusted and confirmed with the simulation results for a varying-load condition. Voltage and current control loops are implemented in dq rotating reference frame to extract the reference voltage of the grid-forming inverter. Thus, the main contributions of this paper are:

(1) A detailed insight of the theoretical analysis for the grid-forming inverter and its control structure.
(2) The validation of the assumed and resultant $P-f$ and $Q-V$ droop characteristics used to drive the grid-forming inverter.

(3) The verification of the power-sharing concept employed by connecting two grid-forming inverters to satisfy a varying load demand.

Five sections make up the remaining text of the paper. Following the Introduction, Section 2 presents a general illustration of the grid-forming inverter and its control schemes. In Section 3, where a thorough examination of the droop control strategy and the construction of the cascaded voltage and current control loops are studied, the system structure and methodology, including the assumed droop control and the voltage and current control loops, are explained. Discussion of the simulation results, including the setting of the assumed $P-f$ and $Q-V$ droops, performance evaluation of the system under a varying load, and the validation of the simulation results as well as the active and reactive power sharing between two grid-forming inverters is in Section 4. In Section 5, the conclusions and future work are outlined.

2. Configuration and Control of Grid-Forming Inverter

Grid forming is an umbrella for any power converter that can deliver constant voltage magnitude and frequency when the main supply grid is not available. It does not need a PLL similar to how grid-following converters work and can work with other grid-forming or -following inverters on the same network. A grid-forming inverter is usually operated in island mode and is regulated to adjust the frequency and voltage forming a local grid. Therefore, it can be shown by connecting a low impedance to an idealized AC voltage source [21]. The use of the grid-forming converter mostly lies in the context of microgrids. In this case, the grid-forming converter can operate in grid-connected mode as supporting inverters or it can operate autonomously, which means in island mode [22].

As mentioned previously, the grid-forming power converter appears as a low-output impedance connected to an ideal ac voltage source, where the voltage level and frequency V_m, δ_m are set by utilizing an appropriate control loop, as indicated in Figure 1. Therefore, the grid-forming converter needs an accurate system of synchronization in order to be able to work in parallel with other grid-forming converters. A real-world illustration of the grid-forming converter is the standby uninterruptible power supply (UPS), where it can form the grid voltage when a grid failure occurs. In the case of grid failure in a grid-connected microgrid system, the grid-forming converter will be employed as a reference if multiple grid-following converters are running on the system.

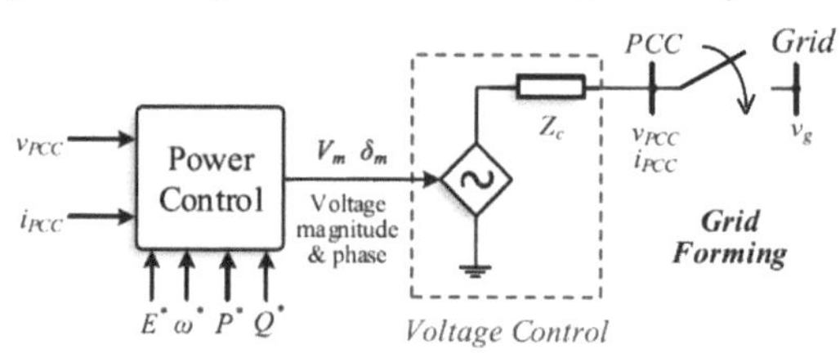

Figure 1. Simplified representation of grid-forming converter.

The following [23] is a classification of certain grid-forming converter control strategies:

- Droop control: this method offers a linear trade-off between the frequency and real power in addition to the voltage and reactive power. This principle is inspired from the steady-state operation of a typical synchronous machine. Therefore, the droop characteristics are expressed as $P-f$ and $Q-V$ relationships. Thus, each distributed generation unit will have the same frequency and it will supply power according to its capacity or, in other words, its droop gain.
- Virtual synchronous machines: this method employs the behavior of a synchronous machine in power systems to be applied in the inverter control. Here, the measurements of the inverter terminals are inserted as inputs into a digital model of a synchronous machine to deliver an inverter output according to the digital model.

To match the machine characteristics, $P - \omega$ and $Q - V$, which are nominated as synchronverters, are usually employed. The virtual machine concept can be implemented considering the detailed electromechanical models or simply the simplified swing dynamics.

- Virtual oscillator controllers: this method emulates the behavior of nonlinear loads where a digital model processes the real-time measurements like a virtual synchronous machine. The major difference is that an oscillator circuit with a natural frequency that matches with AC main grid frequency forms the digital model. Also, by tuning this oscillator circuit, the control bandwidth and nominal voltage can be adjusted. In steady-state conditions, this method will eventually offer $P - \omega$ and $Q - V$ droops.

All the methods described above have analogous properties despite the differences between them. The inverter output of any of the abovementioned grid-forming controllers looks like a voltage source, with amplitude and frequency that changes with the power and load variation of the system. This hierarchy permits grid-forming inverters to instantaneously conform the load demand with the generated output power, control local voltage, and take part in frequency control without the need for a reference voltage from the main grid, as in the case of grid-following inverters [24]. The control methods applied for the grid-forming inverters are depicted in Figure 2. Here, the grid-forming inverter acts like a voltage source, which can be driven by one of the mentioned control methods. In the context of microgrids, grid-following and grid-forming inverters differ significantly in that the grid following always need a PLL as well as voltage reference to extract the real and reactive power.

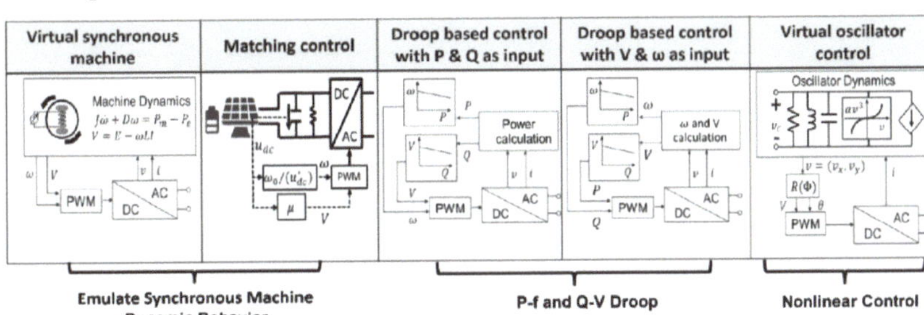

Figure 2. Control methods of the grid-forming inverter [21].

3. System Structure and Methodology

As discussed earlier, the grid-forming inverter should be driven by a proper control methodology to deliver a reference voltage, as the main grid no longer exists. The reference voltage should be regulated by two cascaded voltage and current control loops before being sent to the grid-forming inverter. Figure 3 depicts the structure of the implemented network, which includes one grid-forming inverter and one local load. The grid-forming inverter is represented by an ideal controlled voltage source. The reference voltage V_{ref} is delivered to the inverter after being extracted and controlled by the droop control and sinewave generation as well as the voltage and current control loops, respectively. Those two blue subsystems will be clarified in the subsections that follow.

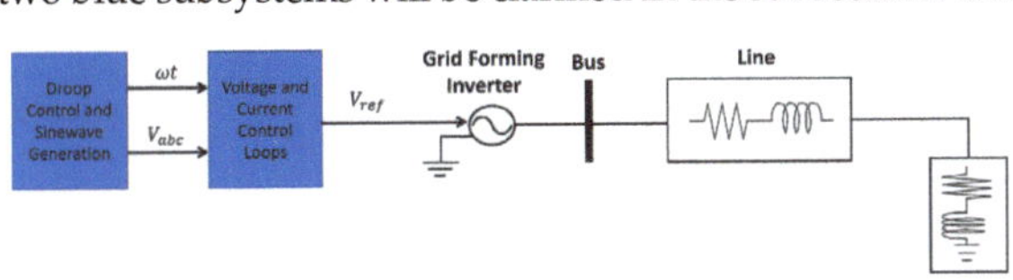

Figure 3. Structure of the proposed network.

3.1. Droop Control and Sinewave Generation

In synchronous machines, the generator rotor speed is decreased if the demand of power is suddenly increased. Thus, the frequency is lowered in order to deliver extra power. This consistency between P and f is introduced as a standalone control in DG systems. The only difference here is that the inverter lacks inertia, unlike the synchronous machine. As a result, the transmission-line impedance plays a major role in determining the droop-control-based active power and frequency $P-f$. Considering the system shown below in Figure 4, which connects a voltage source with a load through a transmission line, the real and reactive powers are given as [25]:

$$P = \left(\frac{EV}{Z}cos\delta - \frac{V^2}{Z}\right)cos\theta + \frac{EV}{Z}sin\delta sin\theta \tag{1}$$

$$Q = \left(\frac{EV}{Z}cos\delta - \frac{V^2}{Z}\right)sin\theta - \frac{EV}{Z}sin\delta cos\theta \tag{2}$$

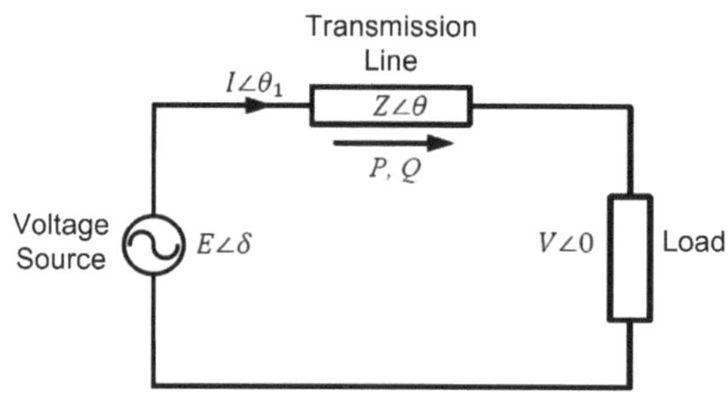

Figure 4. System diagram of a voltage source with load through a transmission line.

When $\theta = 90°$ for an inductive transmission line, the power equations reduce to:

$$P = \frac{EV}{Z}sin\delta \quad \& \quad Q = \frac{EV}{Z}cos\delta - \frac{V^2}{Z} \tag{3}$$

If the phase angle δ is small, $sin\delta \approx \delta$ and further reduction to the power equation yields

$$P \approx \frac{EV}{Z}\delta \quad \& \quad Q = \frac{V}{Z}(E-V) \tag{4}$$

Thus, it can be observed that if the transmission line is inductive, the active power is connected to the phase angle, and the reactive power is coupled to the terminal voltage. However, the frequency is selected in place of the phase angle to regulate the active power for control of DG-based power converters. This is because the initial phase values of the DG systems are unknown in the absence of the grid. According to this logic, the $P-f$ droop can be represented as:

$$f = f_{ref} - k_P\left(P_{ref} - P\right) \tag{5}$$

Also, the terminal voltage and reactive power $Q-V$ droop are characterized by:

$$V = V_{ref} - k_q\left(Q_{ref} - Q\right) \tag{6}$$

where P_{ref} and Q_{ref} are the reference active and reactive power, P and Q are the measured active and reactive power, f_{ref} and V_{ref} are the reference frequency and voltage, and k_P and k_q are the droop gain coefficients.

It is worth mentioning that as the droop gain coefficient is low, the control response will be slow, whereas for large droop gain coefficient, the load sharing becomes faster. This is due to the swap between the droop gain coefficient and stability of the system.

The concept resulting from Equations (5) and (6) has been implemented in the droop control and sinewave generation block depicted in Figure 3. Its representation is presented, as shown in Figure 5. Here, the measured current and voltage at the grid-forming inverter are inserted into a block that calculates the power to extract the measured real and reactive powers, which are then compared with their rated (reference) values.

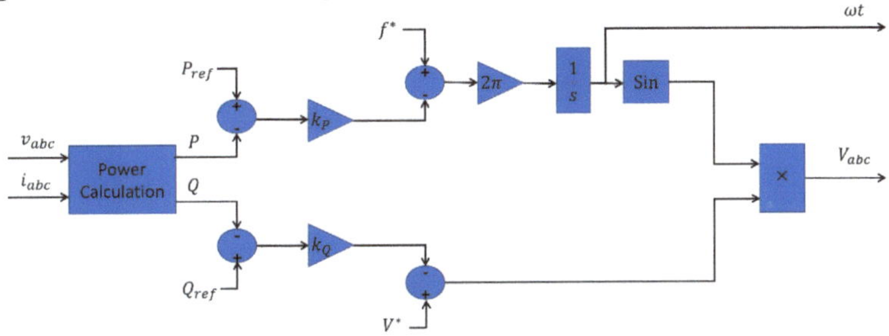

Figure 5. A representation of the droop control and sinewave generation block.

The droop gain coefficients k_P and k_Q are adjusted according to the capacity of the renewable energy source, in other words, the rated values for both active and reactive power of P_{ref} and Q_{ref}. The output of these droop gains is compared with the nominal frequency and terminal voltage values. Thus, if the measured active power is increased/decreased, the frequency will be lowered/raised, respectively. This logic also extends to the measured reactive power and the terminal voltage. That is why the $P-f$ and $Q-V$ droops have inverse characteristics. Finally, the voltage V_{abc} that should be controlled and the phase angle ωt required for dq transformation are extracted and inserted into the control loop block of voltage and current, which will be discussed further.

3.2. Voltage and Current Control Loops

In this section, the sinewave-generated voltage V_{abc} from the droop control block will be regulated by two cascaded control loops, as depicted in Figure 6. The control loop of voltage is in charge of controlling the output voltage, where the distinction between the measured and reference voltage is an input to the control loop of current that delivers the current reference to be injected by the converter. Only when the DG is operating in islanded mode and separated from the main grid does the voltage control loop exist. The current control loop regulates the current provided by the power inverter and tracks, at the same time, the current reference supplied by voltage control loop.

In the block diagram shown below, the ABC sinusoidal voltage is transformed into a synchronous frame by Park's transformation to extract $V_{d,ref}$ and $V_{q,ref}$, such that,

$$\begin{bmatrix} X_d \\ X_q \\ X_0 \end{bmatrix} = \frac{2}{3} \begin{bmatrix} -\cos\omega t & -\cos(\omega t + 2\pi/3) & -\cos(\omega t - 2\pi/3) \\ \sin\omega t & \sin(\omega t + 2\pi/3) & \sin(\omega t - 2\pi/3) \\ 1/2 & 1/2 & 1/2 \end{bmatrix} \times \begin{bmatrix} X_a \\ X_b \\ X_c \end{bmatrix} \quad (7)$$

The measured and reference dq voltages are then compared and passed through PI controllers to deliver the reference dq currents. This part specifically identifies the voltage control loop. Further, the measured and reference dq currents are compared and also inserted to PI controllers to deliver the final voltage reference of the grid-forming converter, which includes the reference voltages V_d^* and V_q^*. This part specifically identifies the current control loop. The voltages V_d^* and V_q^* are transformed into stationary reference frame by the inverse Park's transformation, such that,

$$\begin{bmatrix} X_a \\ X_b \\ X_c \end{bmatrix} = \begin{bmatrix} -\cos\omega t & \sin\omega t & 1/2 \\ -\cos(\omega t - 2\pi/3) & \sin(\omega t - 2\pi/3) & 1/2 \\ -\cos(\omega t + 2\pi/3) & \sin(\omega t + 2\pi/3) & 1/2 \end{bmatrix} \times \begin{bmatrix} X_d \\ X_q \\ X_0 \end{bmatrix} \quad (8)$$

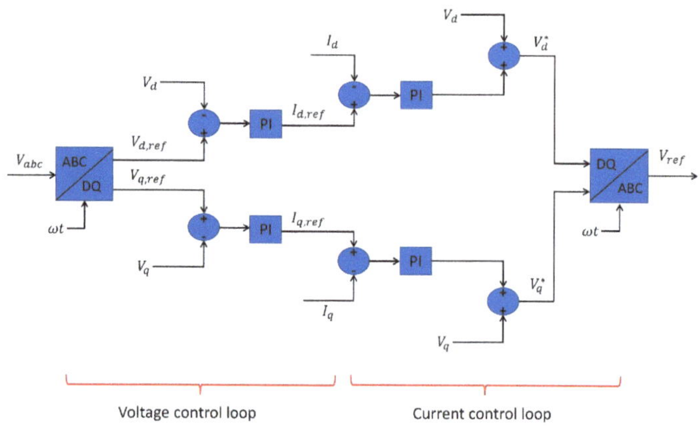

Figure 6. Representation of the voltage and current control loops.

In real practice, the grid-forming converter is fed by a balanced dc source driven by a fuel cell or a battery, for example. This dc source remains disconnected when the system is operating within acceptable limits, and it is connected in case of grid failure to form the grid voltage and frequency.

A flowchart of the overall control process for generating the reference voltage of the grid-forming converter is depicted in Figure 7. In the flowchart, the droop characteristics are adjusted to deliver the sinusoidal signal; then, it is passed through voltage and current control loops to generate the reference voltage.

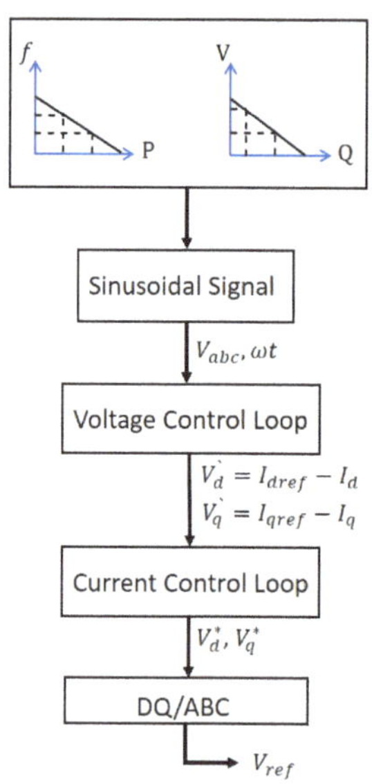

Figure 7. Flowchart of the overall control process.

4. Discussion of Simulation Results

4.1. Setting the Droop Characteristics

The grid-forming power converter's droop characteristics have been tuned so that the voltage V controls the reactive power Q and the frequency f controls the active power P. The $P-f$ and $Q-V$ settings are presented in Figure 8a,b. The permissible limits of the frequency and RMS voltage in the distribution networks are (48 Hz $< f <$ 52 Hz) and (207 V $< V_{rms} <$ 253 V), respectively. The frequency range in the $P-f$ droop has been adjusted between (50 Hz and 52 Hz) as the active power is assumed to be generated to satisfy the load and not to be stored for future usage. Keep in mind that if the active power needs to be stored, the frequency range should be from (48 Hz to 50 Hz). In addition, the voltage range in the $Q-V$ droop has been adjusted between (230 V and 253 V) as the reactive power is assumed to be inductive (load draws + ve VAR). If the reactive power is capacitive, the voltage range should be from (207 V to 230 V). The real and reactive power range in the $P-f$ and $Q-V$ droops was adjusted arbitrarily, as there is no restriction similar to the permitted ranges for voltage and frequency.

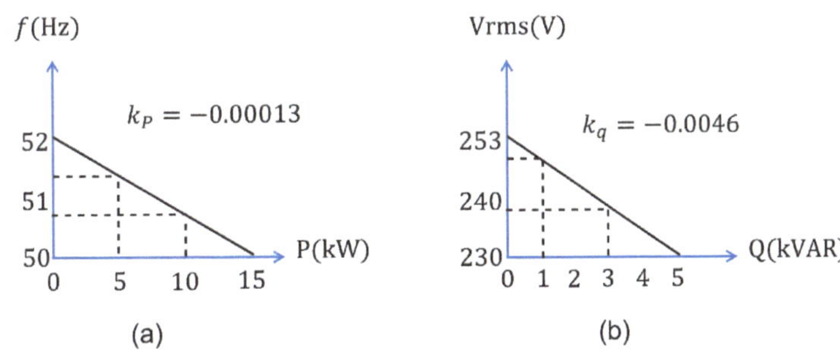

Figure 8. Setting of the droop characteristics. (**a**) $P-f$ droop (**b**) $Q-V$ droop.

In subfigures (a) and (b), the droop gains k_P and k_q are calculated as follows:

$$k_P = \frac{50-52}{15000-0} = -0.00013$$

$$k_q = \frac{230-253}{5000-0} = -0.0046$$

Those droop gains are inserted into the unit of control for the grid-forming power converter, where, if the active power of the load is increased, the frequency is decreased and vice versa. Furthermore, if the load reactive power is increased, the voltage is decreased and vice versa. The references of active and reactive power P_{ref} and Q_{ref} are 15 kW and 5 kVAR, which represent the maximum capacity of the grid-forming inverter, whereas the frequency and voltage references are f_{ref} and V_{ref} are 50 Hz and $\sqrt{2} \times 230$ V, respectively.

4.2. Performance Evaluation under Load Variation

The proposed model structure of Figure 3 was constructed in Matlab/Simulink version R2014b, where the grid-forming power converter was designed as a controlled voltage source. The load is assumed to be a resistive-inductive (PQ) load to examine the reaction of the proposed droop controllers, and the line is an $R-L$ series branch. The Simulink model of the proposed network, including the droop and control structure, is presented in Figure 9. Subfigure (a) shows the network model, while subfigures (b) and (c) present the droop and sinewave generation in addition to the voltage and current control loops. The evaluation of the results was divided into three regions: the first region between (0 and 0.5 s) is when no load is connected to the grid-forming inverter, the second region between (0.5 s and 1 s) is when the load is increased to 50% of the grid-forming inverter capacity, and the third region between (1 s and 1.5 s) is when the load is increased to 100% of the grid-forming inverter capacity. Reactive power vs. voltage $Q - V$ droop and active power versus frequency $P - f$ droop performance under the load variation for the three regions are depicted in Figures 10 and 11. The left y-axis for the two figures represents the active and reactive power, and the right y-axis represents the frequency and RMS voltage, respectively.

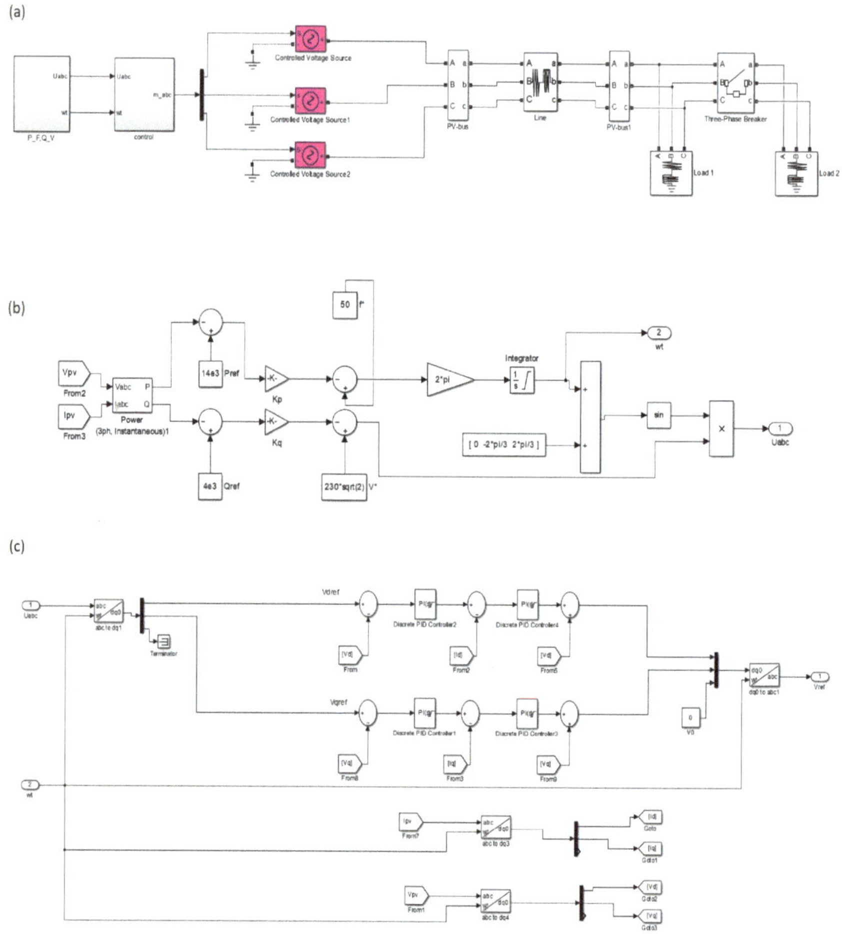

Figure 9. The proposed Simulink model. (**a**) The proposed network; (**b**) the droop characteristics and the sinewave generation; (**c**) the voltage and current control loops.

In no-load conditions (the load active and reactive power are zeros), the grid-forming inverter ran at the highest frequency, 52 Hz, which corresponds to zero active power according to the assumed $P - f$ droop setting. The same condition was applied for the reactive power where the RMS voltage was at the highest permissible value, 253 V, corresponding to the assumed $Q - V$ droop setting.

When the load is increased to 50% of the grid-forming-inverter capacity, the frequency is decreased to 51 Hz according to the assumed $P - f$ droop so that the inverter active power is increased to 7.5 kW to satisfy the active power load demand. Similarly, the RMS voltage is decreased to 241.5 V according to the assumed $Q - V$ droop so that the inverter reactive power is increased to 2.5 kVAR to satisfy the reactive power load demand.

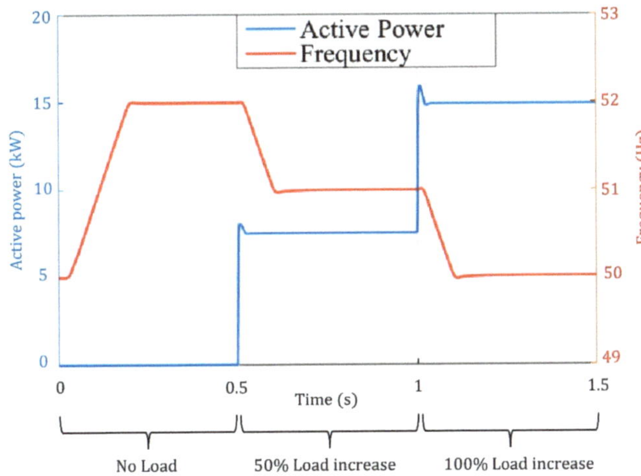

Figure 10. $P - f$ droop performance under load variation.

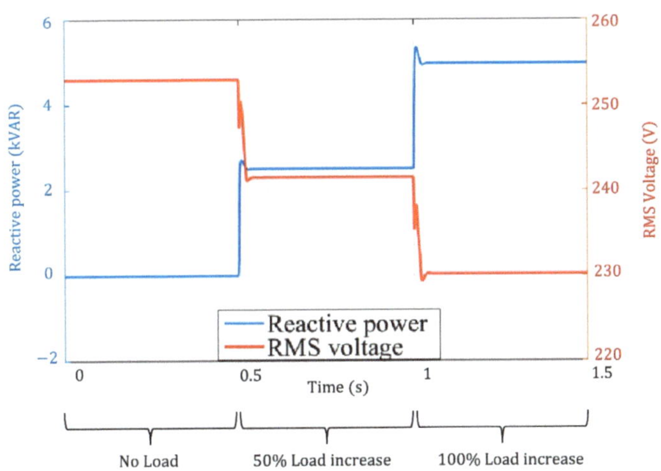

Figure 11. $Q - V$ droop performance under load variation.

Furthermore, when the load is increased to 100% of the grid-forming inverter capacity, the frequency is decreased to 50 Hz (lowest frequency permissible value) according to the assumed $P - f$ droop so that the inverter active power is increased to 15 kW (maximum capacity) to satisfy the active power load demand. At the same time, the RMS voltage is decreased to 230 V (lowest RMS voltage permissible value) according to the assumed $Q - V$ droop so that the inverter reactive power is increased to 5 kVAR (maximum capacity) to satisfy the reactive power load demand. It should be noted that the spike in the RMS voltage is due to the PI controller reaction in the control loop at the moment of 50% and 100% load increase. When the PI controller is subjected to a sudden load increase, it needs a little bit of time to minimize the error difference between the measured and reference quantity. That is why a spike appeared and then settled down to a steady-state response.

The peak voltage and current at the grid-forming inverter bus during the load variation condition are depicted in Figure 12. It can be seen that the current is zero until $t = 0.5$ s as an indication of the zero-load demand while the voltage is oscillating at $\sqrt{2}$ times 253 V. During 0.5 s $< t <$ 1 s, the current increases as the load demand is increased to 50% of the inverter capacity while the voltage is decreased to $\sqrt{2}$ times 241.5 V due to the reaction of the $Q - V$ droop to satisfy the reactive power demand of the load. After $t = 1$ s, the current increases further due to the further increase in the load demand. Thus, the inverter needs to generate more active power according to the $P - f$ droop to satisfy the active power load demand. Also, the voltage is decreased to $\sqrt{2}$ times 230 V due to the reaction of the $Q - V$ droop to satisfy the further increase in the load reactive power demand.

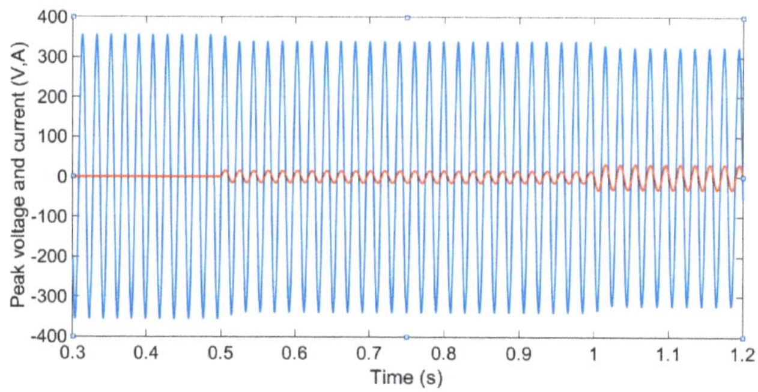

Figure 12. Peak voltage and current at the grid-forming bus under load variation condition. The blue line refers to the voltage and red line to the current.

4.3. Validation of Results

The validation of results is accomplished by monitoring the reference and measured quantities of the droop control loop of the grid-forming inverter, as presented in Figure 13. As shown from subfigure (a), the reference frequency f_{ref} is always 50 Hz, and the measured frequency is changing according to the load demand variation and depending on the assumed $P - f$ droop setting. Notice that the output of K_P in subfigure (b) is changing exactly according to the difference between the measured and reference frequency. Extending the same logic for subfigure (c), the grid-forming inverter active power is changing according to the change in K_P, which represents the assumed $P - f$ droop characteristic.

In subfigure (d), the reference voltage V_{ref} is always 230 V, and the measured voltage is changing according to the load demand variation and depending on the assumed $Q - V$ droop setting. Notice that the output of K_q in subfigure (e) is changing exactly according to the difference between the measured and reference voltage. Extending the same logic for subfigure (f), the grid-forming inverter reactive power is changing according to the change in K_q, which represents the assumed $Q - V$ droop characteristic.

Furthermore, to confirm that the assumed droop control discussed in Section 4.1 has been applied efficiently in the control of the proposed grid-forming inverter, the measured active power and frequency values in addition to the measured reactive power and rms voltage values are plotted as presented in Figure 14. Notice that the resultant $P - F$ droop equals exactly the assumed $P - F$ droop, in which $K_P = -0.00013$ and the resultant $Q - V$ droop equals exactly the assumed $Q - V$ droop, in which $K_q = -0.0046$.

4.4. Active and Reactive Power Sharing

Sharing between active and reactive power among the grid-forming inverters based on their rated capacity is employed to avoid overloading the sources. A microgrid consisting of two grid-forming inverters sharing a single load is represented in Figure 15 to investigate how the droop controllers equally shared the load demand of active and reactive power.

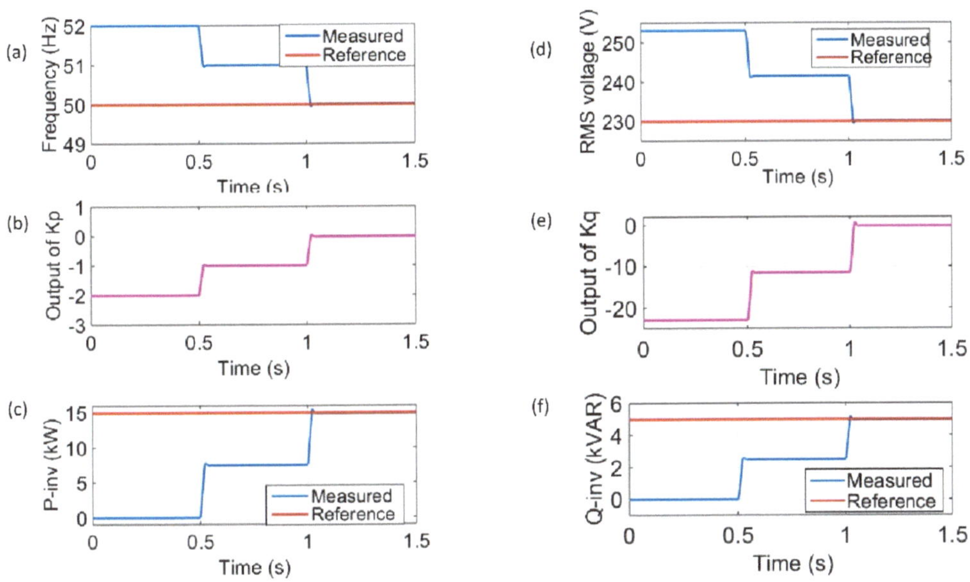

Figure 13. Reference and measured quantities of the droop control loop. (**a**) Reference and measured frequency; (**b**) output of Kp; (**c**) reference and measured active power; (**d**) reference and measured RMS voltage; (**e**) output of Kq (**f**) reference and measured reactive power.

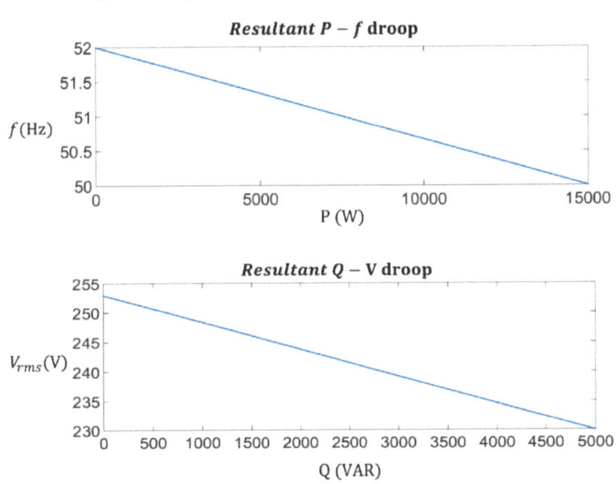

Figure 14. The resultant $P - f$ and $Q - V$ droops.

The droop characteristics of both grid-forming inverters are adjusted to be identical so that they share the same amount of the load demand in terms of real and reactive power. Their settings are as follows:

$$k_P = \frac{50 - 52}{10000 - 0} = -0.0002$$

$$k_q = \frac{230 - 253}{5000 - 0} = -0.0046$$

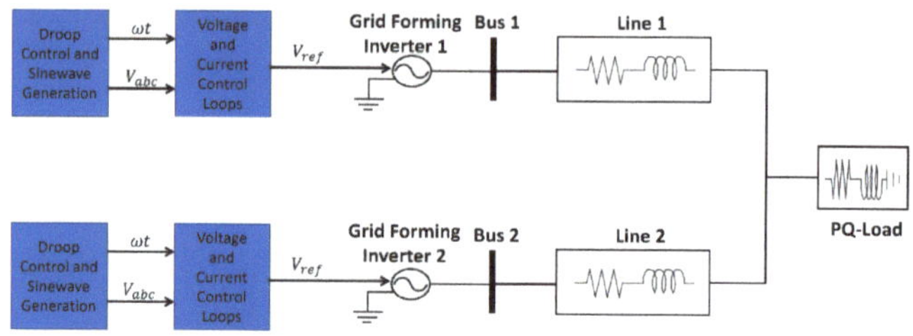

Figure 15. A microgrid consisting of two grid-forming inverters sharing a single load.

The single load in the above model is assumed again as PQ load, where between $0 < t < 0.5$ s, the demand of active power is 12 kW and the demand of reactive power is 4 kVAR. After that the load demand is increased between $0.5\text{ s} < t < 1$ s, the demand of active power becomes 20 kW and the demand of reactive power becomes 10 kVAR.

Figure 16 shows the power sharing in terms of both real and reactive power for both inverters that are grid forming. It can be noticed how both inverters share the same amount of active power and reactive power during the normal load period and when the load is increased. This is due to the fact that both inverters are equipped with the same droop characteristics.

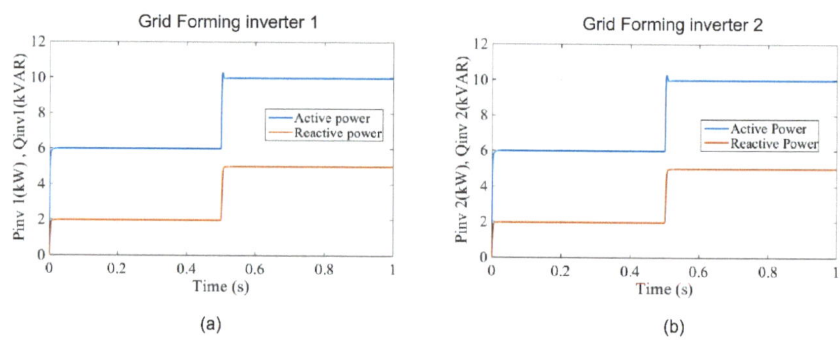

Figure 16. Active and reactive power sharing of both grid-forming inverters during the load variation. (**a**) Active and reactive power of GFM inverter 1; (**b**) active and reactive power of GFM inverter 2.

The frequency and RMS voltage for both grid-forming inverters are presented in Figure 17, where it can be observed that the change in frequency and RMS voltage is the same for both inverters.

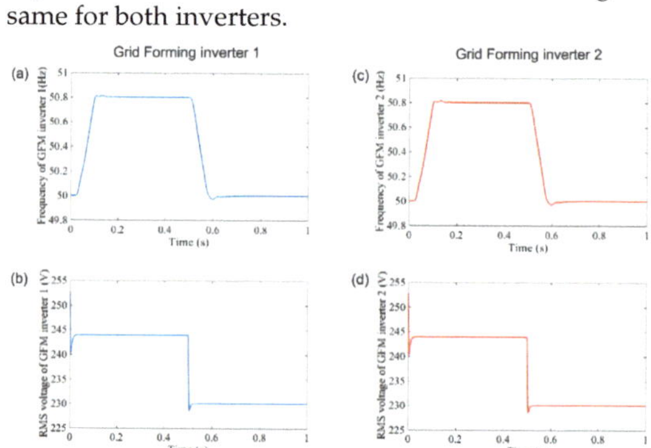

Figure 17. Frequency and RMS voltage of both grid-forming inverters during the load variation. (**a**) Frequency of GFM inverter 1; (**b**) RMS voltage of GFM inverter 1; (**c**) frequency of GFM inverter 2; (**d**) RMS voltage of GFM inverter 2.

Here, the frequency for both inverters was oscillating at 50.8 Hz when the total load demand of active power was 12 kW and then dropped to 50 Hz as the active power load demand increased to 20 kW, which is divided between the maximum capacity of both grid-forming inverters (10 kW for both). The same was the case for the RMS voltage, where the voltage of both inverters was oscillating at approximately 244 V when the total load demand of reactive power was 4 kVAR and then dropped to 230 V as the reactive power load demand increased to 10 kVAR, which is divided between the maximum capacity of both grid-forming inverters 5 kVAR for both.

5. Conclusions and Future Work

Inverters of the grid-forming type have been suggested as a favorable arrangement for the various problems in modern power systems. In this paper, the autonomous operation of the grid-forming inverter with the integration of the droop control as a decentralized method was investigated. The decentralized method used in this work confirms the applicability and accuracy of installing grid-forming inverters in the distribution network without the need for communication. The simulation results confirmed the precision of the assumed droop setting with the resultant droop in response to the PQ load variation in the three regions. It was also clear how the load demand is exactly satisfied with the generation that employs the droop control methodology. Furthermore, the simulation results confirmed the power sharing, where the generation of active and reactive power for both grid-forming inverters was divided equally to satisfy the load demand. Future work can include further investigations into grid-forming inverters connected at the distribution and transmission levels. These investigations can be the hardware implementation, energy storage, system protection, fault ride through capability, stability analysis, economic aspects and economic dispatching of units, and the transition from islanded mode to grid-connected mode. Another path for future work may include a comprehensive analysis of the limitations and constraints inherent in the proposed methodology and the overall study.

Author Contributions: Conceptualization, Q.S.; methodology, Q.S.; software, Q.S.; validation, Q.S. and R.A.; formal analysis, Q.S. and R.A.; writing—original draft preparation, M.K.; writing—review and editing, M.K. and A.A.-Q. All authors have read and agreed to the published version of the manuscript.

Funding: This research received no external funding.

Data Availability Statement: Not applicable.

Acknowledgments: The authors would like to thank PSUT for their support.

Conflicts of Interest: The authors declare no conflict of interest.

References

1. Rosso, R.; Wang, X.; Liserre, M.; Lu, X.; Engelken, S. Grid-Forming Converters: Control Approaches, Grid-Synchronization, and Future Trends—A Review. *IEEE Open J. Ind. Appl.* **2021**, *2*, 93–109. [CrossRef]
2. Lasseter, R.H.; Chen, Z.; Pattabiraman, D. Grid-Forming Inverters: A Critical Asset for the Power Grid. *IEEE J. Emerg. Sel. Top. Power Electron.* **2020**, *8*, 925–935. [CrossRef]
3. Mahamedi, B.; Fletcher, J.E. The Equivalent Models of Grid-Forming Inverters in the Sequence Domain for the Steady-State Analysis of Power Systems. *IEEE Trans. Power Syst.* **2020**, *35*, 2876–2887. [CrossRef]
4. Benzaquen, J.; Miranbeigi, M.; Kandula, P.; Divan, D. Collaborative Autonomous Grid-Connected Inverters: Flexible grid-forming inverter control for the future grid. *IEEE Electrif. Mag.* **2022**, *10*, 22–29. [CrossRef]
5. Unruh, P.; Nuschke, M.; Strauß, P.; Welck, F. Overview on Grid-Forming Inverter Control Methods. *Energies* **2020**, *13*, 2589. [CrossRef]
6. Ratnama, K.S.; Yang, K.P. Future low-inertia power systems: Requirements, issues, and solutions—A review. *Renew. Sustain. Energy Rev.* **2020**, *124*, 109773. [CrossRef]
7. Tamrakar, U.; Shrestha, D.; Maharjan, M.; Bhattarai, B.; Hansen, T.; Tonkoski, R. Virtual Inertia: Current Trends and Future Directions. *Appl. Sci.* **2017**, *7*, 654. [CrossRef]
8. Matevosyan, J.; Badrzadeh, B.; Prevost, T.; Quitmann, E.; Ramasubramanian, D. Grid-Forming Inverters: Are They the Key for High Renewable Penetration? *IEEE Power Energy Mag.* **2019**, *17*, 89–98. [CrossRef]

9. Groß, D.; Colombino, M.; Brouillon, J.-S.; Dörfler, F. The Effect of Transmission-Line Dynamics on Grid-Forming Dispatchable Virtual Oscillator Control. *IEEE Trans. Control Netw. Syst.* **2019**, *6*, 1148–1160. [CrossRef]
10. Zhong, Q.-C.; Weiss, G. Synchronverters: Inverters That Mimic Synchronous Generators. *IEEE Trans. Ind. Electron.* **2011**, *58*, 1259–1267. [CrossRef]
11. Han, H.; Hou, X.; Yang, J.; Wu, J.; Su, M.; Guerrero, J.M. Review of Power Sharing Control Strategies for Islanding Operation of AC Microgrids. *IEEE Trans. Smart Grid* **2016**, *7*, 200–215. [CrossRef]
12. Gursoy, M.; Mirafzal, B. Self-Security for Grid-Interactive Smart Inverters Using Steady-State Reference Model. In Proceedings of the 2021 IEEE 22nd Workshop on Control and Modelling of Power Electronics (COMPEL), Cartagena, Colombia, 2–5 November 2021.
13. Geng, S.; Hiskens, I.A. Unified Grid-Forming/Following Inverter Control. *IEEE Open Access J. Power Energy* **2022**, *9*, 489–500. [CrossRef]
14. Me, S.P.; Zabihi, S.; Blaabjerg, F.; Bahrani, B. Adaptive Virtual Resistance for Postfault Oscillation Damping in Grid-Forming Inverters. *IEEE Trans. Power Electron.* **2022**, *37*, 3813–3824. [CrossRef]
15. Rathnayake, D.B.; Bahrani, B. Multivariable Control Design for Grid-Forming Inverters With Decoupled Active and Reactive Power Loops. *IEEE Trans. Power Electron.* **2023**, *38*, 1635–1649. [CrossRef]
16. Hart, P.J.; Lasseter, R.H.; Jahns, T.M. Coherency Identification and Aggregation in Grid-Forming Droop-Controlled Inverter Networks. *IEEE Trans. Ind. Appl.* **2019**, *55*, 2219–2231. [CrossRef]
17. Huang, X.; Wang, K.; Qiu, J.; Hang, L.; Li, G.; Wang, X. Decentralized Control of Multi-Parallel Grid-Forming DGs in Islanded Microgrids for Enhanced Transient Performance. *IEEE Access* **2019**, *7*, 17958–17968. [CrossRef]
18. Liu, T.; Wang, X. Physical Insight Into Hybrid-Synchronization-Controlled Grid-Forming Inverters Under Large Disturbances. *IEEE Trans. Power Electron.* **2022**, *37*, 11475–11480. [CrossRef]
19. Li, Y.; Gu, Y.; Green, T.C. Revisiting Grid-Forming and Grid-Following Inverters: A Duality Theory. *IEEE Trans. Power Syst.* **2022**, *37*, 4541–4554. [CrossRef]
20. Yazdani, S.; Ferdowsi, M.; Davari, M.; Shamsi, P. Advanced Current-Limiting and Power-Sharing Control in a PV-Based Grid-Forming Inverter Under Unbalanced Grid Conditions. *IEEE J. Emerg. Sel. Top. Power Electron.* **2020**, *8*, 1084–1096. [CrossRef]
21. Lin, Y.; Eto, J.H.; Johnson, B.B.; Flicker, J.D.; Lasseter, R.H.; Pico, H.N.V.; Seo, G.-S.; Pierre, B.J.; Ellis, A. *Research Roadmap on Grid-Forming Inverters*; National Renewable Energy Laboratory: Golden, CO, USA, 2020.
22. Anttila, S.; Döhler, J.; Oliveira, J.; Boström, C. Grid Forming Inverters: A Review of the State of the Art of Key Elements for Microgrid Operation. *Energies* **2022**, *15*, 5517. [CrossRef]
23. Rathnayake, D.B.; Akrami, M.; Phurailatpam, C.; Me, S.P.; Hadavi, S.; Jayasinghe, G.; Zabihi, S.; Bahrani, B. Grid Forming Inverter Modeling, Control, and Applications. *IEEE Access* **2021**, *9*, 114781–114807. [CrossRef]
24. Sharma, D.; Sadeque, F.; Mirafzal, B. Synchronization of Inverters in Grid Forming Mode. *IEEE Access* **2022**, *10*, 41341–41351. [CrossRef]
25. Hossain, M.; Pota, H.; Issa, W.; Hossain, M. Overview of AC Microgrid Controls with Inverter-Interfaced Generations. *Energies* **2017**, *10*, 1300. [CrossRef]

sustainability

MDPI

Article

A Droop-Controlled Interlink Converter for a Dual DC Bus Nanogrid with Decentralized Control

Ahmad M. A. Malkawi [1,*], Ayman AL-Quraan [2] and Luiz A. C. Lopes [3]

1 Mechatronics Engineering Department, Engineering Faculty, The University of Jordan, Amman 11942, Jordan
2 Electrical Power Engineering Department, Hijjawi Faculty for Engineering Technology, Yarmouk University, Irbid 21163, Jordan; aymanqran@yu.edu.jo
3 Department of Electrical and Computer Engineering, Concordia University, Montreal, QC H4B 1R6, Canada; luiz.lopes@concordia.ca
* Correspondence: ah.malkawi@ju.edu.jo

Abstract: This paper proposed a dual DC bus nanogrid with 380 V and 48 V buses and allows the integration of distributed energy resources on two buses. The proposed system employs an interlink converter to enable power sharing between the buses. The integration of distributed energy resources has been found to enhance the reliability of the low-voltage bus in comparison to those that lack such integration. The integration process requires the introduction of a new V-I curve for the interlink converter within a DC nanogrid controlled by DC bus signaling and droop control. Furthermore, selecting a power electronics converter for the interlink converter is essential. This paper employs a dual active bridge with galvanic isolation as an interlink converter and proposes a control strategy for the converter that relies on DC bus signaling and droop control. Moreover, this control methodology serves the purpose of preventing any detrimental impact of the interlink converter on the DC buses through the reprogramming of the V-I curve. Subsequently, the suggested control methodology underwent simulation testing via MATLAB/Simulink, which included two different test categories. Initially, the DAB was evaluated as an interlink converter, followed by a comprehensive assessment of the interlink converter in a complete dual DC bus nanogrid. The results indicate that the DAB has the potential to function as an interlink converter while the suggested control approach effectively manages the power sharing between the two buses.

Keywords: distributed energy resources; dual DC bus nanogrid; nanogrid; interlink; PV system; battery converter; droop control; DC bus signaling

Citation: Malkawi, A.M.A.; AL-Quraan, A.; Lopes, L.A.C. A Droop-Controlled Interlink Converter for a Dual DC Bus Nanogrid with Decentralized Control. *Sustainability* **2023**, *15*, 10394. https://doi.org/10.3390/su151310394

Academic Editor: Elisa Marrasso

Received: 9 May 2023
Revised: 27 June 2023
Accepted: 29 June 2023
Published: 30 June 2023

1. Introduction

Integrating renewable energy sources (RESs) at the distribution levels is a promising alternative for reducing the environmental impact of conventional fossil-fuel-based sources and the need for building new transmission lines to bring power from large plants far away from consumer centers. However, the stochastic nature of RESs and their variable power generation can lead to power quality issues in the distribution system. One way around this is to associate them with energy storage units and controllable loads, which could be controlled as a cluster or a "microgrid". For that, one needs detailed information about the state and operating conditions to be sent timely to the grid interfaces of generation and storage units, which must be fast acting and reliable. This is achieved with sophisticated information and communication technology (ICT) and modern power electronic converters [1–4]. The microgrids study has focused on distribution feeders and neighborhoods with medium-sized power plants, 100's kW [4–6]. The microgrids concept enables the distributed energy resources (DERs), which include power sources, energy storage units, and controllable loads, to operate in islanding mode following a power grid failure [2,5,7]. This is done by dynamically decoupling the microgrids from the utility grid using a full-power bidirectional converter. This grid tie or interface converter can be used as an energy control center [7]. This concept can be extended for smaller units, 10 s of kW, for a residence. The future smart homes are envisioned to be net-zero energy homes (NZEH). These are homes with net-zero energy consumption, meaning that the total energy they use annually is nearly equal to the amount of energy they produce, ideally with RESs [8–10]. To be able to operate autonomously, at least for some time, they should present not only generation as well as storage units and controllable loads which, with a suitable control structure, can be seen as a "nanogrid". Nanogrids can be based on AC or DC distribution [7,11].

In order to connect RESs, plug-in hybrid electric vehicles, and storage units to an AC nanogrid, power electronics converters of the DC to AC type are used. Unidirectional power electronic converters are employed for power sources such as PV and wind turbines. In the case of PV, the required DC–AC converter usually consists of two stages: A step-up (boost) DC–DC converter and a single or three-phase DC–AC converter [7,12,13]. Two-stage converters, now bidirectional, connect energy storage units, usually batteries, to an AC nanogrid [12–14]. In AC nanogrids, the DC–AC converters are responsible for voltage and frequency regulation of the AC bus in islanding (stand-alone) and grid-connecting modes [15,16]. For distributed energy sources, energy storage units, and controllable loads in a nanogrid, a DC bus leads to the most straightforward interconnection scheme [7,17]. Compared to the AC distribution, the DC distribution does not require frequency or phase control. Another aspect of the AC distribution is controlling the reactive power [18]. In addition, DC–DC interfaces have higher efficiency and reliability [7,17,19–21].

Furthermore, DC has lower distribution losses, lower cost of conductors for the same power, and does not use bulky transformers leading to cost and volume reduction [7,17,22]. In addition to that, a DC nanogrid should employ a DC–AC converter to serve as the interface to a conventional AC distribution feeder/neighborhood. It should offer the desirable capability of dynamically decoupling the DC nanogrid from the AC utility grid. Therefore, the future residential electrical system of an NZEH could very well be based on DC distribution [7].

One key aspect that remains unclear is the voltage level that should be adopted. According to the IEC, extra low voltage (ELV) DC presents a magnitude of less than 120 V and a lower risk of electrical shock. However, relatively high currents are required to supply kW loads, leading to high distribution losses and conductors. The 48 V DC is a standard telecom voltage level [7,23,24]. The following higher voltage category is the low voltage (LV) magnitudes between 120 V and 1500 V. Higher voltages lead to lower currents for a given power demand but this might complicate the design of the DC protection system and require power interfaces with significant voltage gains. The 380 V DC level matches the industry-standard intermediate DC voltage [7,25]. One option that industry associations have considered and proposed is using two voltage levels (buses). The Emerge Alliance [26,27] advocates using a 24 V, mainly to comply with current LED technology for lighting, and a 380 V for other loads. However, in order to be able to supply other small loads, the use of a 48 V, which is also considered for mild-hybrid vehicles, will be selected.

In the Emerge Alliance scheme, power sources and storage units are connected to the 380 V bus, and a step-down unidirectional DC–DC converter is used to supply the 24 V lighting bus. One issue of concern is the reliability of the 24 V, which will be de-energized if either the 24 V bus interface fails or the 380 V bus is not operational. The dual DC bus nanogrid considered in this paper presents power sources and energy storage units in both DC buses and a bidirectional "interlink" converter/interface to allow power flow from one bus to the other and, indirectly, from the 48 V bus to the AC utility grid, assuming that a 380 V DC to AC grid interface is present and operational. A simplified Dual DC buses nanogrid is shown in Figure 1. One can see that the RESs (solar energy) and electrical storage systems (batteries) are connected to both DC buses. The first can employ a unidirectional converter but the second requires a bidirectional one. A bidirectional DC–AC converter connects the utility grid to the HV DC bus. A bidirectional interlink converter allows the power flow between the two DC Buses.

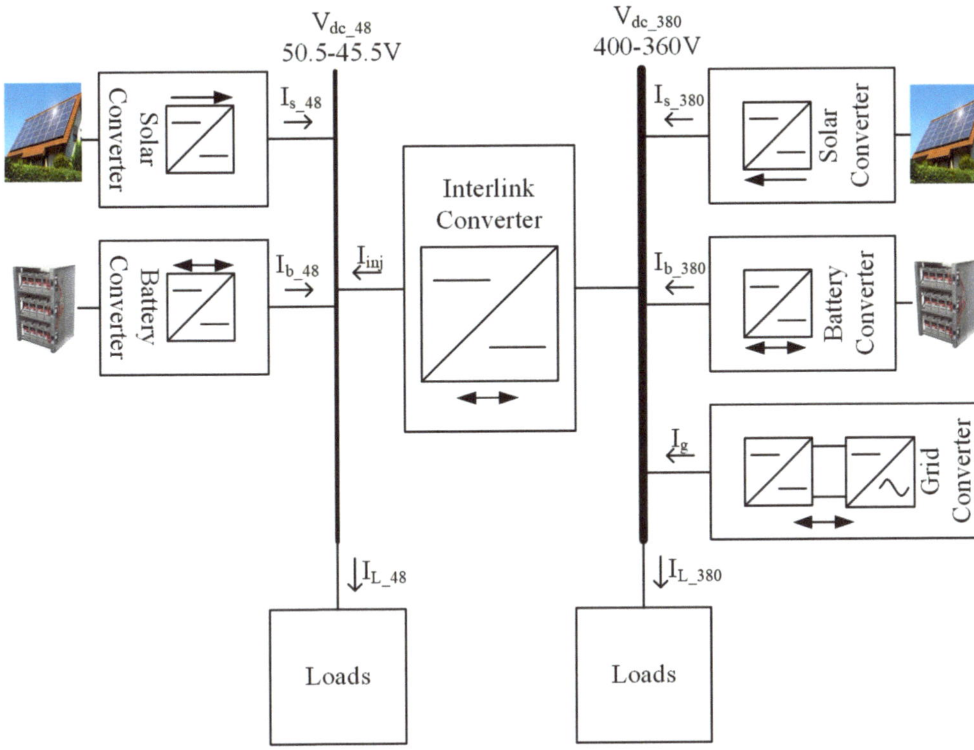

Figure 1. Dual DC Bus Nanogrid.

In terms of research and development of the interlink converter, one should select a suitable power topology with rated voltages of 48 V:380 V, a modulation/control scheme, as well as define an appropriate control strategy (V-I curve) for the interlink converter to operate with DC bus signaling (DBS). The latter should be done in a way as to not interfere with the conventional approach of managing power and energy in a single-bus DC nanogrid. It should be noted that although 48 V and 380 V, as stated before, are ELV and LV, they would be called LV side and HV side, respectively, in this work.

This paper introduces a nanogrid system with dual DC buses, namely a low voltage (48 V) bus and a high voltage (380 V) bus, which are interconnected through an interlink converter responsible for managing the power flow between them. The buses at the booth are equipped with various resources such as renewable energy sources (RESs), storage systems, and load. Furthermore, the paper proposes V-I curves for the power interfaces that manage the power sharing in decentralized control of the dual DC bus nanogrid through DC bus signaling and droop control. The present paper introduces a control strategy for the interlink converter to enhance the power balance of the two buses. Additionally, the authors analyze the small signal model of the dual active bridge in its capacity as an interlink converter. The authors validated the proposed dual DC bus nanogrid through simulation results utilizing MATLAB/Simulink.

2. Power Control and Current Sharing in a Single DC Bus Nanogrid

A conventional single-bus DC nanogrid can be decentralized with a hierarchical structure with DC bus signaling (DBS) and droop control used at the primary level [7,13,28–32]. In addition, adaptive droop control could be used to improve the power sharing for parallel DC–DC converters [33]. This work does not consider the secondary and tertiary levels, which require a communication means for energy management. Thus, with fixed parameters for the primary control level, the DC bus voltage will vary in a small range around the rated voltage, as defined by the V-I curves of the DERs. DBS uses the DC bus voltage as the communication link to coordinate the operation of DERs in a decentralized way. With droop control, the current (power) injected by each DER in the DC nanogrid depends on its threshold voltage (V_{NL}), where the injected current is zero, and its droop slope/factor (R_d) is shown in (1). The latter determines how the injected current varies as a function of grid voltage variations.

$$I_{DC} = (V_{NL} - V_{DC})\frac{1}{R_d} \tag{1}$$

where;

$$R_d = \frac{\Delta V_{DC}}{\Delta I_{DC}} \tag{2}$$

Figure 2 shows the V-I curves of the standard components found in a DC nanogrid with a proposed set of parameters [7,13,30]. The DC–AC grid converter is discussed first. When the DC bus voltage is between 390 V and 370 V, the converter operates in droop mode with a droop constant (slope) R_{dg}. Otherwise, it operates in current-limiting mode, with the DC–AC converter absorbing rated current for V_{DC} > 390 V and supplying for V_{DC} < 370 V. Its threshold, or no-load, voltage is the rated voltage of the DC bus: 380 V. That is to say that at this voltage, the current and power flowing through the converter is zero. The values of the threshold voltage and the DC bus voltage determine the power-flow direction. When the DC bus voltage is higher than 380 V but lower than 390 V, power flows from the DC nanogrid into the AC utility grid (I_g < 0 A). Conversely, for DC bus voltage lower than 380 V but higher than 370 V, power flows to the DC nanogrid from the AC utility grid (I_g > 0 A). The threshold voltage is the main control parameter of this DC–AC converter, which the secondary control level can adjust.

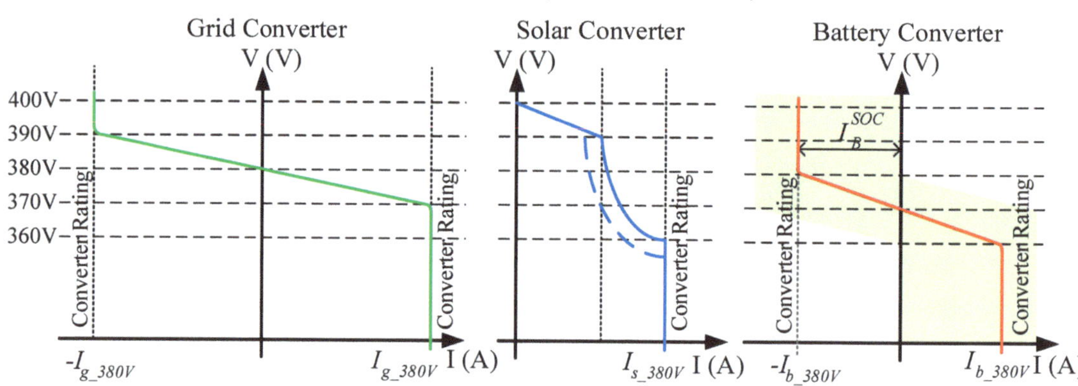

Figure 2. V −I curves of the high voltage DC bus DERs: Grid Converter, Solar Converter, and Battery Converter.

In the case of the RESs, such as the solar converter, the V. vs. I curve usually presents three regions: droop, constant power, and constant current, as shown in Figure 2. It operates with droop constant R_{ds} for a DC bus voltage between 400 V (threshold voltage) and 390 V when the converter starts to operate in the maximum power point tracking (MPPT) mode with the maximum available power injection. In the concept of DC bus signaling, a higher threshold voltage for the solar converter than the grid converter gives the RESs a higher priority to feed the DC nanogrid demand/load when operating in parallel. The following section will discuss the actual power flow between the solar converter, grid converter, battery, and variable load. For DC bus voltages between 390 V and 360 V, the converter operates in the maximum power point tracking (MPPT) mode injecting virtually constant power for constant solar irradiance (W/m^2). The dashed line in this region shows that the solar converter is reprogrammed to operate in MPPT mode, injecting the maximum available power when the solar irradiance decreases from the rated value or the PV panel temperature increases [7]. When the DC bus voltage drops below 360 V, the solar converter operates at the current-limiting mode.

The electrical storage units operate in droop mode, with droop constant R_{db} when the DC bus voltage is between 380 V and 360 V. Otherwise, it operates in the current-limiting mode, absorbing rated current for $V_{DC} > 380$ V and supplying for $V_{DC} < 360$ V. The storage system usually has the third priority, after the RESs and the utility grid, to feed the load. It has a threshold voltage of 370 V, lower than the other system interfaces. This prevents the electrical storage system from discharging through the utility grid. On the other hand, the electrical storage system is discharged at DC bus voltages less than 370 V.

Concerning power-flow control and energy management in the DC nanogrid, the actual value of the threshold voltage of the power interfaces can be adjusted based on the instantaneous and historical price of energy that could be drawn from or supplied to the utility grid. Long- and/or short-term histories of the local renewable energy generation and the nanogrid load profiles play a role in reprograming the threshold voltage. In addition, the history of the storage units' state of charge (SoC) should be considered. To achieve optimal threshold voltage preprogramming, an *optimal energy utilization algorithm* is used [7]. To implement this algorithm, historical data about the system is needed. The maximum charging and discharging current are based on the battery manufacturer and power-converter ratings. Sometimes, when the SoC drops below 20%, the control will stop the discharging operation mode. On the other hand, the battery will not be charged when it has a SoC greater than 90% [34]. The current limits could be a function of the state of charge (SoC) [7,35–37]. This produces the shaded area in the V-I curve, as shown in Figure 2.

Neglecting the voltage drops between the DER interfaces and the equivalent system load, the DC bus voltage (V_{DC}) of a nanogrid operating with DBS and multiple DERs can be determined from the:

$$I_{Load} = \sum_{i=1}^{n} I_{DC\_i} \quad (3)$$

where I_{DC_i} are computed according to the V-I curves of the n DERs, with the value of V_{DC}, one can then calculate the individual contributions of the DERs from their V-I curves.

3. Power Control of the Interlink Converter

As a conventional single-bus DC nanogrid, the dual DC bus nanogrid could be decentralized with a hierarchical structure based on droop control and DC bus signaling [7,28,29]. In principle, the V-I curves and parameters of the DERs in the LV DC and HV DC buses could be the same as in the case of a single-bus DC nanogrid. For instance, it is assumed that the values of threshold voltages, droop slopes, and current limits for the DERs in the LV bus are proportional to those of the HV side, shown in Figure 2. With the addition of the interlink converter, an opportunity for optimizing the operation of both buses arises in terms of voltage regulation and management of power flow and energy management. For that, a suitable strategy for controlling the power flow of the interlink converter is needed. Ideally, it should be based on DBS, considering the voltages of both DC buses, which reflect the power availability in those buses. Considering that the batteries are the weak link in the system since they are not supposed to be over- or undercharged, the control law of the interlink converter must not place unnecessary stress on them. In addition, in terms of managing the system to operate efficiently, having the battery of one bus charge, and the battery of the other should be avoided.

Two control strategies for the interlink converter in a dual DC bus nanogrid, average droop control, and constant-voltage ratio, were presented in [38]. The first determines the power flow between the two buses based on the average value obtained from droop curves of the LV and HV sides. On the other hand, the second employs a PI controller to keep the per-unit voltage ratio constant between the two buses without a V-I curve. Those control strategies can be used for a dual DC bus nanogrid with similar RESs and storage unit participations, but they should not be used when the dual DC nanogrid is connected with the AC utility grid.

This paper presents an alternative approach to control the interlink converter, which is to use, in principle, the same V-I curve of the grid converter of the HV bus. Recall that this is controlled based only on the HV DC bus's voltage, neglecting the AC bus's voltage. This is reasonable since the latter is assumed to be much stronger, with more power availability, than the former and can supply/absorb the matching power for/from the HV DC bus. By making the control of the interlink converter based on the voltage of the LV bus only, one does not compromise the ability of the HV bus to operate without stressing its storage units since the grid converter is there to provide/absorb the required matching power.

Therefore, one can use the V-I curves shown in Figure 3 for the interfaces of the LV DC bus. Note that the voltage levels, in terms of pu, are comparable to those employed in the HV DC bus. The voltage range in normal operating conditions varies by about 5% above and below the rated voltage (48 V).

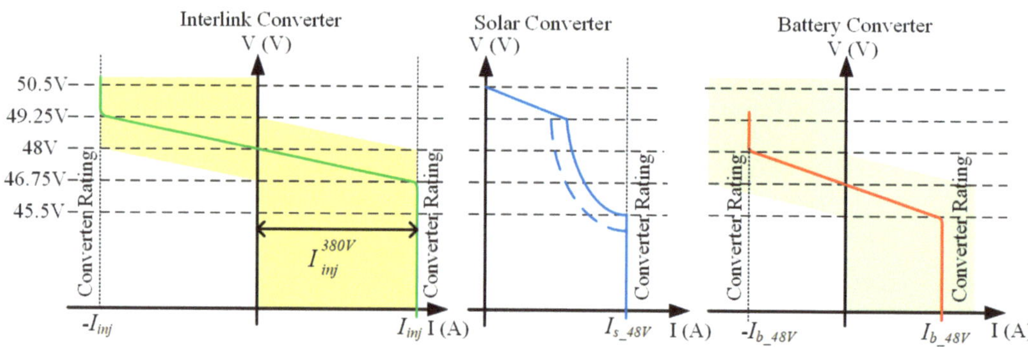

Figure 3. Low voltage 48 V DC bus V-I characteristics: Interlink Converter, Solar Converter, and Battery Converter.

Based on the V-I curves in Figure 3, one can say that the LV DC bus battery will not be discharged into the HV DC bus. The LV DC bus is discharged with DC bus voltages less than 46.75 V, and, at this DC bus voltage, power should flow from the HV DC bus to the LV DC bus, according to the interlink converter V-I characteristic. To avoid power flowing from the LV DC bus to the HV DC bus at light load conditions or when the utility grid cannot absorb the excess power, the interlink converter (current) limits should be reprogrammed. They are changed based on the HV DC bus voltage, as shown in Figure 4. In addition, Figure 4 shows that the interlink converter (current) limits are reprogrammed to prevent the HV DC bus battery from discharging into the LV DC bus.

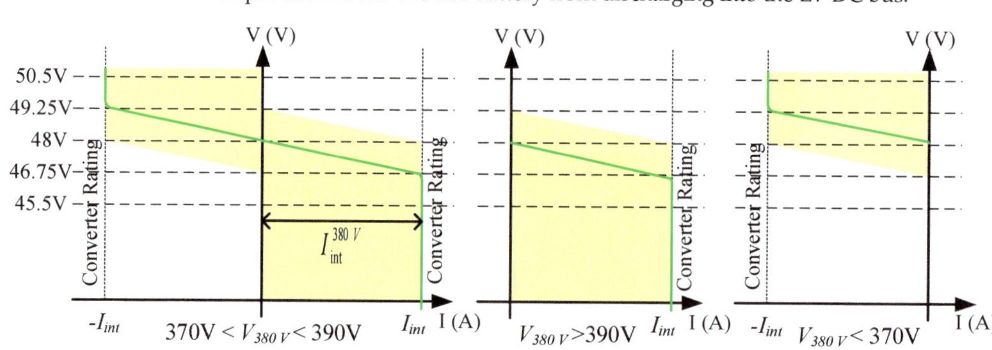

Figure 4. Interlink converter current limit reprogrammed based on the HV DC bus: 370 V < V_{HV} < 390 V, V_{HV} > 390 V, and V_{HV} < 370 V.

4. Interlink Converter Modeling and Control Design in Dual DC Buses Nanogrid

Figure 5 shows the schematic of a full-bridge isolated bidirectional DC–DC converter called a dual active bridge (DAB) [39–41]. It can provide a high voltage gain using a high-frequency transformer with a small volume and weight. Moreover, the DAB provides galvanic isolation between the two DC buses and there is no need for the circuit breakers to isolate the buses at fault conditions in one of them [42]. There, V_{LV} and V_{HV} are the LV and HV DC bus voltages, respectively, $S_1 - S_4$ and $Q_1 - Q_4$ are the controllable switches, n is the transformer's turn ratio, and L is the equivalent leakage inductance of the transformer.

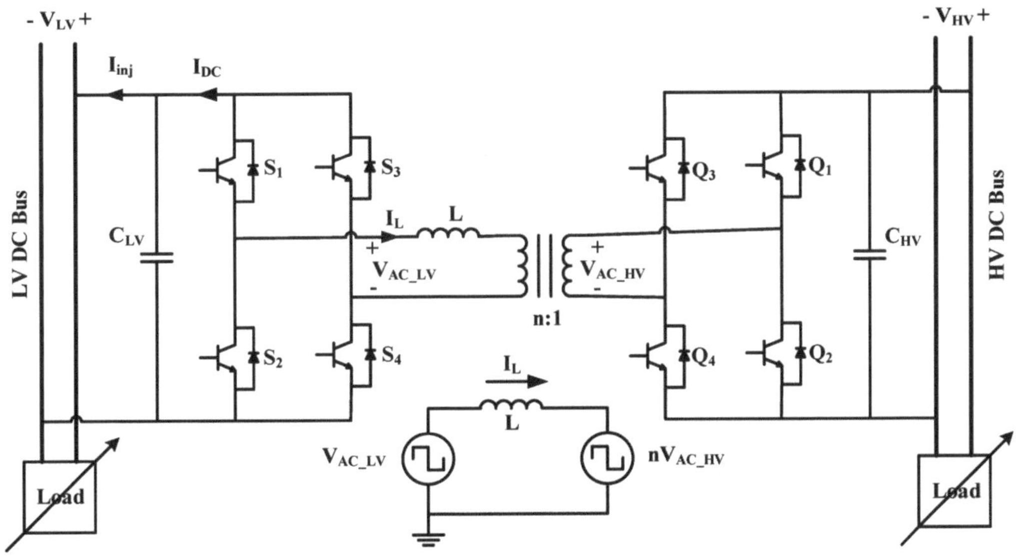

Figure 5. Full-bridge isolated DC–DC converter (DAB).

The DAB DC–DC converter is controlled with phase-shift control in this work. There are many types of phase-shift control in the literature [39–41,43]. However, the most common one (and suitable for this application) is the single-phase-shift (SPS) control shown in Figure 6 [31]. All the switches are gated in this control scheme with a 50% duty cycle. S_1 and S_4 are switched as a pair, complementarily to S_2 and S_3 in the first full bridge. Likewise, Q_1 and Q_4 are switched complementarily to Q_2 and Q_3 in the second HV side bridge. That generates square waveforms, V_{AC_HV} and V_{AC_LV}, at the transformer's terminals. As in traditional AC power-transmission systems, the inductor current's direction and magnitude can be adjusted by changing the phase shift between the two square waves. This allows the control of the direction and magnitude of the power flow [43–45]. This is achieved with a single control parameter, the phase-shift ratio (d), for an actual phase shift (πd) between V_{AC_HV} and V_{AC_LV}.

The power injected into the LV side of the DC–DC converter in Figure 5 can be expressed as follows for SPS control [39,46]:

$$P = \frac{nV_{AC\_HV}V_{AC\_LV}}{2f_sL}d(1 - |d|) \tag{4}$$

where f_s is the switching frequency and V_{AC_HV} and V_{AC_LV} are the transformer's high and low side voltages. By ignoring the voltage loss of the power switch, V_{AC_HV} could equal V_{HV} and $-V_{HV,}$ while V_{AC_LV} could equal V_{LV} and $-V_{LV}$. From Equation (4), the power has a maximum value when the phase-shift ratio (d) is 0.5. Figure 7 shows the power injected by the 48 V bus vs. the phase $-$shift ratio (d). Power flows from the HV side to the

LV side at positive phase shifts ($0 < \pi d < \pi$) and from the LV side to the HV side at negative phase shifts ($-\pi < \pi d < 0$).

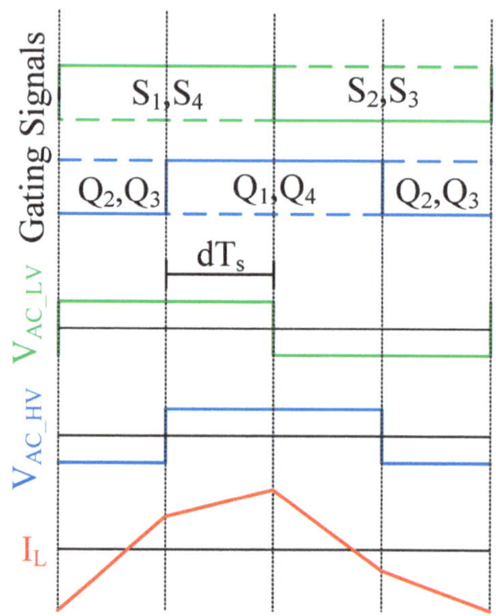

Figure 6. SPS Control waveforms.

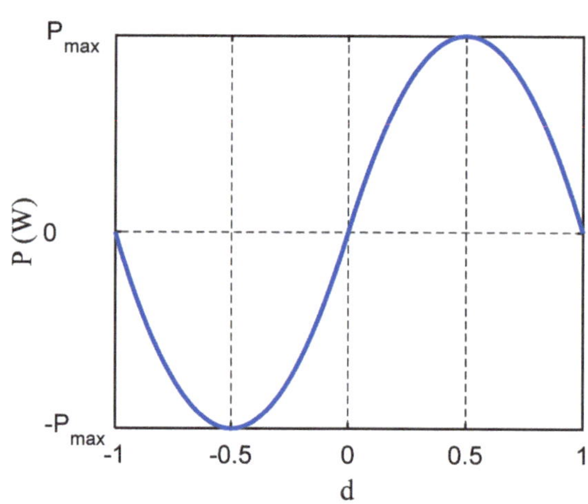

Figure 7. Power injected by the LV bus vs. phase −shift ratio.

Then, the average value of the current at the LV side (I_{DC}) is [46]:

$$I_{DC} = \frac{nV_{AC\_HV}}{2f_sL}d(- |d|) \tag{5}$$

Therefore, the large signal model for the DAB can be obtained by replacing the two full bridges and the HV voltage with a current source of magnitude given in (5) and the large signal equivalent circuit is shown in Figure 8 [46]. Moreover, the LV DC bus could be modeled using Thevenin's equivalent circuit with a threshold voltage V_{NL_LV} and a droop resistance R_{dLV}. Then, the DAB in the dual DC buses nanogrid could be modeled with respect to the LV side in the dual DC nanogrid, as shown in Figure 9.

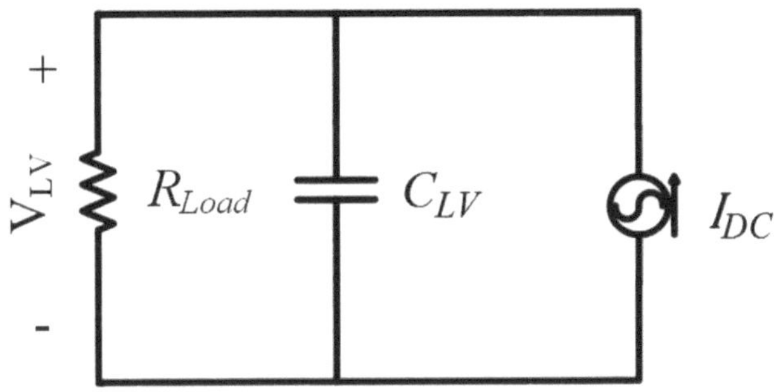

Figure 8. Equivalent circuit of DAB feeding a resistive load.

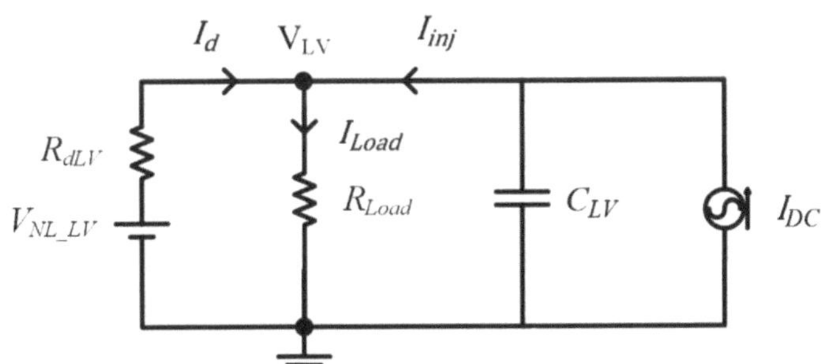

Figure 9. Equivalent circuit of DAB in dual DC buses nanogrid with respect to LV side.

For the study of the dynamic response of the DAB, the expression for the average value of the current injected in the LV DC bus should be linearized around an operating point, leading to a small signal model as in (6) [46]. However, the injected current $\tilde{i}_{inj}(s)$ as a function of the phase-shift ratio $\tilde{d}(s)$ is needed to control the DAB based on the proposed V-I curve. The small signal model for the DAB DC–DC converter with SPS reflected the LV side in the DC nanogrid, assuming that the equivalent threshold voltage of the LV bus (V_{NL_LV}) is constant, is shown in Figure 10.

$$\tilde{i}_{DC}(s) = \frac{nV_{AC\_HV}}{2f_sL}(1-2d)\tilde{d}(s) \tag{6}$$

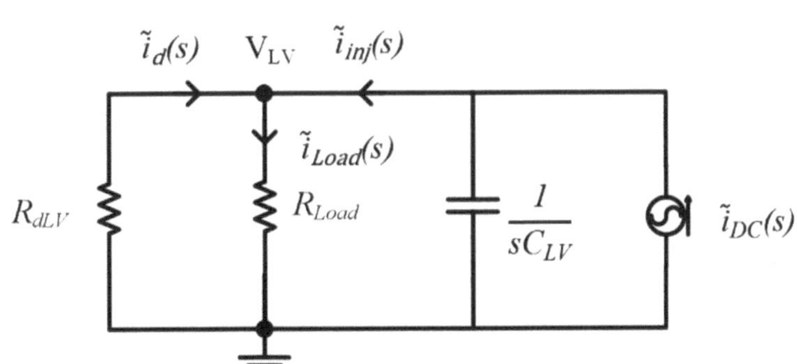

Figure 10. Small signal model of DAB in dual DC buses nanogrid with respect to LV side.

Then, one gets the following expression for the current injected by the interlink converter into the LV DC bus ($\tilde{i}_{inj}(s)$) as a function of variations in the phase-shift ratio ($\tilde{d}(s)$).

$$\tilde{i}_{inj}(s) = \frac{nV_{AC\_HV}(1-2d)(R_{dLV}+R_{Load})}{2f_sL(R_{Load}R_{dLV}C_{LV}s+R_{dLV}+R_{Load})}\tilde{d}(s) \tag{7}$$

The transfer function of the DAB $G_p(s)$ for the design of the current controller becomes.

$$\frac{\tilde{i}_{inj}(s)}{\tilde{d}(s)} = \frac{nV_{AC\_HV}(1-2d)(R_{dLV}+R_{Load})}{2f_sL(R_{Load}R_{dLV}C_{LV}s+R_{dLV}+R_{Load})} \tag{8}$$

5. Case Study

As a case study in this work, a DAB, as in Figure 5, is used as an interlink converter with the following parameters: transformer-turns ratio n = 0.25, leakage inductance L = 300 µH, switching frequency f_s = 20 kHz, and HV and LV capacitors C_{HV} = C_{LV} = 1500 µF. The LV DC bus is modeled by a Thevenin equivalent with the DC voltage V_{NL_LV} = 49.25 V and a droop constant R_{dLV} = 0.289 Ω. In addition, the HV DC bus is modeled by Thevenin's with the DC voltage V_{NL_HV} = 390 V and a droop constant R_{dHV} = 0.289 Ω. Table 1 summarises the case-study parameters.

Table 1. The case-study parameters.

The Parameter	*The Value*
Transformer turns ratio (n)	0.25
Leakage inductance (L)	300 µH
Switching frequency (f_s)	20 kHz
HV and LV capacitors (C_{HV} = C_{LV})	1500 µF
The LV DC bus threshold voltage (V_{NL_LV})	49.25 V
The LV droop constant (R_{dLV})	0.289 Ω
The HV DC bus threshold voltage (V_{NL_HV})	390 V
The HV droop constant (R_{dHV})	0.289 Ω
The load resistance (R_{Load})	10 Ω

For a load resistance of 10 Ω, the transfer function of the plant becomes:

$$\frac{\tilde{i}_{inj}(s)}{\tilde{d}(s)} = \frac{81.5}{0.004335\,s+10.289} \tag{9}$$

Then, a PI type-II compensator (10) is designed for a cut-off frequency of $f_x = 240$ Hz [31] and a phase margin PM = 80°, as illustrated in the following steps:

Step 1: From the Bode plot of the plant, obtain $|G(f_x)| = 6.7 \equiv 16.5$ dB and $\angle(G(f_x)) = -32.4°$.

Step 2: Step 4: Calculate the $\angle(C(f_x))$ and $|C(f_x)|$:

$\angle(C(f_x)) = PM - (180° + \angle(G(f_x))) = -67.6°$
$|C(f_x)| = \frac{1}{G(f_x)} = 0.15 \equiv -16.5$ dB

Step 3: Calculate the controller parameters:

Boost: $\varnothing(\omega_m) = -90° + boost \rightarrow boost = 22.4°$

K factor: $K = \tan\left(\frac{boost}{2} + 45°\right) = 1.5$

Zeros and Poles:

$Z = \frac{1}{\tau} = \frac{2\pi f_x}{K} \rightarrow \tau = 991\ \mu s$
$P = \frac{1}{T_P} = 2\pi K f_x \rightarrow T_P = 443\ \mu s$
Gain: $|C(f_x)| = K_{PI} = 0.15 \equiv -16.5\ \text{dB}$

The compensator parameters are τ = 991 µs, T_P = 443 µs, and K_{PI} = 0.15. Figure 11 shows the Bode plot for the plant (G), the controller (C), and the loop transfer function (CG).

$$C(s) = K_{PI}\frac{1+\tau s}{\tau s(1+T_P s)} \tag{10}$$

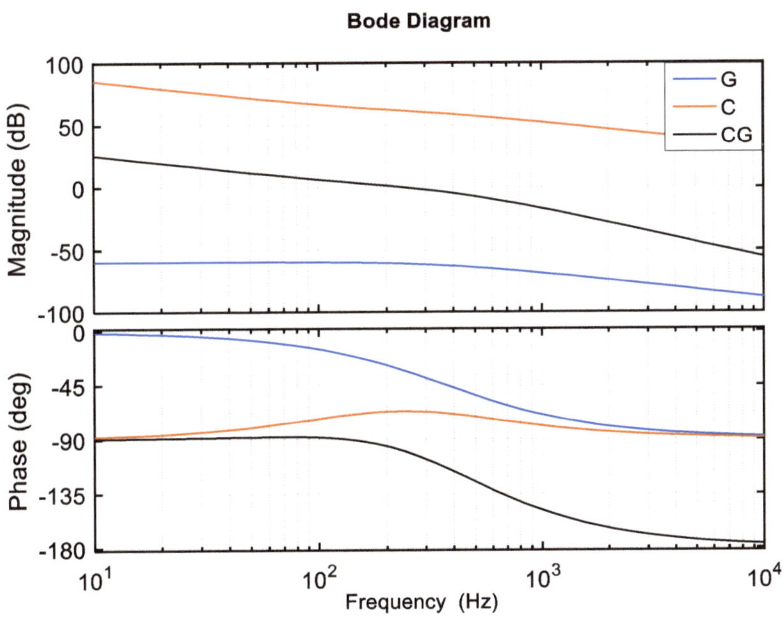

Figure 11. The Bode plot of the model (G), the controller (C), and the combination (CG).

For the current or power sharing, the interlink converter is controlled with the V-I curve in Figure 3, with threshold voltage V_{NL_int} = 48 V and R_{dint} = 0.289 Ω. The injected current (I_{inj}) from DAB is given by the droop equation (1) in the droop mode or constant at 4.325 A in the current-limiting mode. The control block diagram is shown in Figure 12. The phase-shift PWM control used in this work is realized, as shown in Figure 13. The HV side triangular waveform was shifted by 0.25 duty cycle ($\pi/2$) to allow negative delay using a time delay block. The fixed duty cycle is used to be 0.499 instead of 0.5 to implement a small dead time for the switches.

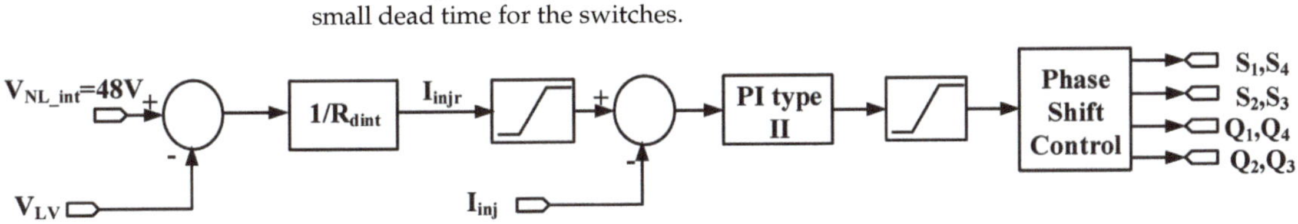

Figure 12. Control block diagram.

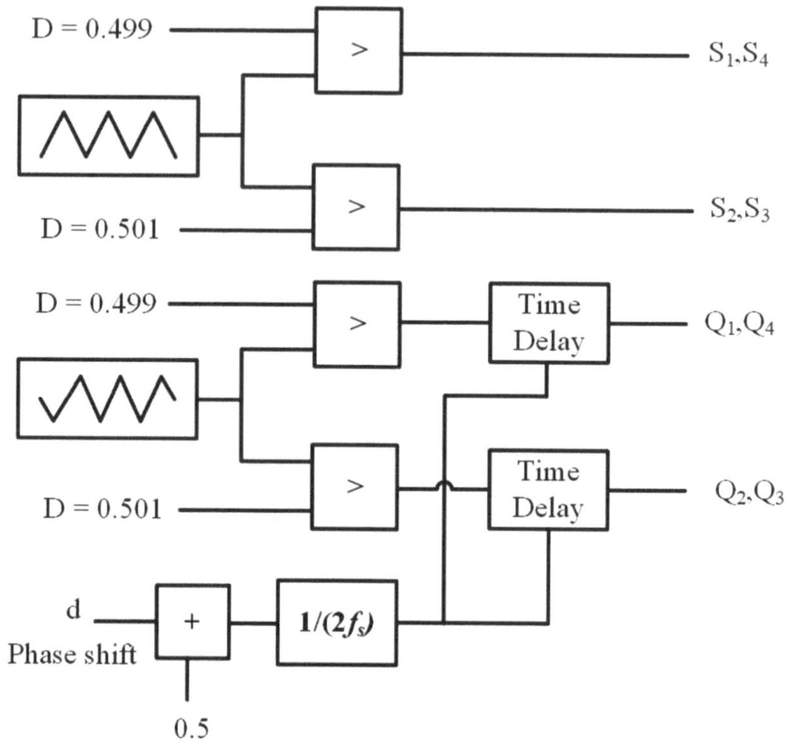

Figure 13. Phase-shift PWM block.

6. Simulation Results

Two simulation results were obtained considering the dual DC bus nanogrid discussed in this paper. The first is to test the DAB as an interlink converter. That can be done by verifying the SPS modulation waveforms described in Figure 6. Then, test the proposed model of the DAB in the DC nanogrid. Finally, test the ability of the interlink converter to direct the power between the two DC buses, as described in Figure 4. The second test is regarding the full dual DC bus nanogrid in Figure 1 and controlling the power and the current sharing with the V-I curve presented in Figures 2 and 3.

6.1. Simulation Results of DAB as Interlink Converter

MATLAB/Simulink is used to simulate the DAB in Figure 5 with the case-study parameter and control loop presented in Section 4. The first is an open loop test to verify the SPS modulation waveforms of the DAB described in Figure 6. Figure 14 shows the gating signals of the LV and HV sides of DAB, the AC voltage at the LV and HV sides of the transformer, and the transformer leakage inductor current at $d = -0.25$. The results show that the phase-shift PWM block in Figure 13 works as expected to perform the required modulation for the DAB.

The second test is to verify the performance of the control loop of the current injected into the LV bus. Again, the voltages of the LV and HV sides are set at 48 V and 380 V, respectively. The reference current is initially at −1.5 A, then changed in a step to 1.5 A. As shown in Figure 15, the injected current I_{inj} follows the reference signal very well, reaching the steady state at 0.007 s with zero error. The negative injected current I_{inj} sign means that the power flow is from the LV to the HV side, achieved with a negative phase shift. Conversely, a positive injected current I_{inj} means the power flow from the HV side to the LV side, requiring a positive phase shift.

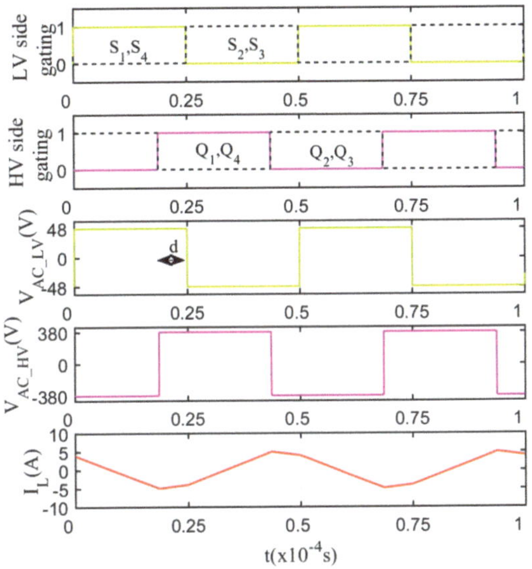

Figure 14. Simulation result of SPS modulation waveforms.

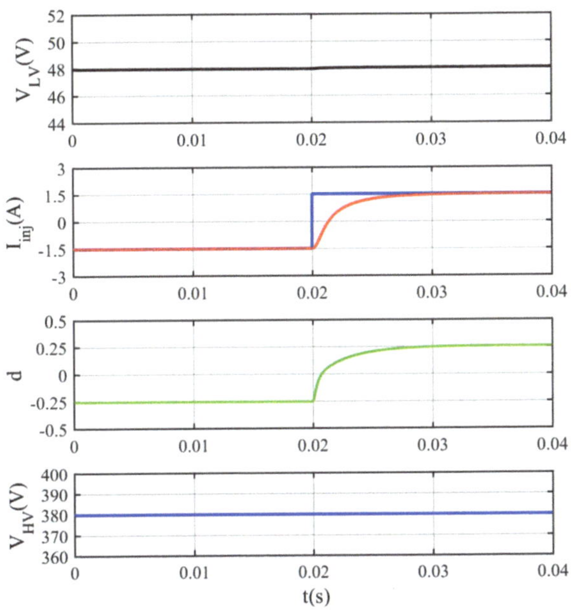

Figure 15. Simulation results for a step change in the injected reference current.

The last test investigates the impact of load variations in both DC buses and the power flow controlled by the interlink converter, according to the proposed droop control scheme, on the LV and HV buses' voltages. Three cases are considered: first with 370 V < V_{HV} < 390 V, then with V_{HV} < 370 V, and finally with V_{HV} > 390 V. As discussed before, the interlink converter is droop-controlled based on the magnitude of the voltage at the LV bus, but for low and high at the HV bus, it should be reprogrammed to prevent worsening the problem of excess surplus or shortage of power in that bus.

Threshold voltages and droop resistances modeled the LV DC bus and HV DC bus. For the LV DC bus, V_{NL_LV} = 49.25 V and R_{dLV} = 0.289 Ω, while for the HV DC bus, V_{NL_HV} = 390 V and R_{dHV} = 0.289 Ω. In this case, the LV DC bus voltage can be calculated by (11) through the droop Equation (1) with an equivalent threshold voltage of V_{NL_eq} = 48.625 V, obtained considering the threshold voltages of the LV DC bus and interlink converter, and a droop constant R_{deq} = 0.1445 Ω, the parallel combination of R_{dLV} and R_{dint}. For a given current injected by the DAB into the LV side, the current drawn from the HV side will be much smaller due to the transformer's turn ratio. The HV DC bus voltage can be calculated by (11) through the droop Equation (1) with a threshold of V_{NL_HV} and a droop constant of R_{dHV}.

The simulation results are shown in Figure 16. Initially, both DC buses are connected to 80 Ω loads, presenting voltages V_{LV} = 48.54 V and V_{HV} = 388.6 V. The value of the current injected into the LV bus (I_{inj}) is −1.87 A, determined by the V-I droop characteristic of the interlink converter. With V_{LV} = 48.54 V higher than V_{NL_int} = 48 V, power is shown to be drawn from the LV bus and sent to the HV bus. At t = 0.02 s, an additional 40 Ω is connected to the LV side, resulting in a decrease in the voltage at that bus, V_{LV} = 48.36 V, and in the current, the interlink converter draws power from the LV bus, I_{inj} = −1.26 A. The change, a reduction, in V_{HV} is minimal due to the small value of the LV side current reflected to the HV side current, and transformer turns ratio. At t = 0.04 s, an additional 40 Ω is connected to the HV side and, since the reference value for the current of the interlink is a function of the LV DC bus only, the injected current does not change; the HV DC bus voltage becomes V_{HV} = 385.8 V, lower than before. At t = 0.06 s, an additional 20 Ω is connected to the LV side, reducing the voltage in the LV bus to V_{LV} = 48 V, which makes $I_{inj} \approx 0$ A. At t = 0.08 s, an additional 20 Ω is connected to the HV side; again, it does not affect the power sharing and the HV DC bus voltage drops to V_{HV} = 380.4 V. At t = 0.1 s, an additional 40 Ω is connected to the LV side, reducing the voltage at this DC bus below 48 V V_{LV} = 47.85 V, and the power starts to flow from the HV DC bus to the LV DC bus with injected current: I_{inj} = 0.52 A. At t = 0.12 s, an additional 40 Ω is connected to the HV side and, again, it does not affect the power sharing, and the HV DC bus voltage drops to V_{HV} = 377.7 V. At t = 0.14 s, an additional 20 Ω is connected to the LV side; therefore, the injected current increases I_{inj} = 1.73 A and this means more power flow from the HV side to the LV side with more drop in the LV DC bus V_{LV} = 47.5 V. Finally, at t = 0.16 s, an additional 20 Ω is connected to the HV side and it does not affect the power sharing, and the HV DC bus voltage drops to V_{HV} = 372.5 V. The results show that the power injected by the interlink converter does not depend on the HV voltage side, as the V-I curve is a function of the LV side. In addition, the interlink converter successfully directs the power between the LV and the HV sides based on the proposed V-I curves.

$$V_{DC} = \frac{R_{Load} V_{NL}}{R_d + R_{Load}} \tag{11}$$

Figure 17 shows simulation results for a case with V_{NL_HV} = 395 V, where the voltage at the HV bus is higher than 390 V, from t = 0 s to 0.6 s. In such a case, the proposed control law for the interlink converter should prevent any power flow to the HV bus, which already has a surplus of power. Initially, both DC buses are connected to 80 Ω loads, presenting voltages V_{LV} = 49.32 V and V_{HV} = 393.6 V. At t = 0.02 s, an additional 40 Ω load is connected to the HV DC bus, V_{HV} decreases to 390.8 V, and there is no power flow from the LV to the HV DC bus. At t = 0.04 s, an additional 40 Ω load is connected to the LV DC bus, V_{LV} = 48.72 V, still higher than 48 V, and there is no power flow from the LV to the HV DC bus because of the voltage at the HV bus. At t = 0.06 s, an additional 40 Ω load is connected to the HV DC bus, V_{HV} drops to 385.26 V, below 390 V, and power starts to flow from the LV to the HV DC bus. Due to that, V_{LV} drops to 48.36 V with I_{inj} = −1.25 A as the V-I curve in Figure 4. At t = 0.08 s, an additional 40 Ω load is connected to the LV DC bus, $V_{LV} \approx 48$ V, which makes $I_{inj} \approx 0$ A. In this case, there is no power flow from the LV DC to the HV DC but now it is due to $V_{LV} \approx 48$ V= V_{NL_int}. After that, power flow from the

HV to the LV DC bus increases at $t = 0.12$ s and $t = 0.16$ s, when loads are added to the LV bus, but not at $t = 0.14$ s, when there is a load increase in the HV DC bus. The results show that the proposed interlink V-I curve could prevent power flow from the LV to the HV side when the HV bus has a voltage higher than 390 V, which means that the HV DC bus cannot supply power to the grid for some reasons.

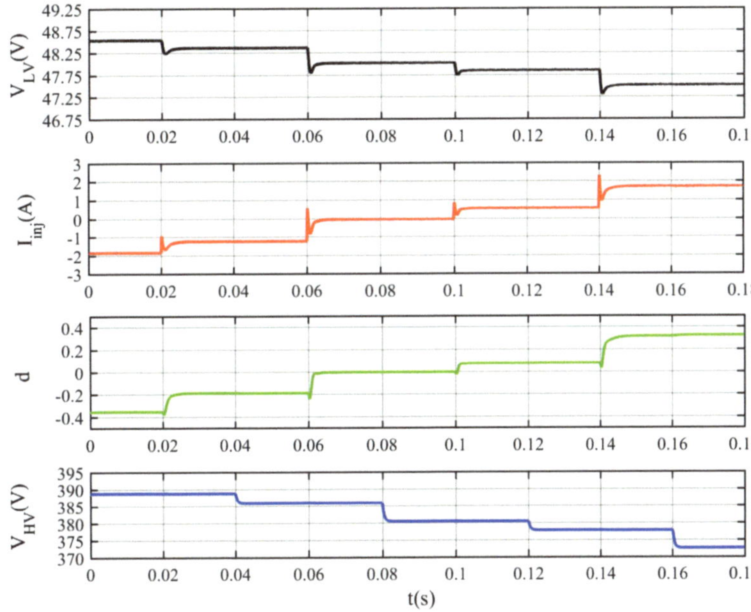

Figure 16. Simulation results considering that 370 V < V_{HV} < 390 V throughout the simulation time.

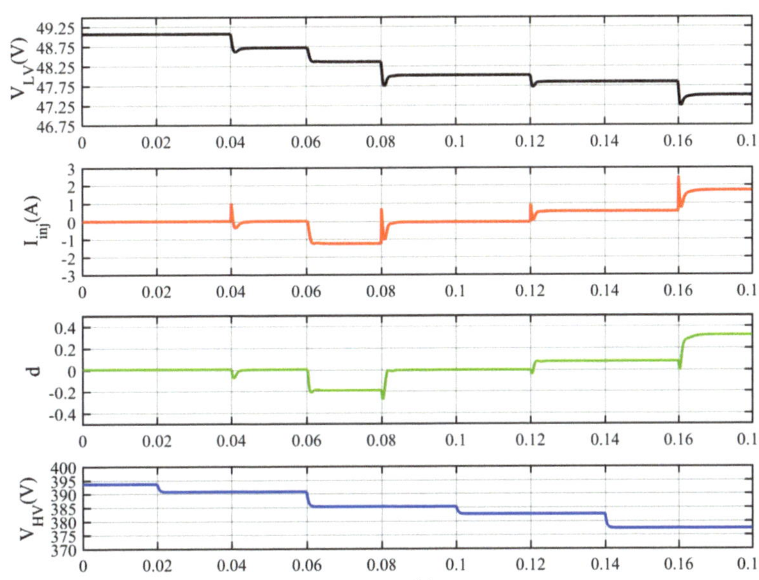

Figure 17. Simulation results for the case where V_{HV} > 390 V for a specific time interval.

Figure 18 shows simulation results for a case with V_{NL_HV} = 380 V, where the HV DC bus is lower than 370 V, from t = 0.1 s to 0.18 s. In such a case, the proposed control logic of the interlink converter prevents the power flow from the HV DC bus to the LV DC bus, achieved by reprogramming the current limit as in Figure 4, I_{inj} = 0. To prevent the HV DC bus batteries from discharging through the LV DC bus, initially, both DC buses are connected to 80 Ω loads, presenting voltages V_{LV} = 48.54 V and V_{HV} = 378.63 V with $I_{inj} = -1.87$ A, meaning that power flows from the LV to the HV side. At t = 0.02 s, an additional 40 Ω load is connected to the HV DC bus and V_{HV} decreases to 375.9 V without changes in the current of the interlink converter. At t = 0.04 s, an additional 40 Ω load is connected to the LV DC bus, and V_{LV} decreases to 48.36 V, indicating a reduction of the power surplus on the LV side, which makes I_{inj} reduce to -1.26 A. At t = 0.06 s, an additional 40 Ω load is connected to the HV DC bus and V_{HV} drops to 370.63 V. At t = 0.08 s, an additional 40 Ω load is connected to the LV DC bus, $V_{LV} \approx 48$ V, which makes $I_{inj} \approx 0$ A. In this case, there is no power flow from the LV DC to the HV DC since $V_{LV} \approx 48$ V = V_{NL_int}. At t = 0.1 s, an additional 40 Ω load is connected to the HV DC bus and V_{HV} drops to 368.05 V, below 370 V, and the interlink should prevent the power flow from the HV DC bus to the LV DC bus as described in Figure 4. Therefore, at t = 0.12 s, when an additional 40 Ω is connected to LV DC, I_{inj} = 0, indicating that power does not flow from the HV DC bus to the LV DC bus, even though V_{LV} = 47.7 V, < V_{NL_int} = 48 V, as discussed in Figure 4. After that, power flow from the HV to the LV DC bus is still prevented at t = 0.14 s and t = 0.16 s when loads are added to the HV bus and LV bus, respectively. The results show that the proposed interlink V-I curve could prevent power flow from the HV to the LV side when the HV side has a voltage less than 370 V, which prevents the HV side batteries from discharging into the LV side.

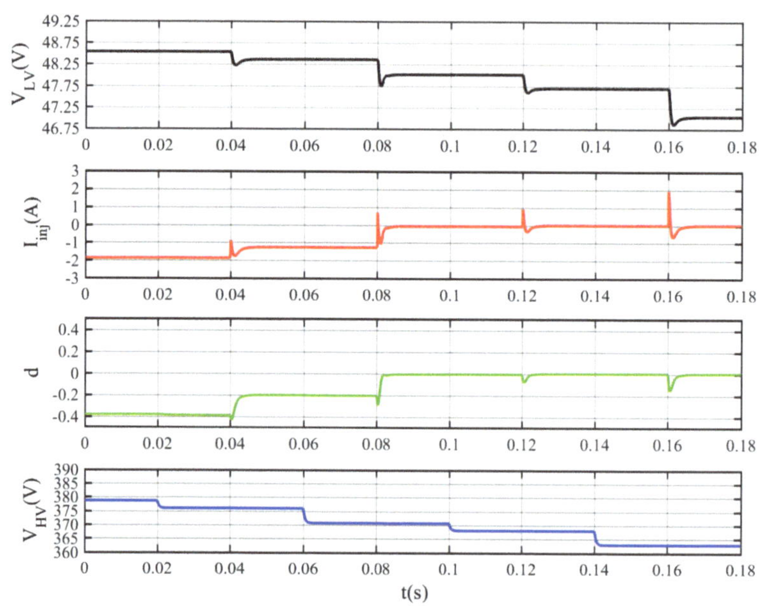

Figure 18. Simulation results for the case where V_{HV} < 370 V for a specific time interval.

6.2. Simulation Results of Dual DC Buses Nanogrid

MATLAB/Simulink is used to simulate the dual DC buses nanogrid in Figure 19 under different operation conditions. A class C DC–DC converter, as in Figure 19, is used as the main power electronics interface of the DERs: solar panels and batteries. A class C DC–DC converter is also used along with a DC–AC converter as the two-stage AC grid interface.

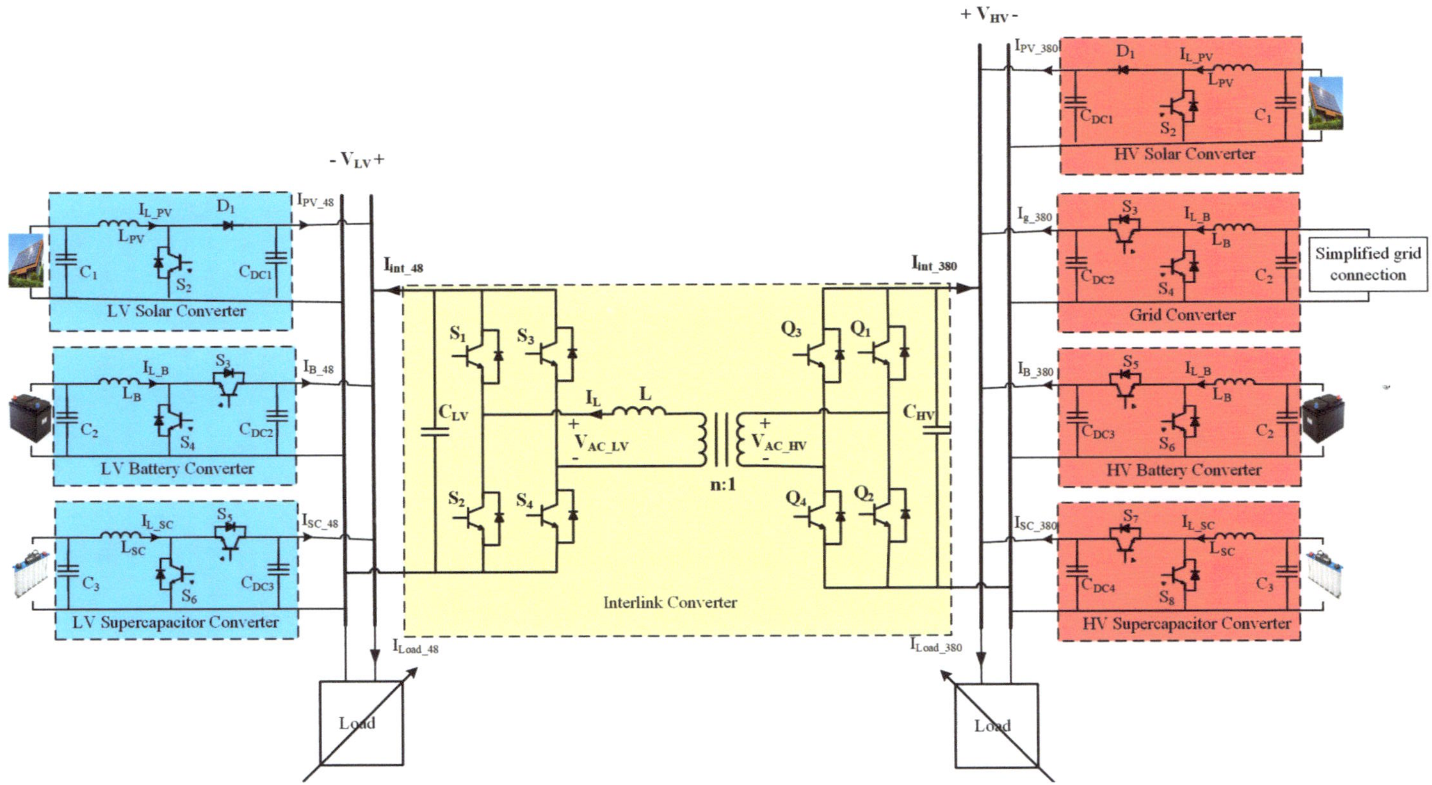

Figure 19. Dual DC buses nanogrid power electronics interfaces.

Regarding the power and the current sharing, the V-I in Figures 2 and 3 were used. For the HV DC bus DERs, a 5 kW PV panel and battery were considered with a droop constant $R_{dS\_380} = R_{dB\_380} = 0.763\ \Omega$, and a 10 kW grid connection was considered with a droop constant $R_{dG} = 0.3815\ \Omega$. They resulted in 14.2 A, 13.11 A, and 26.22 A as current limits for the solar, battery, and grid converters. For the LV DC bus DERs, a 2 kW PV panel, battery, and interlink were considered with a droop constant $R_{dS\_48} = R_{dB\_48} = R_{dint} = 0.0289\ \Omega$. They resulted in a 44 A current limit for the solar converter and a 43.25 A current limit for the battery and interlink converters. The supercapacitors (SC) were controlled to support the dynamics of the battery's current by splitting the storage inductor current into low frequencies for the battery and high frequencies for the SC using a low-pass filter (LPF) with a 30 Hz cut-off frequency.

The same current control loop parameters can be used for all DERs at the same DC bus. A PI type-III controller was designed for the same equivalent plant with a transfer function shown in (12) [47,48]. They were designed for a crossover frequency of $f_x = 2$ kHz (10% of the switching frequency) and a phase margin of $PM = 80°$ for both DC buses. For the LV DC bus converters, the following plant parameters were used: $V_{out} = V_{DC} = 48$ V, $I_{DC} = 41.67$ A, $R = 1.152\ \Omega$, $L = 300\ \mu$H, $C = 1500\ \mu$F, and $D = 0.46$ for $V_{PV} = V_B = V_{SC} = 29$ V. R was selected as the maximum load that could be supplied by the RES alone. Finally, the PI controller parameters are computed as $K_{PI} = 0.0117$, $\tau = 175\ \mu$s, and $T_P = 36\ \mu$s. For the HV DC bus converters, the following plant parameters were used: $V_{out} = V_{DC} = 380$ V, $I_{DC} = 13.16$ A, $R = 27.94\ \Omega$, $L = 3000\ \mu$H, $C = 1500\ \mu$F, and $D = 0.39$ for $V_{PV} = V_{Bat} = V_{SC} = 232$ V. R was selected as the maximum load that could be supplied by the RES alone. Finally, the PI controller parameters are computed as $K_{PI} = 0.0015$, $\tau = 171\ \mu$s, and $T_P = 37\ \mu$s. The interlink converter with the following parameters: transformer turns ratio n = 0.25, leakage inductance L=10.7 μH, switching frequency $f_s = 20$ kHz, and HV and LV capacitors $C_{HV} = C_{LV} = 1500\ \mu$F. Then, a PI type-III compensator is designed for a cut-off frequency of 240 Hz [31] and a phase margin $PM = 80°$. Therefore, the compensator parameters can be calculated as $\tau = 1.4$ ms, $T_P = 318\ \mu$s, and $K_{PI} = 0.049$ at a load resistance of 1.15 Ω.

$$G_{di}(s) = \frac{\tilde{i}_L(s)}{\tilde{d}(s)} = \frac{CV_{out}s + 2I_{DC}}{LCs^2 + \frac{L}{R}s + (1-D)^2} \tag{12}$$

The first test of the dual DC buses nanogrid is with a grid connection, and the batteries are not fully charged. The indication of the load variations in the system with the respective times and the numerical values of the voltage and current waveforms shown in Figures 20 and 21 are summarized in Tables 2 and 3 for the HV DC bus and the LV DC bus, respectively. The figures demonstrate that while the grid, the interlink, and the battery converters operate with droop control, the solar converters operate at the MPPT. The grid converter supports the HV bus and indirectly the LV bus through the interlink converter with $I_{int} = 2$ A when a 305 Ω load is initially attached to the HV bus and 40 Ω to the LV bus. The LV bus load increased to 7 Ω at $t = 0.1$ s and additional support came from the HV bus at $I_{int} = 4.8$ A, which raised the grid current to 5.3 A. At $t = 0.2$, the HV DC bus increased to 63.4 Ω; the grid current changed to 5.2 A to support the HV bus, while the interlink converter current remained constant because it depended on the LV bus. Figure 20 shows that the grid current varied when the HV bus load or LV bus changed, while Figure 21 shows that the interlink converter current changed when the LV bus load changed. Based on these results, the nanogrid operates according to the V-I curves presented in Figures 2 and 3 for the HV DC and LV DC buses, respectively. The RESs operate at maximum power. At the HV DC bus, the battery and the grid converter operate with droop control, the battery is in charging mode, and the grid supports the nanogrid by supplying power. At the LV DC bus, the battery and the interlink converter operate with droop control, the battery is in charging mode, and the interlink supports the LV DC bus. In addition, Figures 20 and 21 show that the PI type III controller successfully controls the dynamic response of the converters where the currents reach the steady state in almost 10 ms. However, the batteries' currents are much slower than other resources due to hybrid energy storage between the supercapacitors and the batteries, which is more efficient for the batteries.

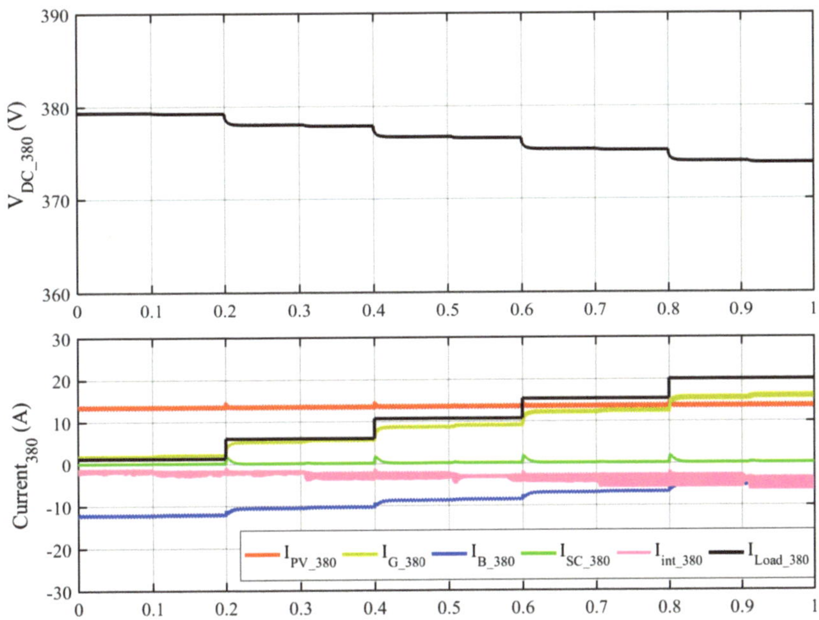

Figure 20. Simulation results of the dual DC buses nanogrid with grid connection—the HV DC bus waveforms.

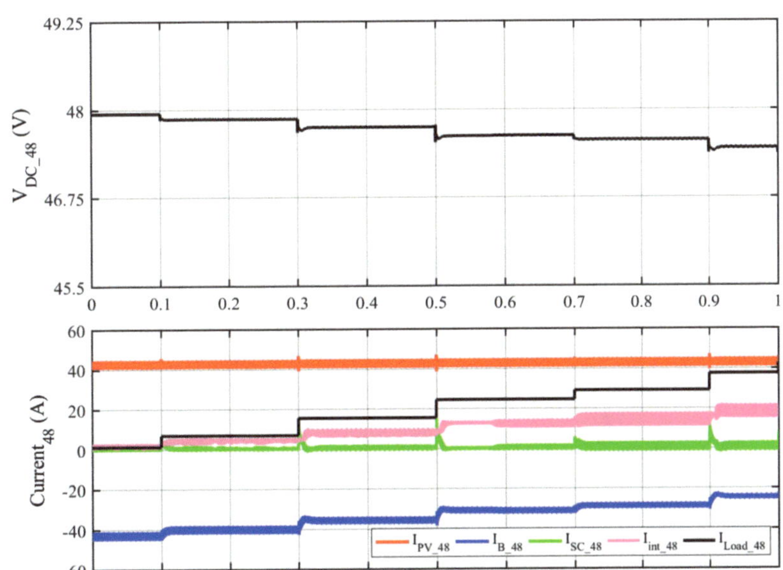

Figure 21. Simulation results of the dual DC buses nanogrid with grid connection—the LV DC bus waveforms.

Table 2. Numerical results of the HV DC bus from Figure 20.

Time (s)	Load (Ω)	V_{HV}(V)	I_{PV_380}(A)	I_G(A)	I_{B_380}(A)	I_{Load_380}(A)
Initially	305	379.3	13.2	1.8	−12.2	1.2
t = 0.2	63.4	378	13.2	5.2	−10.5	6
t = 0.4	35.4	376.7	13.25	8.6	−8.8	10.7
t = 0.6	24.5	375.3	13.3	12.3	−7	15.3
t = 0.8	18.8	374	13.3	15.6	−5.3	20

Table 3. Numerical results of the LV DC bus from Figure 21.

Time (s)	Load (Ω)	V_{LV}(V)	I_{PV_48}(A)	I_{int}(A)	I_{B_48}(A)	I_{Load_48}(A)
Initially	40	47.94	41.7	2	−42	1.2
t = 0.1	7	47.86	41.7	4.8	−40	6.8
t = 0.3	3	47.75	41.7	8.7	−35	16
t = 0.5	1.95	47.62	42	13	−30	24.4
t = 0.7	1.64	47.56	42	15.2	−28	29
t = 0.9	1.27	47.45	42	19	−24	37.5

The second test of the dual DC buses nanogrid is without a grid connection, and the batteries are not fully charged. The indication of the load variations in the system with the respective times and the numerical values of the voltage and current waveforms shown in Figures 22 and 23 are summarized in Tables 4 and 5 for the HV DC bus and the LV DC bus, respectively. Based on these results, the nanogrid operates with the V-I presented in Figures 2 and 3 for HV DC and LV DC buses, respectively. By comparing the cases above, one can say that the grid connection gives the DC nanogrid more power availability; missing this support leads to a higher voltage drop in the DC buses, especially at the HV DC bus. Moreover, the battery started to discharge at t = 0.6 s while in the charging mode with the grid connection.

Regarding the LV DC bus, the effect of missing the grid connection is less once the HV DC bus is higher than 370 V due to the power and current sharing V-I being based on the LV DC bus. Therefore, there is no change in the LV DC bus before t = 0.5 s according to Table 5 compared to Table 3. At t = 0.6 s, the HV DC load was increased to 24.5 Ω, and the HV DC bus dropped below 370 V; the HV battery operates in discharging mode and the interlink converter prevents the power flow from the HV DC to the LV DC bus (I_{int} = 0), so the HV DC bus battery will not discharge through the LV DC bus. After t = 0.6 s, the LV DC bus misses the support from the HV DC bus. Therefore, the voltage drops in the LV DC bus in Table 5 increase compared with the case with grid connection in Table 3 at the same load level. The RESs operate at maximum power at all periods. Moreover, the interlink converter prevents the LV and HV side batteries from discharging to charge each other. In addition, Figures 22 and 23 show that the PI type III controller successfully controls the dynamic response of the converters where the currents reach the steady state in almost 10 ms. However, the batteries' currents are much slower than other resources due to hybrid energy storage between the supercapacitors and the batteries, which is more efficient for the batteries.

Table 4. Numerical results of the HV DC bus from Figure 22.

Time (s)	Load (Ω)	V_{HV}(V)	I_{PV_380}(A)	I_G(A)	I_{B_380}(A)	I_{Load_380}(A)
Initially	305	378	13.2	0	−10.5	1.2
t = 0.2	63.4	374.3	13.25	0	−5.6	5.9

Table 4. *Cont.*

Time (s)	Load (Ω)	V_{HV}(V)	I_{PV_380}(A)	I_G(A)	I_{B_380}(A)	I_{Load_380}(A)
$t = 0.4$	35.4	370.5	13.4	0	−0.5	10.5
$t = 0.5$	No change	370	13.5	0	0	10.5
$t = 0.6$	24.5	368	13.5	0	2.5	15
$t = 0.8$	18.8	364.7	13.6	0	7	19.5

Table 5. Numerical results of the LV DC bus from Figure 23.

Time (s)	Load (Ω)	V_{LV}(V)	I_{PV_48}(A)	I_{int}(A)	I_{B_48}(A)	I_{Load_48}(A)
Initially	40	47.94	41.7	2	−42	1.2
$t = 0.1$	7	47.86	41.7	4.8	−40	6.8
$t = 0.3$	3	47.75	41.7	8.7	−35	16
$t = 0.5$	1.95	47.62	42	13	−30	24.4
$t = 0.6$	No change	47.28	42	0	−18	24.2
$t = 0.7$	1.64	47.16	42	0	−14	28.8
$t = 0.9$	1.27	46.93	42	0	−6	37

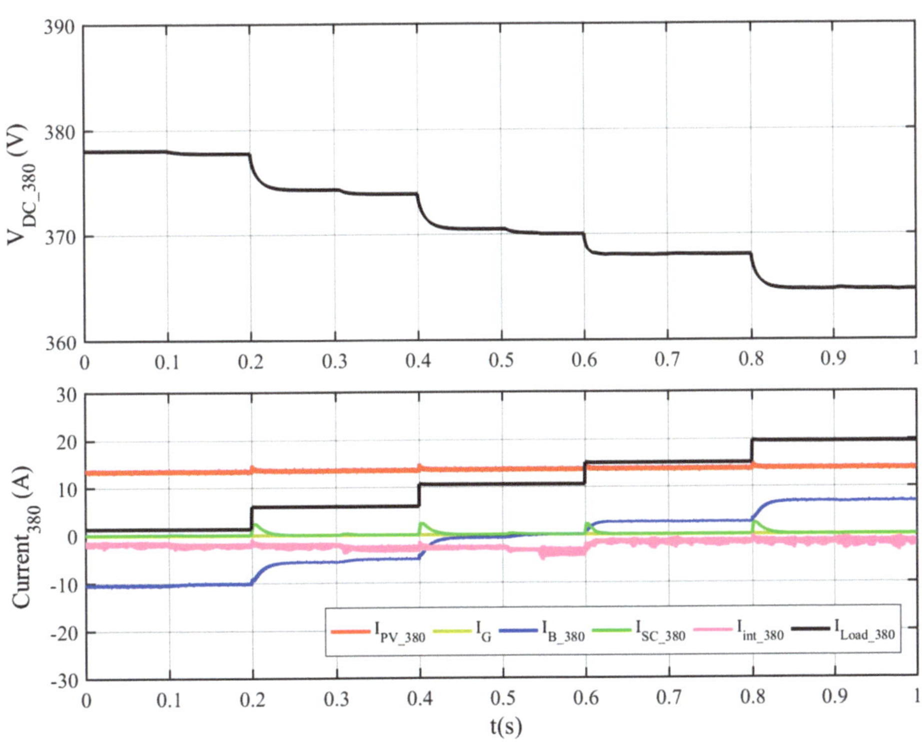

Figure 22. Simulation results of the dual DC buses nanogrid without grid connection—the HV DC bus waveforms.

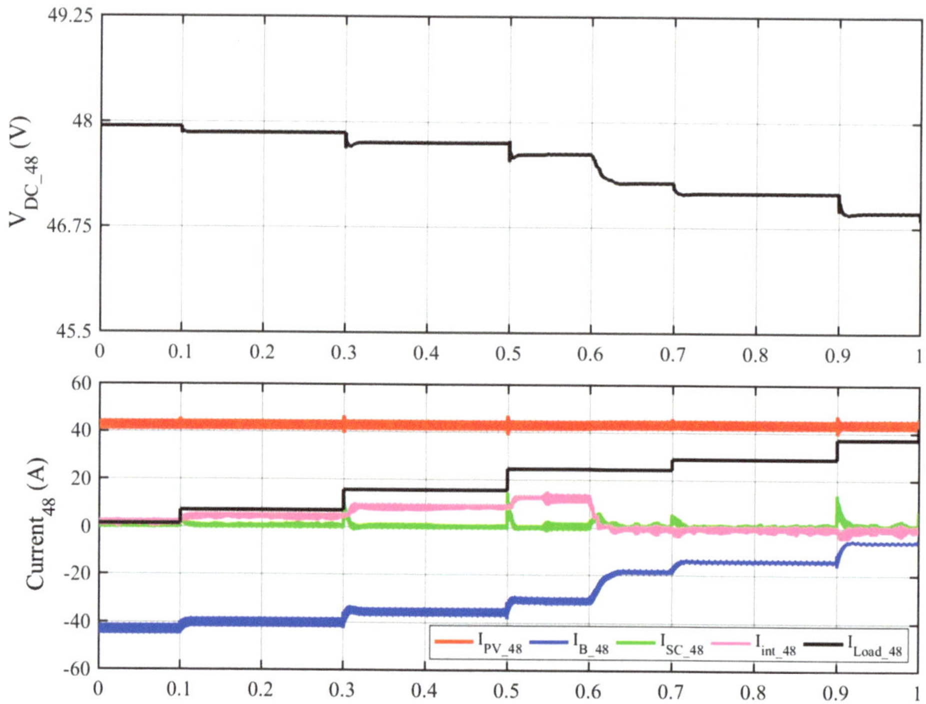

Figure 23. Simulation results of the dual DC buses nanogrid without grid connection—the LV DC bus waveforms.

7. Conclusions

This research presents a nanogrid system with a dual DC bus configuration, including a bidirectional interlink converter capable of directing the power flow between the two buses. The high voltage (HV) bus provides power to high-power appliances, while the low voltage (LV) bus supplies energy to low-power loads. Nanogrid reliability was enhanced by integrating distributed energy resources on the two buses. Furthermore, a control methodology utilizing DC bus signaling (DBS) is suggested, which facilitates the assessment of the impact on the DC buses in the presence of other DERs controlled by DBS. It is commonly observed that the HV bus is more robust than the LV bus. As a result, the voltage at the LV side determines the power flow between DC buses. The above statement is equivalent to the V-I curve of an AC grid converter operating within a single DC bus nanogrid. This implies that the power delivered by the interlink converter is independent of the voltage on the high voltage side since the low voltage side determines the V-I curve.

Furthermore, utilizing the suggested V-I curve, the interlink converter effectively facilitates power transfer between the low voltage (LV) and high voltage (HV) sides. If the high voltage (HV) bus exhibits a significant excess or lack of power, identified by its bus voltage (V_{HV}), the interlink converter is prevented from worsening these circumstances. One potential solution involves implementing a mechanism that prevents power transfer from the low voltage (LV) to the high voltage (HV) side in instances where the HV bus voltage exceeds 390 V. This would effectively prevent the HV DC bus from supplying power to the grid under certain circumstances. Additionally, preventing power flow from the HV to the LV side would be necessary in cases where the HV side voltage falls below 370 V, as this would prevent the discharge of HV side batteries into the LV side. The characteristic above is essential to dual DC bus nanogrids that function continuously or intermittently disconnected from the AC utility grid. In addition, a topology that is considered appropriate for the interlink power converter is chosen. This topology is bidirectional, has high gain, and is equipped with galvanic isolation. It is commonly referred to as the dual active bridge (DAB). A dynamic model has been proposed for the interlink converter. This converter operates using a single phase-shift modulation scheme and behaves within a droop-controlled environment. A closed-loop controller has been developed to regulate the current injected into the LV DC bus based on the voltage of the bus. The performance of the droop-controlled interlink converter in a dual DC bus nanogrid is examined via MATLAB/Simulink simulations. Per the proposed approach, the interlink converter effectively manages power sharing between the HV and LV buses.

Author Contributions: Conceptualization, A.M.A.M.; Methodology, A.M.A.M. and L.A.C.L.; Validation, A.M.A.M.; Formal analysis, A.M.A.M.; Investigation, A.M.A.M.; Writing—review & editing, A.M.A.M. and A.A.-Q.; Visualization, L.A.C.L.; Supervision, L.A.C.L. All authors have read and agreed to the published version of the manuscript.

Funding: This research was funded by the The University of Jordan and Natural Sciences and Engineering Research Council (NSERC) of Canada.

Data Availability Statement: Data are available upon request.

Acknowledgments: The authors appreciate the financial support provided by The University of Jordan, through the Deanship of Scientific Research, and of the Natural Sciences and Engineering Research Council (NSERC) of Canada.

Conflicts of Interest: The authors declare no conflict of interest.

Abbreviations

RES	Renewable Energy Source
ICT	Information and Communication Technology
DER	Distributed Energy Resources
NZEH	Net-Zero Energy Homes
ELV	Extra Low Voltage
LV	Low Voltage
HV	High Voltage
DBS	DC Bus Signaling
MPPT	Maximum Power Point Tracking
SoC	State of Charge
DAB	Dual Active Bridge
SPS	Single-Phase Shift
d	Phase-Shift Ratio
PWM	Pulse Width Modulation

References

1. Alepuz, S.; Busquets-Monge, S.; Bordonau, J.; Gago, J.; Gonzalez, D.; Balcells, J. Interfacing Renewable Energy Sources to the Utility Grid Using a Three-Level Inverter. *IEEE Trans. Ind. Electron.* **2006**, *53*, 1504–1511. [CrossRef]
2. Gungor, V.C.; Sahin, D.; Kocak, T.; Ergut, S.; Buccella, C.; Cecati, C.; Hancke, G.P. Smart Grid Technologies: Communication Technologies and Standards. *IEEE Trans. Ind. Inform.* **2011**, *7*, 529–539. [CrossRef]
3. Bryan, J.; Duke, R.; Round, S. Distributed generation-nanogrid transmission and control options. In Proceedings of the International Power Engineering Conference, Singapore, 27–29 November 2003; pp. 341–346.
4. Leonori, S.; Rizzoni, G.; Mascioli, F.M.F.; Rizzi, A. Intelligent energy flow management of a nanogrid fast charging station equipped with second life batteries. *Int. J. Electr. Power Energy Syst.* **2020**, *127*, 106602. [CrossRef]
5. Rubio, C.; Siddiqui, A. Shape of the microgrid. In Proceedings of the 2001 IEEE Power Engineering Society Winter Meeting, Columbus, OH, USA, 28 January–1 February 2001; pp. 150–153.
6. Lakhina, U.; Elamvazuthi, I.; Badruddin, N.; Jangra, A.; Truong, B.-H.; Guerrero, J.M. A Cost-Effective Multi-Verse Optimization Algorithm for Efficient Power Generation in a Microgrid. *Sustainability* **2023**, *15*, 6358. [CrossRef]
7. Boroyevich, D.; Cvetković, I.; Dong, D.; Burgos, R.; Wang, F.; Lee, F. Future electronic power distribution systems a contemplative view. In Proceedings of the 2010 12th International Conference on Optimization of Electrical and Electronic Equipment, Brasov, Romania, 20–22 May 2010.

8. Sartori, I.; Napolitano, A.; Voss, K. Net zero energy buildings: A consistent definition framework. *Energy Build.* **2012**, *48*, 220–232. [CrossRef]
9. Muhsen, H.; Alkhraibat, A.; Al-Halhouli, A. Real-Time Simulation and Energy Management Attainment of Microgrids. *Sustainability* **2023**, *15*, 2696. [CrossRef]
10. Al-Quraan, A.; Al-Mhairat, B.; Malkawi, A.M.A.; Radaideh, A.; Al-Masri, H.M.K. Optimal Prediction of Wind Energy Resources Based on WOA—A Case Study in Jordan. *Sustainability* **2023**, *15*, 3927. [CrossRef]
11. Hammerstrom, D.J. AC Versus DC Distribution Systemsdid We Get it Right? In Proceedings of the 2007 IEEE Power Engineering Society General Meeting, Tampa, FL, USA, 24–28 June 2007; pp. 1–5.
12. Silveira, J.P.C.; Neto, P.J.D.S.; Barros, T.A.d.S.; Filho, E.R. Power management of energy storage system with modified interlinking converters topology in hybrid AC/DC microgrid. *Int. J. Electr. Power Energy Syst.* **2021**, *130*, 106880. [CrossRef]
13. Malkawi, A.; Lopes, L.A.C. Control of the power electronics interface of a PV source in a smart residential DC nanogrid. In Proceedings of the 2016 IEEE Canadian Conference on Electrical and Computer Engineering (CCECE), Vancouver, BC, Canada, 15–18 May 2016; pp. 1–4. [CrossRef]
14. Committee ISC. *IEEE Standard for Interconnecting Distributed Resources with Electric Power Systems*; Institute of Electrical and Electronics Engineers: New York, NY, USA, 2003.
15. Dong, D.; Thacker, T.; Burgos, R.; Boroyevich, D.; Wang, F.; Giewont, B. Control design and experimental verification of a multi-function single-phase bidirectional PWM converter for renewable energy systems. In Proceedings of the 2009 13th European Conference on Power Electronics and Applications, Barcelona, Spain, 8–10 September 2009; pp. 1–10.
16. Thacker, T.; Burgos, R.; Wang, F.; Boroyevich, D. Single-phase islanding detection based on phase-locked loop stability. In Proceedings of the 2009 IEEE Energy Conversion Congress and Exposition, San Jose, CA, USA, 20–24 September 2009; pp. 3371–3377.
17. Roggia, L.; Schuch, L.; Baggio, J.E.; Rech, C.; Pinheiro, J.R. Integrated Full-Bridge-Forward DC-DC Converter for a Residential Microgrid Application. *IEEE Trans Power Electron* **2013**, *28*, 1728–1740. [CrossRef]
18. Tazy, A.; Samy, M.M.; Ghalib, M.A. Enhancement and Control of Reactive Power Sharing Utilizing Circulating Current in AC Microgrids. In Proceedings of the 2022 23rd International Middle East Power Systems Conference (MEPCON), Cairo, Egypt, 13–15 December 2022; pp. 1–5. [CrossRef]
19. Balog, R.S.; Krein, P.T. Bus Selection in Multibus DC Microgrids. *IEEE Trans. Power Electron.* **2010**, *26*, 860–867. [CrossRef]
20. Chen, J.; Chen, J.; Chen, R.; Zhang, X.; Gong, C. Decoupling control of the non-grid-connected wind power system with the droop strategy based on a DC micro-grid. In Proceedings of the 2009 World Non-Grid-Connected Wind Power and Energy Conference, Nanjing, China, 24–26 September 2009; pp. 1–6. [CrossRef]
21. Blaabjerg, F.; Consoli, A.; Ferreira, J.; Vanwyk, J. Special Feature The Future of Electronic Power Processing and Conversion. *IEEE Trans. Ind. Appl.* **2005**, *41*, 3–8. [CrossRef]
22. Correa, J.M.; Farret, F.A.; Simoes, M.; Ramos, D.B.; Ferrigolo, F.Z. Aspects of the integration of alternative sources of energy for application in distributed generation systems. In Proceedings of the XI Brazilian Power Electronics Conference, Natal, Brazil, 11–15 September 2011; pp. 819–824. [CrossRef]
23. Biczel, P. Power Electronic Converters in DC Microgrid. In Proceedings of the 2007 Compatibility in Power Electronics, Gdansk, Poland, 29 May–1 June 2007; pp. 1–6.
24. Sun, K.; Zhang, L.; Xing, Y.; Guerrero, J.M. A Distributed Control Strategy Based on DC Bus Signaling for Modular Photovoltaic Generation Systems with Battery Energy Storage. *IEEE Trans. Power Electron.* **2011**, *26*, 3032–3045. [CrossRef]
25. Xu, L.; Chen, D. Control and Operation of a DC Microgrid With Variable Generation and Energy Storage. *IEEE Trans. Power Deliv.* **2011**, *26*, 2513–2522. [CrossRef]
26. E-Merge Alliance. Public Overview of the EMerge Alliance Data. Telecom Center Standard Version. 2014. Available online: https://www.emergealliance.org/standards/data-telecom/ (accessed on 26 June 2023).
27. E-Merge Alliance. Public Overview of the EMerge Alliance Occupied Space Standard. 2015. Available online: https://www.emergealliance.org/standards/occupied-space-2/ (accessed on 26 June 2023).
28. Liu, C.; Chau, K.; Diao, C.; Zhong, J.; Zhang, X.; Gao, S.; Wu, D. A new DC micro-grid system using renewable energy and electric vehicles for smart energy delivery. In Proceedings of the 2010 IEEE Vehicle Power and Propulsion Conference, Lille, France, 1–3 September 2010; pp. 1–6. [CrossRef]
29. Lago, J.; Heldwein, M.L. Operation and Control-Oriented Modeling of a Power Converter for Current Balancing and Stability Improvement of DC Active Distribution Networks. *IEEE Trans. Power Electron.* **2011**, *26*, 877–885. [CrossRef]
30. Malkawi, A.; Lopes, L. A novel seamless control algorithm for a single-stage photovoltaic interface employing DC bus signaling. *Int. J. Electr. Power Energy Syst.* **2019**, *113*, 90–103. [CrossRef]
31. Malkawi, A.M.A.; Lopes, L.A.C. Improved Dynamic Voltage Regulation in a Droop Controlled DC Nanogrid Employing Independently Controlled Battery and Supercapacitor Units. *Appl. Sci.* **2018**, *8*, 1525. [CrossRef]
32. Malkawi, A.M.A.; Al-Quraan, A.; Lopes, L.A.C. Extending DC Bus Signaling and Droop Control for Hybrid Storage Units to Improve the Energy Management and Voltage Regulation. *Inventions* **2022**, *7*, 55. [CrossRef]
33. Ghalib, M.A.; Samy, M.M. Magnification of Performance Operation for Low Voltage DC Microgrids Based on Adaptive Droop Control Technique. In Proceedings of the 2021 22nd International Middle East Power Systems Conference (MEPCON), Assiut, Egypt, 14–16 December 2021; pp. 424–429. [CrossRef]

34. Guerrero, J.M.; Vasquez, J.C.; Matas, J.; de Vicuna, L.G.; Castilla, M. Hierarchical Control of Droop-Controlled AC and DC Microgrids—A General Approach Toward Standardization. *IEEE Trans. Ind. Electron.* **2011**, *58*, 158–172. [CrossRef]
35. Noroozian, R.; Abedi, M.; Gharehpetian, G.; Hosseini, S. Combined operation of DC isolated distribution and PV systems for supplying unbalanced AC loads. *Renew. Energy* **2009**, *34*, 899–908. [CrossRef]
36. Justo, J.J.; Mwasilu, F.; Lee, J.; Jung, J. AC-microgrids versus DC-microgrids with distributed energy resources: A review. *Renew. Sustain. Energy Rev.* **2013**, *24*, 387–405. [CrossRef]
37. Kakigano, H.; Nishino, A.; Ise, T. Distribution voltage control for DC microgrid with fuzzy control and gain-scheduling control. In Proceedings of the 8th International Conference on Power Electronics–ECCE Asia, Jeju, Republic of Korea, 30 May–3 June 2011; pp. 256–263. [CrossRef]
38. Chedid, R.; Rahman, S. Unit sizing and control of hybrid wind-solar power systems. *IEEE Trans. Energy Convers.* **1997**, *12*, 79–85. [CrossRef]
39. Dalbon, W.; Roscia, M.; Zaninelli, D.; Leva, S. Hybrid photovoltaic system control for enhancing sustainable energy. In Proceedings of the IEEE Power Engineering Society Summer Meeting, Chicago, IL, USA, 21–25 July 2002. [CrossRef]
40. Tsikalakis, A.G.; Hatziargyriou, N.D. Centralized control for optimizing microgrids operation. In Proceedings of the 2011 IEEE Power and Energy Society General Meeting, Detroit, MI, USA, 24–28 July 2011; pp. 1–8. [CrossRef]
41. Bhaskara, S.N.; Chowdhury, B.H. Microgrids—A review of modeling, control, protection, simulation and future potential. In Proceedings of the 2012 IEEE Power and Energy Society General Meeting, San Diego, CA, USA, 22–26 July 2012; pp. 1–7.
42. Zhang, L.; Wang, Y.; Cheng, L.; Kang, W. A Three-Parameter Adaptive Virtual DC Motor Control Strategy for a Dual Active Bridge DC–DC Converter. *Electronics* **2023**, *12*, 1412. [CrossRef]
43. Zhang, W.; Lee, F.C.; Huang, P.-Y. Energy management system control and experiment for future home. In Proceedings of the 2014 IEEE Energy Conversion Congress and Exposition (ECCE), Pittsburgh, PA, USA, 14–18 September 2014; pp. 3317–3324. [CrossRef]
44. Karlsson, P.; Svensson, J. DC bus voltage control for a distributed power system. *IEEE Trans. Power Electron.* **2003**, *18*, 1405–1412. [CrossRef]
45. Kim, J.-W.; Choi, H.-S.; Cho, B.H. A novel droop method for converter parallel operation. *IEEE Trans. Power Electron.* **2002**, *17*, 25–32. [CrossRef]
46. Kurohane, K.; Senjyu, T.; Yona, A.; Urasaki, N.; Muhando, E.B.; Funabashi, T. A high quality power supply system with DC smart grid. In Proceedings of the IEEE PES T&D 2010, New Orleans, LA, USA, 19–22 April 2010; pp. 1–6. [CrossRef]
47. Kollimalla, S.K.; Mishra, M.K.; Narasamma, N.L. Design and Analysis of Novel Control Strategy for Battery and Supercapacitor Storage System. *IEEE Trans Sustain. Energy* **2014**, *5*, 1137–1144. [CrossRef]
48. Kolluri, S.; Narasamma, N.L. Analysis, modeling, design and implementation of average current mode control for interleaved boost converter. In Proceedings of the 2013 IEEE 10th International Conference on Power Electronics and Drive Systems (PEDS), Kitakyushu, Japan, 22–25 April 2013; pp. 280–285. [CrossRef]

sustainability

Article

A Cost-Effective Multi-Verse Optimization Algorithm for Efficient Power Generation in a Microgrid

Upasana Lakhina [1], Irraivan Elamvazuthi [1], Nasreen Badruddin [1,*], Ajay Jangra [2], Bao-Huy Truong [3] and Joseph M. Guerrero [4]

1 Department of Electrical and Electronics Engineering, Institute of Health and Analytics, Universiti Teknologi PETRONAS, Seri Iskandar 32610, Perak, Malaysia
2 University Institute of Engineering and Technology, Kurukshetra University, Kurukshetra 136119, India
3 Institute of Engineering and Technology, Thu Dau Mot University, Thu Dau Mot 7500, Vietnam
4 Centre of Research on Microgrids, Department of Energy Technology, Aalborg University, 9220 Aalborg, Denmark
* Correspondence: nasreen.b@utp.edu.my

Abstract: Renewable energy sources (RESs) are a great source of power generation for microgrids with expeditious urbanization and increase in demand in the energy sector. One of the significant challenges in deploying RESs with microgrids is efficient energy management. Optimizing the power allocation among various available generation units to serve the load is the best way to achieve efficient energy management. This paper proposes a cost-effective multi-verse optimizer algorithm (CMVO) to solve this optimization problem. CMVO focuses on the optimal sharing of generated power in a microgrid between different available sources to reduce the generation cost. The proposed algorithm is analyzed for two different scale microgrids (IEEE 37-node test system and IEEE 141-node test system) using IEEE test feeder standards to assess its performance. The results show that CMVO outperforms multi-verse optimizer (MVO), particle swarm optimization (PSO), artificial hummingbird algorithm (AHA), and genetic algorithm (GA). The simulation results emphasize the cost reduction and execution time improvement in both IEEE test systems compared with other meta-heuristic algorithms.

Keywords: cost optimization; energy management; microgrid; multi-verse optimizer; renewable energy sources (RESs)

Citation: Lakhina, U.; Elamvazuthi, I.; Badruddin, N.; Jangra, A.; Truong, B.-H.; Guerrero, J.M. A Cost-Effective Multi-Verse Optimization Algorithm for Efficient Power Generation in a Microgrid. *Sustainability* **2023**, *15*, 6358. https://doi.org/10.3390/su15086358

Academic Editors: Ayman Al-Quraan and Ahmad M.A. Malkawi

Received: 23 December 2022
Revised: 21 February 2023
Accepted: 28 February 2023
Published: 7 April 2023

1. Introduction

Microgrids are reliable low-voltage networks that supply energy to consumers efficiently [1,2]. With the rapid growth in power demand in the energy sector, it is essential to use renewable energy resources (RERs) to meet demand. RERs help to fulfill the energy demand and carry out efficient execution of microgrids. Microgrids consist of various distributed energy resources (DERs) such as wind power plants and solar power plants, among others, as well as storage devices and loads [3,4]. Figure 1 describes a microgrid consisting of an energy management system, DERs, storage system, and load. Loads are generally categorized as residential, industrial, and commercial loads. Microgrids are usually operated in islanded mode or grid-connected mode.

A microgrid in islanded mode is a standalone, independent system that operates for different communities. It works totally on distributed energy resources, whereas the microgrid in grid-connected mode means saving fossil fuels and works as a regular microgrid connected to the utility grid for power generation [5]. Energy management in a microgrid is essential for its smooth operation in a real-time environment. It is observed that to solve the optimization problem and perform energy management in a microgrid is a challenging task [6]. In previous studies, many known meta-heuristic algorithms have been applied to optimize various performance attributes of a microgrid.

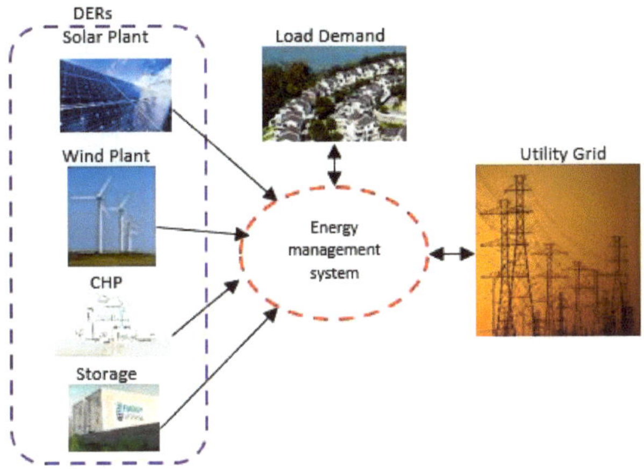

Figure 1. Structure of a microgrid.

Many optimization methods have been reported in the literature for optimizing different parameters in microgrid energy management. The authors in [7] proposed a robust stochastic for hybrid energy systems to perform optimization. The suggested method intends to decrease system losses and the total operating cost of renewable energy resources and was tested on the IEEE 37 node distribution system. Another study [8] presented an integrated method using multi-objective particle swarm optimization to minimize power supply probability loss, levelized energy cost, and greenhouse emissions. The authors in [9] proposed a reduction in cost by optimally placing a capacitor to alleviate the power loss in a radial distribution network. The proposed work is then tested over two different IEEE standard networks. In [10], a real-time energy management system was proposed for the performance optimization of a microgrid. The proposed binary particle swarm optimization was analyzed on the IEEE 14-bus system and focused on minimizing energy costs and CO_2 emissions.

Furthermore, in various studies, many optimizing algorithms have been used for scheduling of renewable energy sources. In [11], the authors proposed quantum-teaching-based learning optimization for optimal energy management using day-ahead optimum power scheduling in microgrids. The authors in [12] used the artificial fish swarm algorithm for cost minimization, whereas in [13], optimal scheduling of dispatchable distributed generations has been studied to minimize the fuel cost in microgrids. Moreover, in [14], the authors presented a multi-objective economic dispatch problem using pareto concavity elimination to minimize the cost in microgrids. Hence, the prime aim of a microgrid energy management system is to effectively schedule the power flow and coordinate between various available generation units and load demand to optimize operational cost [15,16]. It tries to maintain stability between generated and demanded power based on the idea that DERs can supply the requested capacity every hour. The multi-verse optimizer algorithm (MVO) is a nature-derived meta-heuristic algorithm that has improved the ability to exploit and explore search space without getting trapped in local minima. In contrast, it has lower accuracy and slow convergence while solving a problem. Hence, many researchers have proposed different versions of MVO to handle these disadvantages. The authors in [17] have published an improved MVO for feature selection for phishing, spam, and denial of service attacks. Moreover, in [18], the improved multi-verse optimizer is used for text document clustering. Similarly, in [19], a multi-verse optimization algorithm for stochastic bi-objective disassembly sequence planning is proposed. It achieved minimum energy consumption and maximized disassembly profit. It is noted that MVO has proved to be effective in addressing different optimization problems, whereas it is less explored for optimization in microgrids.

Scheduling power among various units to have minimum generation cost, the energy management system must have the capability to allocate power sharing in generous proportions. When RERs are not able to serve the load with generated power, then power is obtained from other storage sources, virtual power plants (VPPs), or utility grids [20,21]. In addition to the above, it is observed in the literature that various other work [22–28] has been performed for optimization using meta-heuristic algorithms, as there has been a sharp increase in demand for optimization in various research areas as well as in the energy sector. Thus, the importance of energy optimization is rapidly increasing.

The contributions of this paper are summarized below.

This paper proposes an enhanced multi-verse optimizer algorithm called 'cost-effective multi-verse optimization algorithm (CMVO)' for optimal power scheduling among available generation units in a microgrid to minimize generation cost. It is observed that modification performed in the original version ensures more balanced exploitation and exploration. It has improved the ability to discover rugged search spaces and avoid local optima stagnation. The CMVO algorithm solves the power scheduling problem for microgrids to minimize the generation cost. It focuses on optimal power sharing among available generation units such that demand is satisfied with minimum cost. The convergence speed of the proposed algorithm is also improved in a way that it explores the search space without getting trapped into local minima and converges quickly. The calculated mean and standard deviation of all of the algorithms illustrate that the proposed algorithm is most stable among all investigated algorithms.

The remaining structure of the paper is organized in a manner where the literature on cost optimization by meta-heuristic algorithms is discussed through related work in Section 2. Section 3 presents the problem formulation for the research conducted and explains various functions and equality constraints. Further, the methodology for the proposed algorithm is described in Section 4, followed by the results and analysis in Section 5.

2. Related Work

This section investigates several aspects of the previous work carried out to optimize microgrids. The best way to optimize microgrids is efficient energy management, so that a perfect balance is maintained between demand and supply. In smart microgrids, the load is first served with energy from renewable energy resources when it works in islanded mode. If renewable energy cannot meet the load demand, the grid works in a grid-connected way, in which they buy the deficient amount of energy from the utility grid [29,30]. In the literature, researchers have operated microgrids in islanded mode and grid-connected mode. Energy management is carried out in microgrids in many ways by solving different kinds of optimization problems such as the optimal allocation of generation sources, optimal dispatch problem, and power scheduling problem. In the optimal generation allocation of conventional sources or optimal dispatch problem, the sizing and siting of the sources is improved, which optimizes various parameters of microgrids, whereas in the power scheduling problem, researchers focused on scheduling the demanded power in microgrid generators in such a way that it optimizes different parameters such as power losses, generation cost, total operation cost, and so on. The authors in [31–33] have discussed the optimal allocation problem considering the sizing and siting of various sources to minimize cost, power losses, emissions, and more in microgrids. The authors in [34] proposed a stochastic multi-objective optimization model to reduce the voltage deviation and operational cost in grid-connected mode for energy management. The proposed idea is applied to an amended IEEE 34-bus test system that consists of diesel generators, solar units, wind turbine units, EVs, and a battery system. It ensures that the malfunctioning of a microgrid is not possible because operators can trade between the high power quality and operational cost. Modified PSO is implemented in grid-connected mode in [35] for real-time energy management. The authors work for efficient demand side management by optimally controlling battery operations. The proposed idea successfully reduced operational costs by 12% over a time horizon of 96 h.

In [36], ANN-based binary particle swarm optimization and ANN-based tracking search algorithm were implemented to schedule a few microgrids in virtual power plants, aiming for optimal scheduling with less fuel consumption, reducing CO_2, and increasing system efficiency. They assessed the system under different system scenarios to evaluate the performance of the system under variable conditions. They used actual load data for trained and untrained models to assess the algorithm's performance, and the results were then compared to previous works based on several parameters. The results demonstrate that the hybrid algorithm was better than the available algorithms. In [11], the authors applied a quantum-based algorithm for better microgrid energy management. A quantum-teaching-learning-based optimization algorithm was employed to optimize the microgrid's energy flow. They studied four scenarios considering seasonal variations to deal with uncertainties related to power generation. It focuses on day-ahead scheduling according to the availability of DERs and shows a significant reduction in operational costs, especially during times of high market prices. It also improved the convergence graph and optimized power scheduling compared with other algorithms. This paper focuses on implementing an optimization algorithm for microgrids working in islanded mode. In previous works, various algorithms have been used to optimize microgrids by concentrating on different performance attributes.

The authors in [37] focused on optimizing renewable energy microgrids for the rural areas of the south Philippines using multi-objective particle swarm optimization and proposed a multi-case power management strategy. An optimized microgrid design was presented, considering various variables like load size, renewable energy sources, and different objective functions. It minimized the loss of supply, the level of energy, and greenhouse emission cost, and maximized reliability. A standalone microgrid was proposed in [38] with renewable energy resources for rural communities. The proposed microgrid performance was evaluated using differential evolution, PSO, and GA to find affordable energy for the community. The studies showed that differential evolution was suitable for the least energy cost compared with PSO. An improved mayfly optimization algorithm was applied for microgrid optimization in [39] for economic emission dispatch. The microgrid worked in islanded mode, utilizing solar power, wind power, and thermal power. The simulation was carried out for 24 h with varying load and supply demands. The algorithm was implemented for various scenarios, and it was observed that it performed better than other algorithms and could reduce total operational cost and emission level. It achieved a better system cost and emission reduction than mayfly and other metaheuristic algorithms.

The authors in [40] implemented a lightning search algorithm for energy management in microgrids on IEEE 14-bus system for a 24 h data. It aims to optimize the microgrid with renewable energy by minimizing operation costs, reducing emissions, and maximizing the usage of renewable energy. In [41], particle swarm optimization and the rainflow algorithm were implemented on a community microgrid for power scheduling with a battery for different scenarios. They focused on day-ahead battery scheduling, considering degradation costs owing to charging and discharging cycles. Moreover, after considering several uncertainties, the proposed technique could minimize operational costs by 40%. The variations in power generation and demand do not impact the battery energy schedule, whereas a fluctuation in electricity price affects the schedule. In another study [42], the PV microgrid was optimized using a mixed integer linear programming model considering social, technical, and economic aspects. It focuses on designing systems using location sizing and microgrid configurations. The results show that it successfully minimized the design cost of a microgrid with several uncertainties in geographic contents and improved its versatility. Several methods have been used in [43–46] for the optimization of various problems of energy management in microgrids.

In [47], the Markov decision process was employed for power scheduling in a renewable-energy-based microgrid. A rollout algorithm was used for decision space and the large state of MDP. A memory-based genetic algorithm [45] was carried out on a microgrid consisting of solar, wind, and a combined heat and power plant (CHP). It focuses on minimizing the cost through optimal energy distribution among available generation sources, whereas the authors of [48] optimize energy, heat, and demand using a mathematical model based on MILP to minimize the operational cost. In [49], the author proposed an artificial hummingbird algorithm for optimal operation of a microgrid. It tends to solve the deterministic incentive DR program that reduces overall cost, taking into account the load demand. It worked in grid-connected mode and the proposed algorithm was tested for two different case studies. The authors in [50] focused on solving numerical optimization problems with the proposed cost-effective multi-verse optimizer algorithm. They modified the updated position mechanism in standard MVO and combined it with a sine cosine algorithm for balanced exploration and exploitation. It is observed that the proposed algorithm achieves much better results for optimization, and the proposed method was evaluated on 27 benchmark functions. Similarly, in [51], multi-verse optimization is used for power scheduling for loss minimization, and validation was conducted on the IEEE 30-bus test system.

Previous studies have analyzed power scheduling from various aspects for better energy management. Among existing algorithms that have been implemented for optimization, it is observed that they do not explore and exploit the search space effectively and escape the local minima, whereas the multi-verse optimization algorithm has been applied to various research areas and shows promising results; it explores and exploits the search space efficiently. It is observed that it has also been used by various research scholars to optimize various parameters of microgrids while addressing different problems like unit commitment, economical dispatch, demand response problem, optimal allocation, power scheduling, and so on. This gives us the motivation to solve the power scheduling problem using MVO. The previously implemented work lacks an efficient optimization algorithm to solve power scheduling among different generation units to minimize the generation cost in microgrids. This gives us the motivation to solve the power scheduling problem using MVO. The proposed cost-effective optimization algorithm is a modified multi-verse optimization algorithm. As compared with the original algorithm, in this modified version, the best outcome of each iteration is stored in a temporary variable. After n number of iterations, they replace the current universes as an input to the (n + 1)th iteration. Here, n represents the total number of initialized populations. It enhances the operation and improves the solution searching capability of the original version in a search space concerning the challenges and drawbacks of available traditional algorithms. It aims to solve the power scheduling problem considering the intermittent nature of the renewable sources and minimize the generation cost for power generation. The proposed idea effectively fills this research gap by optimally scheduling the power between various generation sources.

3. Problem Formulation

3.1. Cost Function

The microgrid consists of variable load and intermittent generation power, including wind plants, PV, CHP, and so on. The demand at each hour and generated power are different, and the main aim is to supply power to satisfy the demand load. There are many techniques for energy allocation among DERs. The best energy management method in a microgrid is to reduce the generation cost. The commonly used quadratic cost function used to achieve this goal from the literature on microgrid [52] optimization is as follows:

$$C_i = \alpha_i \times P_i^2 + \beta_i \times P_i + \gamma_i \tag{1}$$

where α_i, β_i, and γ_i are the cost coefficients; C_i represents the total cost in dollars; and P_i denotes the power of ith generation units in MW per hour. This quadratic cost function is used for each generation unit to minimize generation cost. The total cost of each hour is a summation of all the DERs used at that hour. The main aim is to satisfy the load through power generation with minimum cost. It is considered that generation will always be equal to load.

3.2. Equality Constraint

For smooth implementation, it is observed that, at any instant, generated power should always be greater than or equal to the demand power. If DERs cannot meet demand, the needed capacity is taken from the utility grid. In this research, it is considered that load will always satisfy the load and it is not necessary to get energy from the utility grid.

This can be formulated as follows:

$$\sum_{g=1}^{DER} P_{g=}P_l \tag{2}$$

where P_g is the total generated power of the available DERs, P_l is the demanded power at a particular hour of the day, and DER represents the total number of DERs in a microgrid. Generated power for every hour is the sum of the power of all of the generation units. In this paper, we have considered two different microgrids, and both microgrids have a different number of generation units. Each generation unit should be operated between its limits. This can be represented as follows:

$$P_{g\ min} \leq P_g(t) \leq P_{g\ max} \quad where \ g = 1,2,\ldots N_{DER}t = 1,2,\ 3\ldots 24. \tag{3}$$

This equation states that, at time t, power from any generation unit should always be in this power range. Here, $P_{g\ min}$ is the minimum power generated by any generation unit, and it is supposed to be zero, whereas $P_{g\ max}$ is the maximum power produced depending on the rated power capacity. These also define the lower and upper bound and form the generation vector's boundary.

3.3. Objective Function

Each generation unit is considered a decision variable to solve the cost minimization problem. Thus, there are two different scale microgrids, and the decision variables differ. In microgrid 1, there are six decision variables; they are three wind power plants, two PV plants, and a CHP. Similarly, for microgrid 2, there are 15 decision variables, which are eight wind power plants, six PV plants, and a CHP. The vector solution for microgrid 1 [45] is represented as $x = [\ P_{wp1}, P_{wp2}, P_{wp3}, P_{pv1}, P_{pv2}, P_{CHP}]$ and that for microgrid 2 [46] is $x = [\ P_{wp1}, P_{wp2}, P_{wp3}, P_{wp4}, P_{wp5}, P_{wp6}, P_{wp7}, P_{wp8}, P_{pv1}, P_{pv2}, P_{pv3}, P_{pv4}, P_{pv5}, P_{pv6}, P_{CHP}]$. The total generation cost is the sum of all of the costs of generation units used at that time. Using the equality constraint, the energy management system should balance the generation power with the demanded load. Thus, the main aim of the optimization algorithm is to use the generated power from all of the available generation units so that the cost of generation is minimized. To solve this power scheduling problem, the following optimization function [45] should be solved:

$$MIN\ OF = \sum_{i=1}^{DER} C(i) \tag{4}$$

where

$$C(i) = \sum_{i=1}^{DER} \left[\alpha_i \times P_i^2 + \beta_i \times P_i + \gamma_i\right] \tag{5}$$

Subject to

$$\sum_{g=1}^{DER} P_{g=}P_l \tag{6}$$

As discussed, the generated power should be in the power range, and handling this issue in an optimization problem is challenging. Thus, the best way to address this problem is to introduce a penalty function. The penalty function helps to implement the optimized process in a balanced manner. The equation for the objective function with the penalty function [45] is given below:

$$C(i) = \left[\sum_{i=1}^{DER} \left[\alpha_i \times P_i^2 + \beta_i \times P_i + \gamma_i \right] \right] + P_f \left| \sum_{i=1}^{DER} P_i - P_l \right| \quad (7)$$

Here, P_f is the penalty factor that maintains the balance equation.

4. Methodology

This section explains the multi-verse optimizer algorithm and how it helps to find an optimized result for the problem. Further, a cost-effective version of the multi-verse optimizer algorithm is discussed and implemented in different scenarios for cost optimization.

4.1. Multi-Verse Optimizer Algorithm

The multi-verse algorithm is a nature-inspired population-based stochastic optimization algorithm. The optimization process for any problem starts by initializing some random set of solutions. These initial solutions are explored over a definite time for some defined step known as iterations or generations. The basic concept for all population-based algorithms is the same, but what makes them different is the process of moving or evolving toward an optimized solution. A population-based algorithm uses two processes during searching: exploration and exploitation [53]. The better the exploration and exploitation, the more optimized the solution is expected to be. The multi-verse optimization algorithm is based on the concept of the multi-verse theory, which came into existence after the big bang theory. This theory implies that a massive eruption led to the universe's presence, whereas the multi-verse theory states that many explosions led to the emergence of several parallel universes that collide and interact with each other, and every existing universe has distinct properties.

MVO works on the principle of the multi-verse theory of white holes, black holes, and wormholes. It is believed that white holes have never been found in the universe, but many physicists consider them a collision between several existing parallel universes. However, black holes attract everything towards them, including light, because of their great gravitational force. Wormholes are considered time/space travel tunnels from which objects travel within the universe from one corner to another (from one universe to another). Universe expansion through space depends on its inflation rate. The creation of stars, different planets, asteroids, wormholes, white holes, black holes, physical laws, and the sustainability of life depends on the inflation speed of the universe. It is argued that universes interact with each other through white holes, black holes, and wormholes to achieve a stable point. This is the motivation for a multi-verse optimization algorithm where universes interact with each other to achieve an optimized solution. In a multi-verse optimization algorithm, a few rules are employed in the universes, which are described below:

1. The inflation rate is directly proportional to the probability of having white holes.
2. The inflation rate is inversely proportional to the probability of having black holes.
3. The universe with more white holes and a high inflation rate tends to send more objects to another universe.
4. The universe with more black holes and a lower inflation rate tends to receive more objects from another.
5. The objects in different universes tend to randomly transfer toward the best universe through wormholes, regardless of the inflation rate.

In a multi-verse optimization algorithm, solutions are called universes and every variable in a solution is a variable in a universe. Moreover, each universe is associated with an inflation rate that is known as the fitness value of that universe. For the exploration process, white/black hole tunnels are used, whereas for the exploitation process, the objects are sent through wormholes. The universes with high inflation rates have more white holes, whereas the universes with lower inflation rates have more black holes. As explained above, the object travels from the source universe of white holes to the destination universe of black holes through white/black tunnels. With this process, the overall average inflation rate of all of the universes increases by the movement of objects from universes with higher inflation rates to universes with lower inflation rates. This process is also explained in Figure 2. To mathematically model this process, the following steps are followed:

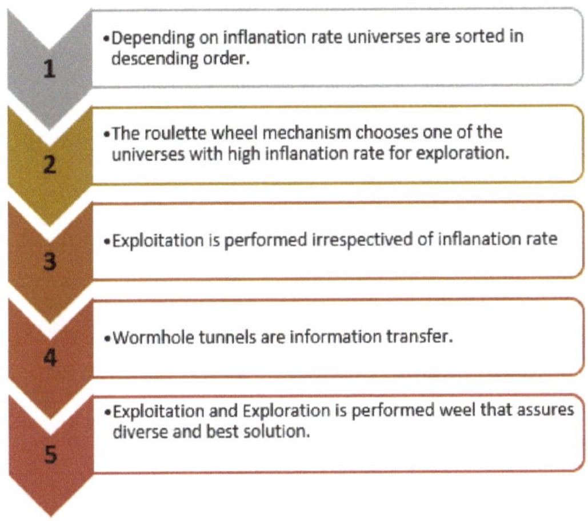

Figure 2. Multi-verse optimizer process.

Step 1: All of the universes are sorted by their inflation rate so that the universes with more white holes, i.e., with a high inflation rate, are kept in front, and the universes with more black holes, i.e., with a lower inflation rate, are kept at the rear.

Step 2: The roulette wheel mechanism chooses one of the universes with a white hole. With this implementation, exploration occurs as the universes interchange objects to explore the search space.

Step 3: After exploration, exploitation is performed using wormholes in universes. Each universe's objects exploit the search space by transforming the objects through the search space regardless of the inflation rate.

Step 4: Wormhole tunnels are formed between a universe and the best universe created.

Step 5: This mechanism assures the diversity of solutions and is expected to expand the local search and enhance the universe's overall inflation rate.

Initially, all of the parameters are defined, such as the objective function, problem dimension, population size, maximum iterations, and upper and lower limits. After explaining all of the parameters, the positions of universes are initialized using random solutions.

Each universe has d number of variables in a solution (here, it is several generation units). Universe i is shown by the vector, $x^i = [x_1{}^i\ x_2{}^i\ x_3{}^i \ldots \ldots\ x_d{}^i]$. The matrix of the universe is shown as follows:

$$U_i = \begin{bmatrix} x_1^1 & x_1^2 & \cdots & \cdots & x_1^d \\ x_2^1 & x_2^2 & \cdots & \cdots & x_2^d \\ \vdots & \vdots & \vdots & \vdots & \vdots \\ x_n^1 & x_n^2 & \cdots & \cdots & x_n^d \end{bmatrix} \tag{8}$$

Here, d is the number of decision variables and n is the number of universes (solutions). Then, the inflation rates are calculated and tend to find the best solution using the following equation:

$$x_i^j = \begin{cases} x_k^j & r1 < NI(U_i\ ) \\ x_i^j & r1 < NI(U_i\ ) \end{cases} \tag{9}$$

where x_i^j indicates the jth parameter of the ith universe, U_i indicates the ith universe, $NI(U_i)$ is a normalized inflation rate of the ith universe, $r1$ is a random number in [0, 1], and x_k^j indicates the jth parameter of the kth universe selected by the roulette wheel selection mechanism. These universes are then sorted in a series where the ones with a high inflation rate, i.e., with a more significant number of white holes, are in front, and those with low inflation rates are kept at the rear. After sorting the universes, the wormhole existence probability (WEP) and traveling distance rate (TDR) are calculated using the formula for both coefficients mentioned below:

$$\mathrm{WEP} = \min + l \times \left(\frac{\max - \min}{L}\right) \tag{10}$$

where min stands for minimum and equals 0.2, max stands for maximum and the value is set to 1, l shows the current iteration, and L indicates the total number of iterations.

$$TDR = 1 - \frac{l^{\frac{1}{p}}}{L^{\frac{1}{p}}} \tag{11}$$

Here, p shows the exploitation accuracy over iterations and is set to 6. The speed and accuracy of the local search or exploitation rate are directly proportional to p. Now, the positions of the universes are updated using the following equation and the current best solution.

$$x_i^j = \begin{cases} X_j + TDR \times \left(\left(ub_j - lb_j\right) \times r4 + lb_j\right) \; r3 < 0.5 \; and \; r2 < WEP \\ X_j + TDR \times \left(\left(ub_j - lb_j\right) \times r4 + lb_j\right) \; r3 < 0.5 \; and \; r2 < WEP, \text{ otherwise} \\ x_i^j \quad r2 \geq WEP \end{cases} \tag{12}$$

where X_j indicates the jth parameter of the best universe formed so far; TDR is a coefficient; WEP is another coefficient; lb_j shows the lower bound of the jth variable; ub_j is the upper bound of the jth variable; x_i^j indicates the jth parameter of the ith universe; and $r2$, $r3$, and $r4$ are random numbers in [0, 1]. This is repeated until the maximum number of iterations is reached and the best-optimized result is outputted.

4.2. Cost-Effective Multi-Verse Optimizer Algorithm

The existing algorithm has yielded promising results, but converges prematurely without proper exploitation in search space. The improved version of the multi-verse optimizer algorithm enhances the optimization results with an improved exploitation process for cost optimization problems. As shown in Figure 3, some modifications in the optimization process using a multi-verse optimizer algorithm ensure better results for cost optimization problems and are labelled as 1 and 2. Initially, all of the parameters are defined as the objective function, dimension of the problem, population size, maximum iterations, and upper and lower limits. Then, the positions of universes are initialized using random solutions. The following steps are achieved by initializing the population of N universes. Inflation rates are calculated and the best solution is decided among all of the universes. Further, sorting of universes is performed, in which a high inflation rate universe is supposed to consist of more white holes stored in front and others in the rear. After the sorting is complete, the best results so far are stored at every consecutive iteration. Then, the WEP and TDR are calculated using the given equations. The positions of the universe are held using the current locations, and the best solution at every iteration is stored followed by the evaluation of inflation rates for new universes. After this, it is checked whether the stored solutions reach the population size. Once the stored solutions are equal to population size, the previous population is replaced, and the new solution serves as the new population for the rest of the implementation. For every n population size, there will be a new population after every $2n$ iterations. This process is followed for N number of iterations and outputs promising results for this problem (Algorithm 1).

Algorithm 1: Algorithm for proposed cost-effective multi-verse optimizer algorithm

Input: Number of decision variables, boundaries, loads, cost coefficients
Output: Generation cost at each hour
Step 1: Start, Initial parameters No. of universes = 50, maximum no. of iteration = 1000, WEP = 0.2 and TDR = 1.
Step 2: Initialize Universe positions with random solutions using a matrix.
Step 3: Calculate the inflation rate using $x_i^j = \begin{cases} x_k^j & r1 < NI(U_i) \\ x_i^j & r1 < NI(U_i) \end{cases}$ And find the best solution using Roulette wheel selection.
Step 4: Calculate WEP using WEP $= \min + l \times \left(\frac{\max - \min}{L}\right)$ and TDR by $TDR = 1 - \frac{l^{1/p}}{L^{1/p}}$
Step 5: Update the inflation rates using.

$$x_i^j = \begin{cases} \begin{cases} X_j + TDR \times \left(\left(ub_j - lb_j\right) \times r4 + lb_j\right) & r3 < 0.5 \\ X_j - TDR \times \left(\left(ub_j - lb_j\right) \times r4 + lb_j\right) & r3 \geq 0.5 \end{cases} & r2 < WEP \\ x_i^j & r2 \geq WEP \end{cases}$$

Step 6: After every iteration, store the best solution.
Step 7: If stored solutions are equal to the number of universes, Replace the population with new stored solutions.
Step 8: Repeat the process until the maximum number of iterations.
Step 9: Calculate the Best cost.
Step 10: End

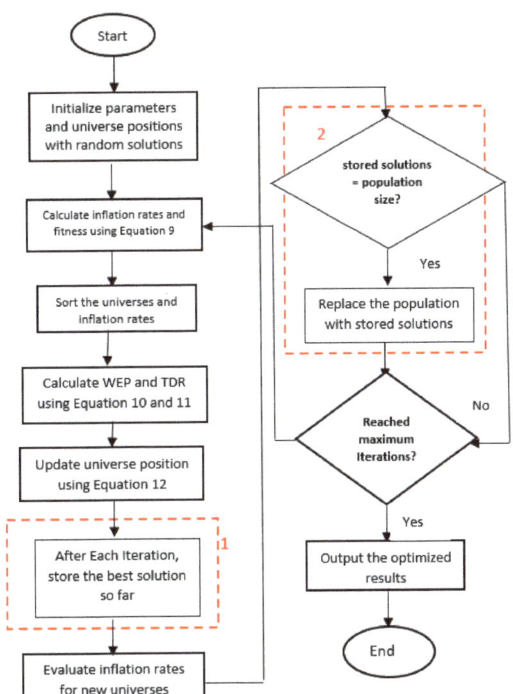

Figure 3. Flowchart of the modified version of MVO.

5. Implementation and Results

In this section, the experimental setup and dataset description are provided. Moreover, the performance of a proposed algorithm is evaluated by implementing it on different scale microgrid models.

5.1. Experimental Setup and Data Set Description

The algorithms were implemented using MATLAB and the experiments were executed on a computer with a Windows 10 64-bit operating system specification, Intel (R) core (TM) i5, and 8 GB RAM. The proposed algorithm is implemented on two test systems and the results are compared to those of other algorithms. According to the classification based on generation power, demands, and electrical lengths, the microgrids are categorized as small-scale, medium-scale, and large-scale microgrids.

In this paper, the performance of the proposed algorithm is evaluated on two microgrid models, i.e., the IEEE 37-node test feeder and the IEEE 141-node test feeder. These two different models are categorized as medium-scale and large-scale microgrids. In this research, we will consider the microgrids in islanded mode, assuming the generated power from all of the available generation sources will satisfy the load at each hour. Islanded mode depicts that the microgrid will serve the demand with the available renewable sources and CHP. A cost-effective multi-verse optimization algorithm optimizes demand power among the various available DERs. It focuses on optimal power sharing among the available sources in such a way that generation cost is optimized and the load is served.

The proposed CMVO and other algorithms were implemented by the authors for both medium-scale and large-scale microgrids. The generation data were obtained from [52] and all of the algorithms were implemented by the authors for the same dataset and then compared in terms of cost. It is shown that the proposed algorithm performs better than the other investigated meta-heuristic algorithms. These algorithms are run 30 times for a 24 h load dataset, and the best results are reported for a fair evaluation. All of the algorithms are executed for the same population size and the maximum number of iterations for the comparison. Other parameters for each algorithm are varied accordingly. The parameter settings of all algorithms are presented in Table 1.

Table 1. Parameter settings of the algorithms.

Algorithm	Parameter	Value
CMVO	Universe size	50
	Number of iterations	1000
	WEP	0.2
	TDR	1
MVO	Universe size	50
	Number of iterations	1000
	Min	0.2
	Max	1
	p	6
GA	Population size	50
	Number of iterations	1000
	Crossover probability	0.9
	Mutation probability	0.5
PSO	Swarm size	50
	Number of iterations	1000
	Learning factor	2
	Inertia weight	0.9 to 0.4
AHA	Population size (n)	50
	Number of iterations	1000
	Migration coefficient	2n

Two different small-scale and large-scale microgrid test systems are used for the implementation. Table 2 represents the configuration of these microgrids and they are discussed in detail below.

Table 2. Configurations of microgrids.

Microgrid	1	2
IEEE Test System	IEEE 37-NODE Test Feeder	IEEE 141-NODE Test Feeder
Scale	Medium	Large
No. of solar plant (PV)	2	6
No. of wind plant	3	8
No. Of CHP	1	1

5.1.1. Microgrid 1 (Medium-Scale Microgrid)

This microgrid consists of three wind plants, two PV plants, and one CHP, whereas the load area is represented by the IEEE 37-bus test system, as seen in Figure 4. The generation and load data for this microgrid are adopted from [45] and the rated capacity of the wind turbine, solar plant, and CHP is 750 KW, 200 KW, and 1000 KW, respectively [45]. Wind and PV plants are intermittent and provide different power at each hour, whereas CHP will offer the same power for a whole day. It is assumed that the generation sources do not run out, and the microgrid works in islanded mode without relying on the grid.

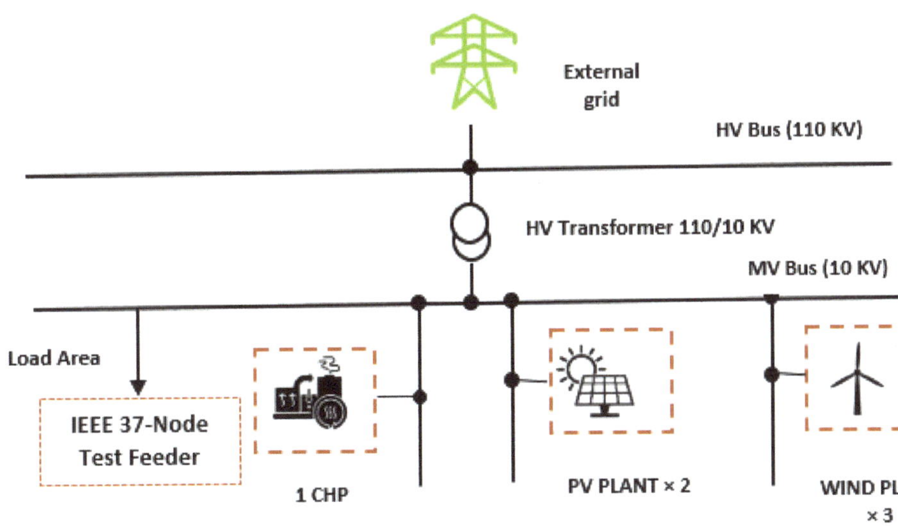

Figure 4. Representation of microgrid 1.

The energy management system will be informed about the demand and generated power for every hour. It evaluates the data and communicates back to generation units that aim to allocate the energy to achieve the minimum generation cost. The power generation of each renewable energy source per hour is shown in Figure 5. Table 3 shows the load for each microgrid [45,54], whereas Table 4 shows the cost coefficient for each generation unit in microgrid 1 [45,46].

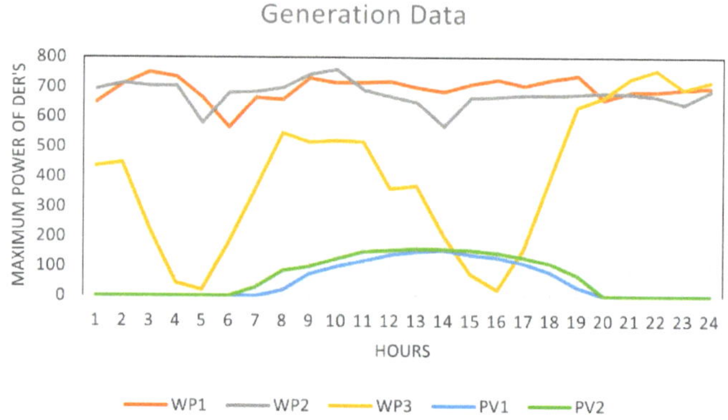

Figure 5. Generation data for each hour.

Table 3. Load data for microgrid 1 and microgrid 2 [45,54].

Hour	Load (KW) IEEE 37	Load (KW) IEEE 141
1	1471	3482
2	1325	2946
3	1263	2761
4	1229	2558
5	1229	2541
6	1321	2616
7	1509	3635
8	1663	4339
9	1657	4748
10	1643	5100
11	1643	5231
12	1652	5306
13	1666	5454
14	1639	5215
15	1642	5363
16	1640	5383
17	1676	5198
18	1920	5051
19	2214	4496
20	2382	5275
21	2382	5479
22	2327	5536
23	2174	5370
24	1903	4611

Table 4. Cost coefficients of DERs in microgrid 1.

Plant	α	β	Υ
WP1	0.0027	17.83	4.46
WP2	0.0028	17.54	4.45
WP3	0.0026	17.23	4.44
PV1	0.0055	29.30	4.45
PV2	0.0055	29.58	4.46
CHP	0.0083	75.73	5.21

5.1.2. Microgrid 2 (Large-Scale Microgrid)

This microgrid consists of eight wind plants, six PV plants, and one CHP, whereas the load area is represented by the IEEE 141-bus test system, as seen in Figure 6. The load data for this microgrid are adopted from [45], and the rated capacity of the wind turbine, solar plant, and CHP is 750 KW, 200 KW, and 1000 KW, respectively. Figure 6 represents microgrid 2, and cost coefficients for microgrid 2 are listed in Table 5 [46].

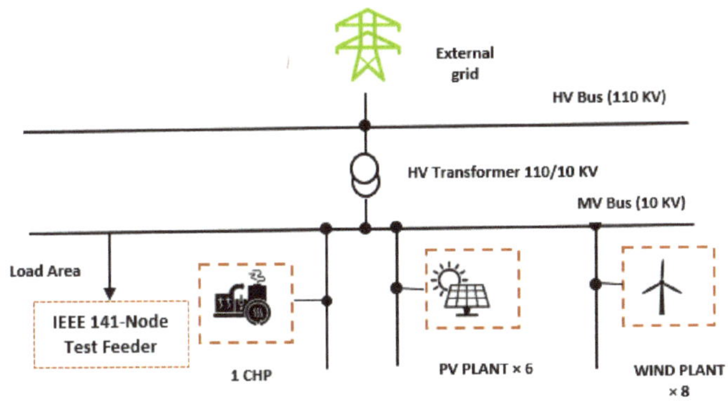

Figure 6. Representation of microgrid 2.

Table 5. Cost coefficient of DERs in microgrid 2.

Plant	α	β	Υ
WP1	0.0027	17.83	4.46
WP2	0.0028	17.54	4.45
WP3	0.0026	17.23	4.44
WP4	0.0027	17.83	4.46
WP5	0.0028	17.54	4.45
WP6	0.0028	17.54	4.45
WP7	0.0026	17.23	4.44
WP8	0.0026	17.23	4.44
PV1	0.0055	29.30	4.45
PV2	0.0055	29.58	4.46
PV3	0.0055	29.30	4.45
PV4	0.0055	29.58	4.46
PV5	0.0055	29.30	4.45
PV6	0.0055	29.58	4.46
CHP	0.0083	75.73	5.21

Wind and PV plants are intermittent and provide different amounts of power at each hour, whereas CHP will offer the same amount of power for a whole day. For this implementation, it is assumed that the generation sources do not run out and the microgrid works in islanded mode without relying on the grid. The load dataset for microgrid 2 is presented in Table 3.

5.2. Experiment Results

In the experiment results, the results for both microgrids are explained by various algorithms. The results are obtained for different microgrids with their respective datasets using initialized parameters. All of the algorithms were run 30 times, and the best results were reported for fair evaluation.

5.2.1. Microgrid 1 (IEEE 37 NODE TEST FEEDER)

The optimization results for microgrid 1 are explained in this section. Five of the algorithms are applied to the given dataset. It is assumed that the generated power will always satisfy the demanded power, so the equality constraint is satisfied. Table 6 shows the generation power of all of the DERs every hour by applying a genetic algorithm (GA). For this microgrid, we have six DERs as WP1, WP2, WP3, PV1, PV2, and CHP. The load data for this microgrid are presented in Table 1. After applying the genetic algorithm, the total generation is $1600.35. Table 7 shows the result for the generation power of all of the DERs using the particle swarm optimization algorithm. The entire generation cost for PSO is $1183.45. Table 8 shows the results performed by the artificial hummingbird algorithm AHA [55], respectively. The total generation cost for AHA is $1353.74. Similarly, Tables 9 and 10 represent the generation power for the multi-verse optimizer algorithm (MVO) and cost-effective multi-verse optimizer (CMVO) algorithm, respectively. The total generation cost for MVO is $1177.20, whereas it is $1167.35 for CMVO. These tables represent various algorithms for optimal power scheduling of all available generation units at the hour.

Table 6. Generation power (KW) by GA for microgrid 1.

No.	WP1	WP2	WP3	PV1	PV2	CHP	Cost ($)
1	383.73	574.293	395.83	0	0	117.134	51.17
2	534.689	399.98	354.819	0	0	35.502	43.91
3	506.64	541.73	168.308	0	0	46.316	43.51
4	604.402	521.86	20.91	0	0	81.82	45.05
5	644.34	460.329	11.67	0	0	112.64	46.86
6	509.143	587.81	161.23	0	0	62.8	45.48
7	643.22	544.58	90.8	0	12.55	217.82	62.48
8	643.23	457.34	453.9	8.2	57.38	42.85	59.97
9	697.466	460.85	359.08	45.26	48.89	45.42	60.39
10	681.39	273.42	434.45	45.86	61.61	146.245	66.14
11	433.81	503.18	493.97	115.71	95.25	1.06	58.83
12	683.71	639.69	154.38	108.2	60.22	5.6	58.93
13	583.55	552.691	287.59	77.51	99.25	65.377	62.69
14	504.43	454.82	203.9	95.97	101.11	278.75	74.87
15	565.73	622.26	25.91	86.12	221.53	320.43	76.35
16	603.22	559.95	10.227	120.11	57.968	288.51	75.31
17	664.6	610.49	123.11	104.07	60.73	112.96	65.55
18	466.86	559.42	361.755	62.14	33.61	436.19	87.69
19	696.422	617.59	501.47	2.64	22.6	373.25	88.38
20	610.1	635.03	474.8	0	0	662.05	98.90
21	505.89	576.14	663.899	0	0	636.06	97.30
22	623.55	648.58	591.11	0	0	463.74	86.36
23	597.81	454.71	647.62	0	0	472.83	84.17
24	573.63	609.4	579.94	0	0	140.01	60.08

Table 11 represents the total generation cost by all of the algorithms after optimal power scheduling among various available DERs. Based on Table 8, it is found that the cost-effective multi-verse optimizer algorithm has improved results. It performs optimal scheduling at a lower cost than all other algorithms. In comparison, the total costs produced by CMVO, MVO, PSO, AHA, and GA are $1167.35, $1177.20, $1183.45, $1353.74, and $1600.35, respectively. It is seen that the daily cost reduction varies from $10.05 for MVO to $433 for GA, whereas the average time taken by CMVO, MVO, PSO, AHA, and GA for each hour is 0.18 s, 0.19 s, 1.90 s, 0.26 s, and 1.05 s, respectively. The mean and standard deviation for each algorithm are presented in Table 11. Here, AHA has a better standard deviation than CMVO, but has a difference of $186, which is more significant. So, the proposed algorithm is more stable than the other investigated algorithms.

Table 7. Generation power (KW) by PSO for microgrid 1.

No.	WP1	WP2	WP3	PV1	PV2	CHP	Cost ($)
1	340.11	693.8836	437	0	0	0	39.12
2	610	715	0	0	0	0	32.33
3	558	705	0	0	0	0	31.23
4	571.28	657.71	0	0	0	0	30.63
5	649	580	0	0	0	0	30.66
6	459	682	180	0	0	0	36.60
7	665	484	360	0	0	0	39.90
8	660	536.89	466.10	0	0	0	42.57
9	478.81	639.15	514.99	0	24.03	0	47.14
10	703.40	453.24	486.34	0	0	0	42.22
11	717	454.65	433.65	0	37.68	0	47.16
12	640.19	667.97	335.13	8.68	0	0	47.01
13	648.60	647.39	370	0	0	0	42.65
14	685	570	210	138.28	35.71	0	53.20
15	710	665	0	112	155	0	50.01
16	725	668.06	0	111.74	135.18	0	49.74
17	705	673.53	159.66	90.68	47.11	0	53.45
18	725	674	395	77.82	48.17	0	57.52
19	740	677.72	634.99	0	65.78	95.49	68.22
20	660	682	667	0	0	373	82.03
21	685	680	730	0	0	287	77.02
22	687	670	758	0	0	212	71.68
23	695	645	695	0	0	138	64.69
24	495	688	720	0	0	0	46.65

Table 8. Generation power (KW) by AHA for microgrid 1.

No.	WP1	WP2	WP3	PV1	PV2	CHP	Cost ($)
1	339.07	694.95	436.96	0	0	0	44.33
2	162.01	714.98	447.99	0	0	0	41.71
3	328.00	704.99	229.99	0	0	0	40.74
4	480.00	704.99	43.99	0	0	0	40.24
5	629.00	579.99	19.99	0	0	0	40.29
6	459.01	681.98	179.99	0	0	0	41.81
7	467.68	682.36	358.92	0	0.029607	0	49.51
8	643.28	505.41	513.19	0.68	0.201648	0.22	56.61
9	729.23	358.55	514.65	54.08	0.466118	0	56.52
10	372.73	754.57	515.38	0.01	0.209078	0	56.25
11	663.77	476.72	501.45	0.32	0.282455	0.43	56.26
12	717.29	617.67	313.45	2.63	0.511972	0.42	56.53
13	699.73	647.00	170.74	148.13	0.382843	0	56.79
14	677.34	568.27	207.93	107.10	78.05455	0.28	58.43
15	709.50	664.99	73.86	39.23	150.4662	3.92	58.75
16	721.78	668.68	0.35	105.60	143.4948	0	59.15
17	704.22	673.99	159.78	89.42	48.06352	0.50	58.66
18	722.77	673.50	393.77	64.56	65.37054	0	62.74
19	738.62	676.32	626.67	29.18	65.55676	77.62	71.15
20	659.99	681.99	666.99	0	0	373.	82.03
21	684.99	679.98	729.99	0	0	287.02	77.02
22	687	670	758	0	0	212	71.68
23	695	645	695	0	0	138	64.69
24	495.00	687.99	719.99	0	0	0	51.86

Table 9. Generation power (KW) by MVO for microgrid 1.

No.	WP1	WP2	WP3	PV1	PV2	CHP	Cost ($)
1	361.18	672.81	437	0	0	0	39.13
2	609.99	715	0	0	0	0	32.34
3	582.44	680.55	0	0	0	0	31.23
4	587.82	641.17	0	0	0	0	30.64
5	648.99	580	0	0	0	0	30.66
6	459.85	682	179.14	0	0	0	36.60
7	464.05	685	359.94	0	0	0	39.85
8	519.48	596.52	547	0	0	0	42.52
9	429.49	745	482.50	0	0	0	42.43
10	445.41	685.22	512.37	0	0	0	42.15
11	717	503.75	398.39	23.85	0	0	46.99
12	720	587.56	324.60	19.83	0	0	47.13
13	694.32	643.01	328.65	0	0	0	42.68
14	685	570	210	155	18.99	0	53.20
15	710	665	0	131.28	135.71	0	50.01
16	725	669	0	104.95	141.03	0	49.74
17	705	674	160	104.21	32.78	0	53.44
18	725	674	395	71.91	54.08	0	57.52
19	740	677.87	634.98	0	0	161.14	66.84
20	660	682	667	0	0	373	82.03
21	685	680	730	0	0	287	77.02
22	687	670	758	0	0	211.99	71.68
23	695	645	695	0	0	137.99	64.70
24	500.83	682.16	720	0	0	0	46.66

Table 10. Generation power (KW) by CMVO for microgrid 1.

No.	WP1	WP2	WP3	PV1	PV2	CHP	Cost ($)
1	353.16	693.51	424.32	0	0	0	39.14
2	609.99	715	0	0	0	0	32.33
3	563.36	699.64	0	0	0	0	31.25
4	558.78	670.21	0	0	0	0	30.64
5	648.99	580	0	0	0	0	30.66
6	470.87	670.12	180	0	0	0	36.61
7	630.78	518.40	359.80	0	0	0	39.90
8	437.34	699.34	526.31	0	0	0	42.49
9	409.15	744.67	503.17	0	0	0	42.39
10	599.26	760	283.73	0	0	0	42.28
11	600.05	526.10	516.83	0	0	0	42.19
12	689.93	669.82	292.24	0	0	0	42.45
13	676.69	622.62	366.68	0	0	0	42.71
14	685	570	209.9	155	19.01	0	53.20
15	710	664.97	0	131.06	135.95	0	50.02
16	725	669	0	113.94	132.05	0	49.74
17	705	674	160	109.61	27.38	0	53.44
18	725	673.99	395	78.130	47.87	0	57.52
19	739.40	677.96	635	0	0	161.62	66.83
20	660	682	667	0	0	373	82.03
21	685	680	729.981	0	0	287.01	77.02
22	687	670	758	0	0	211.99	71.68
23	695	645	695	0	0	138.00	64.70
24	498.23	687.93	716.82	0	0	0	46.66

Table 11. Total generation cost and time of the invested algorithm for microgrid 1.

Algorithm	Total Cost ($)	Total Time (s)	Mean ($)	Standard Deviation
Proposed CMVO	1167.35	0.18	1261.11	17.22
MVO	1177.20	0.19	1265.85	19.87
PSO	1183.45	1.90	1346.91	41.56
AHA	1353.74	0.26	1361.16	16.67
GA	1600.35	1.05	1958.76	32.34

Figure 7 represents the convergence graph for hour 10 and Figure 8 represents the graph for hour 13. These convergences graphs are selected to show the best outcome from all of the running hours. It is observed from the figure that the proposed algorithm that explores the search space is converged efficiently. The convergence graph represents the best solution versus the generation (iteration number). In these graphs at hour number 10, it is observed that, compared with other algorithms, the proposed algorithm finds a promising region in search space for initial generations and rapidly converges to the optimal solution. It performed better than other algorithms in terms of efficiency and accuracy. Similarly, in another hour, hour 13, it is observed that the proposed algorithm converges better as it finds a promising region in fewer generations and achieves better results compared with other existing algorithms.

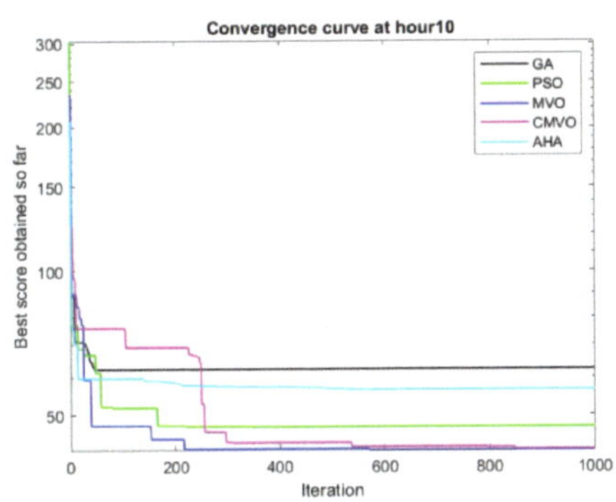

Figure 7. The convergence graphs for microgrid 1 for hour 10.

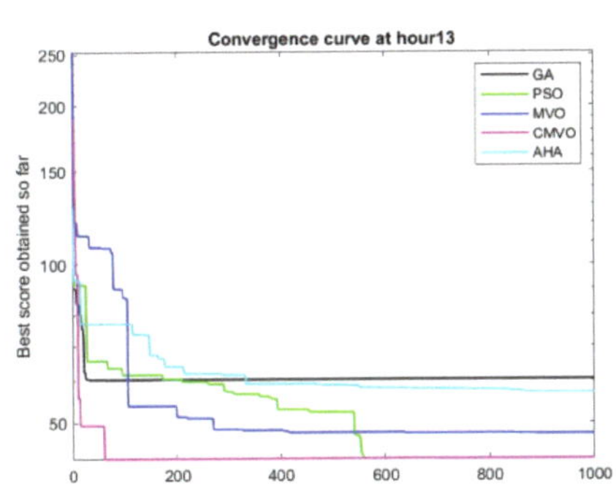

Figure 8. The convergence graphs for microgrid 1 for hour 13.

5.2.2. Microgrid 2 (IEEE 141 NODE TEST FEEDER)

The optimization results for microgrid 2 are explained in this section. Five of the algorithms are applied to the given dataset. It is assumed that the generated power will always satisfy the demanded power, so the equality constraint is satisfied. The generation power of all DERs at every hour is different and, for this microgrid, we have 14 DERs as WP1, WP2, WP3, WP4, WP5, WP6, WP7, WP8, PV1, PV2, PV3, PV4, PV5, PV6, and CHP. As shown in Figure 5, the data for WP1, WP2, and WP3 are the same, and WP4 has the same data as WP1, WP5 and WP6 have the same data as WP2, and the data of WP7 and WP8 are the same as those of WP3. Similarly, for solar datasets, PV1, PV3 and PV5 have the same data as PV1 in Figure 4, and PV2, PV4, and PV6 have the same data as PV2 in Figure 4. The load data for this microgrid are presented in Table 3. Assume CHP has a constant production value of 1000 KW for the entire 24 h. Table 12 shows that the total generation obtained by genetic algorithm is $4711.06. It describes the optimal power scheduling carried out by GA for this microgrid. Each row in the table shows the total power generated by each renewable resource, and it is observed that, at every hour, the demand is satisfied by generated power. Similarly, Tables 13 and 14 describe the optimal power scheduling carried out by PSO and AHA, respectively. The total cost generated by PSO for microgrid 2 with 15 DERs is $3252.05, which shows an improvement in generation cost compared with GA, which is $4711.06, whereas that of AHA is $3598.03. Tables 15 and 16 represent the results of MVO and CMVO, where the total generation cost produced by MVO is $3211.43. Compared with all of these algorithms, CMVO achieves many promising results, and the entire generation cost is $3178.30. It is observed that the daily cost reduction varies from $33.3 for MVO to $1532.96 for GA.

Table 12. Generation power (KW) by GA for microgrid 2.

No.	WP1	WP2	WP3	WP4	WP5	WP6	WP7	WP8	PV1	PV2	PV3	PV4	PV5	PV6	CHP	Cost ($)
1	266.55	562.88	304.41	482.98	642.50	589.01	301.68	331.94	0	0	0	0	0	0	0	96.60
2	362.10	475.85	333.35	296.96	404.02	394.55	322.03	357.10	0	0	0	0	0	0	0	87.14
3	193.03	563.42	191.78	583.54	383.54	651.13	151.11	43.417	0	0	0	0	0	0	0	84.13
4	414.91	611.88	26.44	321.47	566.87	570.09	18.43	22.46	0	0	0	0	0	0	0	80.68
5	458.20	480.10	0	439.96	547.06	431.25	0	16.08	0	0	0	0	0	0	7.89	77.92
6	462.17	483.09	67.89	240.32	634.785	506.96	132.02	88.73	0	0	0	0	0	0	0	81.59
7	572.38	501.92	285.42	594.31	537.65	464.53	299.55	319.80	0	12.35	0	17.038	0	19.89	10.05	119.18
8	522.87	595.10	435.21	484.16	607.71	630.73	389.28	427.95	6.39	36.05	7.32	47.40	4.9578	38.28	105.48	151.37
9	573.92	671.20	344.79	553.14	607.52	635.48	372.39	480.34	45.02	48.05	38.70	58.24	30.99	27.01	5.14	156.70
10	407.38	693.91	399.74	636.00	649.37	500.92	415.77	478.07	80.75	53.73	48.87	60.83	50.55	63.78	560.23	193.76
11	566.47	617.54	427.35	659.78	581.30	466.78	304.23	429.12	89.26	120.08	47.64	85.58	88.28	96.67	650.84	203.45
12	596.61	560.44	217.73	636.78	634.30	574.24	270.83	316.55	104.28	94.62	82.66	106.05	110.42	112.52	887.91	219.65
13	646.29	561.12	289.20	654.15	519.63	562.11	285.74	265.26	115.35	109.49	99.59	85.51	94.11	100.46	937.53	258.77
14	653.23	473.21	133.19	628.34	461.28	488.50	156.81	134.84	112.96	121.97	97.45	122.7	107.30	111.7	849.69	624.76
15	622.73	592.20	61.39	687.11	541.72	585.69	48.81	34.19	114.65	130.77	96.43	131.25	116.53	117.98	939.28	417.22
16	701.51	633.34	19.95	693.21	577.54	628.39	6.94	11.32	118.43	40.96	111.38	57.18	117.81	66.06	899.45	446.78
17	614.11	632.59	146.81	624.06	547.03	638.26	134.89	138.28	52.78	59.93	75.86	89.75	85.87	108.29	812.61	253.12
18	680.39	630.80	271.10	685.45	653.45	611.39	264.69	297.53	62.45	73.12	62.99	37.89	70.00	68.9	540.15	194.58
19	618.00	562.14	497.95	605.59	511.91	521.03	478.84	528.09	12.25	25.82	11.05	46.09	17.00	25.16	34.95	149.96
20	555.56	591.59	610.00	554.79	616.86	588.71	576.92	512.74	0	0	0	0	0	0	468.09	172.64
21	580.69	582.53	610.78	604.42	574.14	625.57	595.95	551.74	0	0	0	0	0	0	753.13	180.54
22	543.86	582.12	621.52	607.25	616.43	564.16	684.28	664.83	0	0	0	0	0	0	651.5	175.55
23	655.44	432.84	543.51	638.87	693.77	607.04	560.54	618.42	0	0	0	0	0	0	149.49	161.62
24	484.30	588.78	641.39	560.66	535.92	600.20	638.19	528.33	0	0	0	0	0	0	33.14	123.36

Figure 9 represents the convergence graph for hour 7 and Figure 10 represents the graph for hour 23. These convergence graphs are selected to show the best outcome from all of the running hours. The convergence graph represents the best solution versus iteration. In these graphs, at hour 7, it is observed that, compared with other algorithms, the proposed algorithm finds a promising region in search space for initial generations and quickly converges to the optimal solution. It performed better than different algorithms in terms of efficiency and accuracy. Similarly, in another hour, hour 23, it is observed that the proposed algorithm converges better as it finds a suitable region in fewer generations and achieves the optimal result compared with other existing algorithms.

Table 13. Generation power (KW) by PSO for microgrid 2.

No.	WP1	WP2	WP3	WP4	WP5	WP6	WP7	WP8	PV1	PV2	PV3	PV4	PV5	PV6	CHP	Cost ($)
1	650	695	352.1	526.13	694.39	564.36	0	0	0	0	0	0	0	0	0	88.02
2	667.00	711.30	448	0	498.38	451.97	0	169.33	0	0	0	0	0	0	0	78.37
3	741.81	0	227.39	0	705	705	166.57	215.21	0	0	0	0	0	0	0	75.14
4	638.69	689.28	0	0	588.0	589.73	43.99	8.27	0	0	0	0	0	0	0	71.73
5	665	580	5.1356	412.81	422.17	455.87	0	0	0	0	0	0	0	0	0	71.59
6	0	520.19	180	528.64	670.65	682	0	34.51	0	0	0	0	0	0	0	72.67
7	307.19	633.48	357.00	278.06	658.41	676.23	359.05	357.46	0	8.09	0	0	0	0	0	103.75
8	660	642.63	547	556.12	347.62	700	547	306.22	6.31	13.36	0	0	12.70	0	0	125.37
9	730	255.29	497.5	730	745	745	398.66	515	0	0	54.16	0	11.331	66.04	0	133.79
10	565.94	678.14	448.38	694.5	741.08	748.66	437.58	519.95	63.98	0	42.93	59.26	0	99.49	0	145.97
11	701.94	692	470.78	717	633.73	691.01	514.98	517	119.50	119.18	0	53.84	0	0	0	144.15
12	720	670	360	720	669.81	670	360	352.7	139.45	155	0	81.57	137.70	155	114.64	170.89
13	700	650	370	699.97	650	650	368.45	369.84	0	160	149.63	145.60	111.39	141.96	287.11	184.00
14	685	570	205	684.78	570	570	205	190.60	155	158	155	84.83	0	155.75	826.02	211.28
15	710	665	0	710	665	665	74	0	140	155	140	144.07	140	154.92	1000	221.72
16	725	669	20	725	669	669	20	18	130	145	130	145	130	145	1000	229.63
17	705	674	149.6	705	674	674	160	106.31	110	130	110	0	0	0	1000	207.97
18	692.87	674	390.51	725	674	672.70	395	395	80	0	0	110	0	0	241.91	154.70
19	586.83	612.30	588.76	548.27	546.84	583.95	444.79	510.93	19.324	46.60	0	7.357	0	0	0	128.55
20	590.17	662.58	667	660	682	681.97	666.97	664.29	0	0	0	0	0	0	0	127.87
21	685	680	730	685	680	679	670	670	0	0	0	0	0	0	0	131.46
22	657.47	670	758	687	577.52	670	758	758	0	0	0	0	0	0	0	132.39
23	561.57	645	695	689.14	744.279	645	695	695	0	0	0	0	0	0	0	129.51
24	0	647.23	720	674.74	628.80	530.02	720	690.1	0	0	0	0	0	0	0	111.55

Table 14. Generation power (KW) by AHA for microgrid 2.

No.	WP1	WP2	WP3	WP4	WP5	WP6	WP7	WP8	PV1	PV2	PV3	PV4	PV5	PV6	CHP	Cost ($)
1	88.88	516.7	310.1	419.8	694.98	695.0	328.0	428.4	0	0	0	0	0	0	0	101.65
2	709.9	554.9	12.73	400.5	234.23	714.9	216.4	102.2	0	0	0	0	0	0	0	92.18
3	638.2	702.7	151.8	0.15	438.05	517.0	190.9	121.9	0	0	0	0	0	0	0	89.06
4	372.4	630.2	35.43	323.0	448.55	705.0	42.89	0.32	0	0	0	0	0	0	0	85.78
5	472.7	309.9	12.92	656.9	482.19	579.7	19.81	6.78	0	0	0	0	0	0	0	85.61
6	344.8	511.2	156.8	251.4	410.46	681.9	79.18	180.0	0	0	0	0	0	0	0	86.64
7	439.1	468.0	352.9	228.5	561.20	679.8	359.5	181.3	0	5.38	0	3.72	0	4.28	350.9	118.10
8	659.7	566.3	411.4	516.3	699.47	542.1	516.	351.6	6.38	1.21	18.20	0.89	7.96	40.30	0.30	144.51
9	712.7	653.9	433.3	690.6	335.18	631.9	432.0	502.9	67.39	3.04	73.95	93.00	49.78	67.99	0.04	152.21
10	646.8	749.5	456.2	596.0	748.50	565.3	4.85	476.8	0.01	101.6	75.08	69.67	44.47	31.90	532.9	160.86
11	716.3	636.6	495.0	469.3	663.85	683.6	516.9	455.4	93.54	0.30	56.64	46.68	25.47	0.13	370.9	164.86
12	619.3	571.4	359.8	638.8	646.03	571.5	333.9	359.8	30.02	91.25	0.29	126.5	140.0	87.56	729.5	189.03
13	698.6	646.2	202.8	684.6	634.66	605.5	354.0	351.6	121.0	75.45	118.8	0.63	0.06	127.5	832.1	201.26
14	646.8	568.3	173.9	594.4	559.59	563.9	152.1	178.1	100.9	155.5	154.2	153.5	149.2	153.0	910.8	218.45
15	673.9	664.7	73.98	686.7	664.97	664.3	69.51	73.68	138.3	145.6	124.5	145.0	124.0	142.6	970.9	225.75
16	724.9	668.9	19.84	724.8	668.79	668.9	19.99	19.67	129.9	145.0	129.7	144.8	129.9	144.9	999.5	229.54
17	688.4	668.7	155.5	662.1	672.05	670.2	130.3	156.2	72.03	74.38	2.28	129.4	90.68	108.0	917.4	212.53
18	715.0	540.8	255.6	692.6	460.89	672.8	247.8	301.6	72.58	3.50	78.98	65.19	53.74	0.05	889.5	176.66
19	721.6	579.9	628.9	463.8	474.76	344.9	312.9	629.7	0.52	16.77	2.35	34.16	10.14	56.17	219.0	147.18
20	438.6	606.8	661.9	540.7	662.02	680.9	309.0	663.9	0	0	0	0	0	0	710.8	151.45
21	566.1	670.7	722.1	684.0	548.94	677.5	592.5	565.9	0	0	0	0	0	0	450.3	154.03
22	686.9	630.4	758.0	686.8	387.71	642.1	555.6	518.9	0	0	0	0	0	0	669.3	154.77
23	491.8	644.6	690.4	694.4	741.58	539.8	428.9	695.0	0	0	0	0	0	0	443.2	134.74
24	699.9	550.0	466.7	699.7	688.00	609.7	83.35	624.4	0	0	0	0	0	0	188.9	121.27

Table 15. Generation power (KW) by MVO for microgrid 2.

No.	WP1	WP2	WP3	WP4	WP5	WP6	WP7	WP8	PV1	PV2	PV3	PV4	PV5	PV6	CHP	Cost ($)
1	650	695	399.20	0	690.46	612.90	434.41	0	0	0	0	0	0	0	0	87.73
2	641.79	692.46	437.38	580.07	594.27	0	0	0	0	0	0	0	0	0	0	74.23
3	670.85	679.14	0	508.41	688.40	0	214.18	0	0	0	0	0	0	0	0	70.97
4	719.64	463.36	0	0	661.67	705	0	8.30	0	0	0	0	0	0	0	67.38
5	432.42	580	0	367.75	572.77	580	0	8.05	0	0	0	0	0	0	0	71.52
6	0	681.56	0	250.12	679.84	681.85	180	142.60	0	0	0	0	0	0	0	72.60
7	636.42	374.36	355.04	646.64	604.91	683.16	334.46	0	0	0	0	0	0	0	0	95.25
8	647.63	555.82	293.82	426.29	599.70	699.65	545.70	495.92	0	16.72	0	57.71	0	0	0	121.46
9	719.18	697.76	437.77	628.16	519.27	709.55	435.10	502.59	0	0	47.62	0	38.90	12.05	0	133.41
10	684.71	676.63	465.77	709.11	750	689.11	424.89	434.5	95.61	73.72	0	0	0	95.90	0	141.73
11	659.38	691.28	500.48	716.90	691.57	689.12	501.29	516.40	114.51	0	0	149.98	0	0	0	139.67

Table 15. *Cont.*

No.	WP1	WP2	WP3	WP4	WP5	WP6	WP7	WP8	PV1	PV2	PV3	PV4	PV5	PV6	CHP	Cost ($)
12	719.52	669.09	346.24	719.87	669.83	670	357.35	359.67	118.58	154.23	116.45	145.22	131.49	120.12	8.2720	170.56
13	699.79	649.12	369.96	700	649.97	649.94	370	369.91	149.97	156.63	89.151	134.45	117.17	78.706	269.16	187.59
14	684.65	568.44	204.81	685	568.40	568.73	197.33	202.31	140.90	151.20	153.56	119.93	153.80	96.17	719.70	210.81
15	710	665	74	710	665	664.61	0	0	129.39	155	140	155	140	155	1000	221.82
16	725	668.88	19.89	725	669	669	20	20	130	145	130	144.87	129.09	145	999.26	229.59
17	704.71	674	159.74	677.09	674	674	158.80	139.1	0	130	109.43	92.21	109.31	111.63	783.87	206.74
18	708.98	657.65	394.70	725	674	674	395	394.97	0	63.15	80	106.75	78.00	98.763	0	151.62
19	675.79	459.15	577.06	310.79	642.88	626.26	574.08	588.61	22.14	19.24	0	0	0	0	0	123.92
20	568.78	682	667	659.2	682	682	667	667	0	0	0	0	0	0	0	127.86
21	684.89	680	729.63	685	680	679.47	670	670	0	0	0	0	0	0	0	131.46
22	593.42	670	758	676.25	652.52	669.82	757.96	758	0	0	0	0	0	0	0	132.37
23	584.93	644.90	694.65	692.44	745	643.14	688.58	676.33	0	0	0	0	0	0	0	129.53
24	700	684.33	649.19	0	688	687.75	720	481.71	0	0	0	0	0	0	0	111.65

Table 16. Generation power (KW) by CMVO for microgrid 2.

No.	WP1	WP2	WP3	WP4	WP5	WP6	WP7	WP8	PV1	PV2	PV3	PV4	PV5	PV6	CHP	Cost ($)
1	523.03	695	437.00	0	695	695	0	437.00	0	0	0	0	0	0	0	87.95
2	0	715	359.59	0	715	715	0	441.40	0	0	0	0	0	0	0	73.76
3	750.00	705	0	600.84	705	0	0	0	0	0	0	0	0	0	0	68.21
4	484.92	632.85	0	735	0	705	0	0	0	0	0	0	0	0	0	65.25
5	202.97	513.15	0	665	580	580	0	0	0	0	0	0	0	0	0	68.33
6	0	682	5.02	565	682	682	0	0	0	0	0	0	0	0	0	68.50
7	478.51	685	274.38	665	685	685	162.13	0	0	0	0	0	0	0	0	95.36
8	0	700	547.00	660	700	700	423.50	547	20	41.46	0	0	0	0	0	117.06
9	730	679.27	515.00	730	604.12	381.26	515	515	75	0	0	3.48	0	0	0	130.09
10	715	750	520.00	349.94	750	750	520	520	0	125	100	0	0	0	0	137.07
11	717	692	517.00	717	692	692	517	460.17	0	0	120	0	106.86	0	0	139.16
12	720.	670	360.00	720	670	648.71	357.75	360	140	155	140	155	54.28	155	0	167.55
13	700	650	370.00	700	650	601.12	370	370	150	160	150	160	139.91	96.83	185.92	186.46
14	685	570	205.00	685	570	570	205	205	155	158	155	158	155	63.70	675.50	210.57
15	710	665	0.00	710	665	665	74	0	140	143.99	140	155	140	155	1000	221.86
16	725	669	20.00	725	669	669	20	20	130	145	130	145	128	145	1000	229.61
17	705	674	160	705	674	674	147.18	160	110	0	109.72	0	0	79.11	1000	207.50
18	725	674	395	725	674	674	395	395	0	110	80	94.41	0	110	0	150.87
19	325.37	678	635	733.47	678	678	635	117.33	15.92	0	0	0	0	0	0	119.82
20	660	682	575	660	682	682	667	667	0	0	0	0	0	0	0	127.92
21	684	680	730	685	680	680	670	670	0	0	0	0	0	0	0	131.49
22	687	670	758	687	670	548	758	758	0	0	0	0	0	0	0	132.41
23	695	645	555	695	745	645	695	695	0	0	0	0	0	0	0	129.63
24	0	688	720	470.47	646.99	645.52	720	720	0	0	0	0	0	0	0	111.70

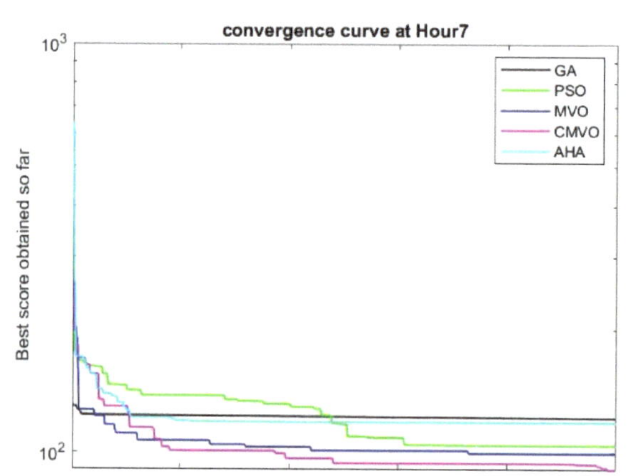

Figure 9. Convergence graphs for microgrid 2 for hour 7.

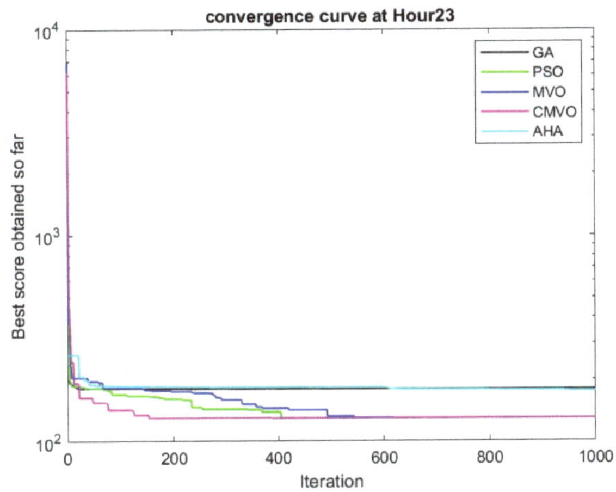

Figure 10. Convergence graphs for microgrid 2 for hour 23.

It is observed that CMVO produces the minimum generation cost for the available power compared with the other algorithms, and the time taken by the proposed algorithm is improved. The average time taken by the investigated algorithms, i.e., CMVO, MVO, PSO, AHA, and GA, is 0.41 s, 0.43 s, 2.66 s, 0.29 s, and 1.6 s, respectively. The mean and standard deviation for each algorithm are provided in Table 17. It is observed that the proposed algorithm is more stable than other investigated algorithms and has the minimum mean and standard deviation.

Table 17. Total generation cost and time of the investigated algorithm for microgrid 2.

Algorithm	Total Cost ($)	Total Time (s)	Mean ($)	Standard Deviation
Proposed CMVO	3178.30	0.27	3353.21	26.78
MVO	3211.43	0.29	3413.56	31.58
PSO	3253.05	2.66	4253.48	885.34
AHA	3598.03	0.29	3846.11	40.57
GA	4711.06	1.6	5348.16	91.60

6. Conclusions

This paper proposed a cost-effective multi-verse optimizer algorithm for optimal power sharing among different generation units. The proposed optimization algorithm, CMVO, enhances the local and global search capacity of MVO and results in cost optimization. It is observed that the proposed algorithm minimizes the generation cost and provides the most cost-effective solution for the power scheduling problem, ensuring its stability and efficacy. The optimization results provide an optimal energy management strategy in renewable-energy-based microgrids. This paper compares the different optimization algorithms for two different scale microgrids, i.e., the IEEE 37 node and 141 node systems, where the proposed algorithm outperforms other implemented algorithms.

Furthermore, it is seen that, in both cases, CMVO outperforms and provides the most effective solution and can be highly desirable for power-sharing problems. It optimally shares the generated power among different DERs, supplies energy at the minimum cost compared with another algorithms, and satisfies the equality constraints. In the future, the proposed algorithm can be investigated for different microgrid systems for various scenarios. Moreover, this algorithm can be tested under uncertain generation and load data in the islanded or grid-connected mode with energy storage devices. With the availability of large datasets for demand and load, this power scheduling problem can also be solved using machine learning or other available methods.

Author Contributions: U.L. carried out the research and participated in drafting the manuscript. N.B. and I.E. supervised, analyzed the results, and reviewed the manuscript. A.J., B.-H.T. and J.M.G. gave critical revision of the manuscript. All authors have read and agreed to the published version of the manuscript.

Funding: The authors are grateful to the sponsors who provided UTP GA and Research, and ICRF Grant (015ME0-255) for this project.

Institutional Review Board Statement: Not applicable.

Informed Consent Statement: Not applicable.

Data Availability Statement: Not applicable.

Acknowledgments: The authors would like to thank Universiti Teknologi PETRONAS (UTP) Malaysia; University Institute of Engineering and Technology, Kurukshetra University, India; and Institute of Engineering and Technology, Thu Dau Mot University, Thu Dau Mot VN-57, Vietnam, Centre of Research on Microgrids, Department of Energy Technology, Aalborg University, 9220 Aalborg, Denmark for their support.

Conflicts of Interest: The authors declare no conflict of interest.

References

1. Barik, A.K.; Jaiswal, S.; Das, D.C. Recent trends and development in hybrid microgrid: A review on energy resource planning and control. *Int. J. Sustain. Energy* **2021**, *41*, 308–322. [CrossRef]
2. Al-Ismail, F.S. DC Microgrid Planning, Operation, and Control: A Comprehensive Review. *IEEE Access* **2021**, *9*, 36154–36172. [CrossRef]
3. Bihari, S.P.; Sadhu, P.K.; Sarita, K.; Khan, B.; Saket, R.K.; Kothari, D.P. A comprehensive Review of Microgrid Control Mechanism and Impact Assessment for Hybrid Renewable Energy Integration. *IEEE Access* **2021**, *9*, 88942–88958. [CrossRef]
4. Kajela, D.; Manshahia, M.S. Optimization of Renewable Energy Systems: A Review. *Int. J. Sci. Res. Sci. Technol.* **2017**, *3*, 765–795.
5. Xing, X.; Xie, L.; Meng, H. Cooperative energy management optimization based on distributed MPC in grid-connected microgrids community. *Electr. Power Energy Syst.* **2018**, *107*, 186–199. [CrossRef]
6. Vera, Y.E.G.; Lopez, R.D.; Agustin, J.L.B. Energy management in microgrids with renewable energy sources: A literature review. *Appl. Sci.* **2019**, *9*, 3854. [CrossRef]
7. Reddy, S.S. Optimization of Renewable Energy Resources in Hybrid Energy Systems. *J. Green Eng.* **2017**, *7*, 43–60. [CrossRef]
8. Sadeghi, D.; Naghshbandy, A.H.; Bahramara, S. Optimal sizing of hybrid renewable energy systems in the presence of electric vehicles using multi-objective particle swarm optimization. *Energy* **2020**, *209*, 118471. [CrossRef]
9. Bilal, M.; Shahzad, M.; Arif, M.; Ullah, B.; Hisham, S.B.; Ali, S.S.A. Annual cost and loss minimization in a radial distribution network by capacitor allocation using pso. *Appl. Sci.* **2021**, *11*, 11840. [CrossRef]
10. Hannan, M.A.; Abdolrasol, M.G.M.; Faisal, M.; Ker, P.J.; Begum, R.A.; Hussain, A. Binary Particle Swarm Optimization for Scheduling MG Integrated Virtual Power Plant towards Energy Saving. *IEEE Access* **2021**, *7*, 107937–107951. [CrossRef]
11. Raghav, L.P.; Kumar, R.S.; Raju, D.K.; Singh, A.R. Optimal Energy Management of Microgrids Using Quantum Teaching Learning Based Algorithm. *IEEE Trans. Smart Grid* **2021**, *12*, 4834–4842. [CrossRef]
12. Kumar, K.P.; Saravanan, B.; Swarup, K.S. Optimization of renewable energy sources in a microgrid using artificial fish swarm algorithm. *Energy Procedia* **2016**, *90*, 107–113. [CrossRef]
13. Maulik, A.; Das, D. Optimal operation of microgrid using four different optimization techniques. *Sustain. Energy Technol. Assess.* **2017**, *21*, 100–120. [CrossRef]
14. Sarfi, V.; Member, S.; Livani, H. An economic–reliability security constrained optimal dispatch for microgrids. *IEEE Trans. Power Syst.* **2018**, *33*, 6777–6786. [CrossRef]
15. Aoun, A.; Ibrahim, H.; Ghandour, M.; Ilinca, A. Supply Side Management vs. Demand Side Management of a Residential Microgrid Equipped with an Electric Vehicle in a Dual Tariff Scheme. *Energies* **2019**, *12*, 4351. [CrossRef]
16. Zunnurain, I.; Maruf, N.I.; Rahman, M.; Shafiullah, G. Implementation of Advanced Demand Side Management for Microgrid Incorporating Demand Response and Home Energy Management System. *Infrastructures* **2018**, *3*, 50. [CrossRef]
17. Alzaqebah, M.; Jawarneh, S.; Mohammad, R.M.A.; Alsmadi, M.K.; Almarashdeh, I. Improved Multi-Verse Optimizer Feature Selection Technique with Application to Phishing, Spam, and Denial of Service Attacks. *Int. J. Commun. Netw. Inf. Secur.* **2021**, *13*, 76–81. [CrossRef]
18. Abasi, K.; Khader, A.T.; Al-Betar, M.A. An Improved Multi-Verse Optimizer for Text Documents Clustering. *Kufa J. Eng.* **2022**, *13*, 28–42. [CrossRef]

19. Fu, Y.; Zhou, M.; Guo, X.; Qi, L.; Sedraoui, K. Multiverse Optimization Algorithm for Stochastic Biobjective Disassembly Sequence Planning Subject to Operation Failures. *IEEE Trans. Syst. Man Cybern. Syst.* **2021**, *52*, 1041–1051. [CrossRef]
20. Emara, D.; Ezzat, M.; Abdelaziz, A.Y.; Mahmoud, K.; Lehtonen, M. Novel Control Strategy for Enhancing Microgrid Operation Connected to Photovoltaic Generation and Energy Storage Systems. *Electronics* **2021**, *10*, 1261. [CrossRef]
21. Taheri, S.I.; Salles, M.B.C.; Costa, E.C.M. Optimal Cost Management of Distributed Generation Units and Microgrids for Virtual Power Plant Scheduling. *IEEE Access* **2020**, *8*, 208449–208461. [CrossRef]
22. Andrew, T.G.; Elamvazuthi, I.; Vasant, P. Solving engineering optimization problems with the Karush-Kuhn-Tucker hopfield neural networks. *Int. Rev. Mech. Eng.* **2011**, *5*, 1333–1339.
23. Mariam, L.; Basu, M.; Conlon, M.F. Microgrid: Architecture, Policy and Future Trends. *Renew. Sustain. Energy Rev.* **2016**, *64*, 477–489. [CrossRef]
24. Vasant, P.; Ganesan, T.; Elamvazuthi, I.; Vo, D.N. Improved Tabu Search Recursive Fuzzy Method for Crude Oil Industry. *Int. J. Model. Simul. Sci. Comput.* **2012**, *3*, 1150002. [CrossRef]
25. Sechilariu, M.; Locment, F.; Wang, B. Photovoltaic Electricity for Sustainable Building Efficiency and Energy Cost Reduction for Isolated DC Microgrid. *Energies* **2015**, *8*, 7945–7967. [CrossRef]
26. Elamvazuthi, I.; Ganesan, T.; Vasant, P. A Comparative Study of HNN And Hybrid HNN-PSO Techniques in the Optimization of Distributed Generation (DG) Power Systems. In Proceedings of the International Conference on Advance Computer Science and Information System, Vancouver, BC, Canada, 18 December 2011.
27. Bayindir, R.; Hossain, E.; Kabalci, E.; Perez, R. A Comprehensive Study on Microgrid Technology. *Int. J. Renew. Energy Res.* **2014**, *4*, 1094–1107.
28. Fayek, H.M.; Elamvazuthi, I.; Nallagownden, P.; Venkatesh, B. A controller based on Optimal Type-2 Fuzzy Logic: Systematic design, optimization, and real-time implementation. *ISA Trans.* **2014**, *53*, 1583–1591. [CrossRef]
29. Mansouri, S.A.; Ahmarinejad, A.; Nematbakhsh, E.; Javadi, M.S.; Jordehi, A.R.; Catalao, J.P. Energy management in microgrids including smart homes: A multi-objective approach. *Sustain. Cities Soc.* **2021**, *69*, 102852. [CrossRef]
30. Shah, P.; Mehta, B. Microgrid Optimal Scheduling with Renewable Energy Sources Considering Islanding Constraints. *Iran. J. Sci. Technol.* **2019**, *44*, 805–819. [CrossRef]
31. Rehman, R.M.A.A.; Alsharabi, N.; Rabbani, S.; Shakir, M.; Malik, N.; Rehman, A.; Khan, A.A. Optimal locating and sizing of DG in radial distribution system using modified Frog leaping algorithm. *IET J. Res.* **2022**, *69*, 1–18. [CrossRef]
32. Shahzad, M.; Akram, W.; Arif, M.; Khan, U.; Ullah, B. Optimal siting and sizing of distributed generators by strawberry plant propagation algorithm. *Energies* **2021**, *14*, 1744. [CrossRef]
33. Shahzad, M.; Shafiullah, Q.; Akram, W.; Arif, M.; Ullah, B. Reactive power support in radial distribution network using mine blast algorithm. *Elektron. IR Elektrotechnika* **2021**, *27*, 33–40. [CrossRef]
34. Zandrazavi, S.F.; Guzman, C.P.; Pozos, A.T.; Quiros-Tortos, J.; Franco, J.F. Stochastic multi-objective optimal energy management of grid-connected unbalanced microgrids with renewable energy generation and plug-in electric vehicles. *Energy* **2021**, *241*, 122884. [CrossRef]
35. Hossain, M.A.; Pota, H.R.; Squartini, S.; Abdou, A.F. Modified PSO Algorithm for Real-time Energy Management in Grid-connected Microgrids. *Renew. Energy* **2019**, *136*, 746–757. [CrossRef]
36. Abdolrasol, M.G.M.; Hannan, M.A.; Hussain, S.M.S.; Ustun, T.S.; Sarker, M.R.; Ker, P.J. Energy Management Scheduling for Microgrids in the Virtual Power Plant System Using Artificial Neural Networks. *Energies* **2021**, *14*, 6507. [CrossRef]
37. Tarife, R.; Nakanishi, Y.; Chen, Y.; Zhou, Y.; Estoperez, N.; Tahud, A. Optimization of Hybrid Renewable Energy Microgrid for Rural Agricultural Area in Southern Philippines. *Energies* **2022**, *15*, 2251. [CrossRef]
38. Kamal, M.M.; Ashraf, I.; Fernandez, E. Planning and optimization of microgrid for rural electrification with the integration of renewable energy resources. *J. Energy Storage* **2022**, *52*, 104782. [CrossRef]
39. Nagarajan, K.; Rajagopalan, A.; Angalaeswari, S.; Natrayan, L.; Mammo, W.D. Combined Economic Emission Dispatch of Microgrid with the Incorporation of Renewable Energy Sources Using Improved Mayfly Optimization Algorithm. *Comput. Intell. Neurosci.* **2022**, *2022*, 6461690. [CrossRef]
40. Roslan, M.; Hannan, M.; Ker, P.J.; Begum, R.; Mahlia, T.I.; Dong, Z. Scheduling controller for microgrids energy management system using optimization algorithm in achieving cost saving and emission reduction. *Appl. Energy* **2021**, *292*, 116883. [CrossRef]
41. Hossain, M.A.; Pota, H.R.; Squartini, S.; Zaman, F.; Guerrero, J.M. Energy scheduling of community microgrid with battery cost using particle swarm optimization. *Appl. Energy* **2019**, *254*, 113723. [CrossRef]
42. Domenech, B.; Ferrer-Martí, L.; García, F.; Hidalgo, G.; Pastor, R.; Ponsich, A. Optimizing PV Microgrid Isolated Electrification Projects—A Case Study in Ecuador. *Mathematics* **2022**, *10*, 1226. [CrossRef]
43. Gholami, K.; Dehnavi, E. A modified particle swarm optimization algorithm for scheduling renewable generation in a micro-grid under load uncertainty. *Appl. Soft Comput. J.* **2019**, *78*, 496–514. [CrossRef]
44. Abaeifar, A.; Barati, H.; Tavakoli, A.R. Inertia-weight local-search-based TLBO algorithm for energy management. *Int. J. Electr. Power Energy Syst.* **2021**, *137*, 107877. [CrossRef]
45. Askarzadeh, A. A Memory-Based Genetic Algorithm for Optimization of Power Generation in a Microgrid. *IEEE Trans. Sustain. Energy* **2018**, *9*, 1081–1089. [CrossRef]
46. Ramli, M.A.; Bouchekara, H.R.E.H.; Alghamdi, A.S. Efficient Energy Management in a Microgrid with Intermittent Renewable Energy and Storage Sources. *Sustainability* **2019**, *11*, 3839. [CrossRef]

47. Lan, Y.; Guan, X.; Wu, J. Rollout strategies for real-time multi-energy scheduling in microgrid with storage system. *IET Gener. Transm. Distrib.* **2015**, *10*, 688–696. [CrossRef]
48. Silvente, J.; Papageorgiou, L.G. A MILP formulation for the optimal management of microgrids with task interruptions. *Appl. Energy* **2017**, *206*, 1131–1146. [CrossRef]
49. Alamir, N.; Kamel, S.; Megahed, T.; Hori, M.; Abdelkader, S. Developing an artificial hummingbird algorithm for probabilistic energy management of microgrids considering demand response. *Front. Energy Res.* **2022**, *10*, 876. [CrossRef]
50. Jui, J.J.; Ahmad, M.A.; Rashid, M.I.M. Modified Multi-Verse Optimizer for Solving Numerical Optimization Problems. In Proceedings of the 2020 IEEE International Conference on Automatic Control and Intelligent Systems, Shah Alam, Malaysia, 20 June 2020; pp. 81–86.
51. Lokman, H.; Musirin, I.; Suliman, S.I.; Suyono, H.; Nur, R.; Mustafa, S.A.S.; Zellagui, M. Multi-verse optimization based evolutionary programming technique for power scheduling in loss minimization scheme. *Int. J. Artif. Intell.* **2019**, *8*, 292–298.
52. Crisostomi, E.; Liu, M.; Raugi, M.; Shorten, R. Plug-and-Play Distributed Algorithms for Optimized Power Generation in a Microgrid. *IEEE Trans. Smart Grid* **2014**, *5*, 2145–2154. [CrossRef]
53. Mirjalili, S.; Mirjalili, S.M.; Hatamlou, A. Multi-Verse Optimizer: A nature-inspired algorithm for global optimization. *Neural Comput. Appl.* **2015**, *27*, 495–513. [CrossRef]
54. Khodr, H.; Olsina, F.; Jesus, P.D.O.-D.; Yusta, J. Maximum savings approach for location and sizing capacitors in distribution systems. *Electr. Power Syst. Res.* **2008**, *78*, 1192–1203. [CrossRef]
55. Zhao, W.; Wang, L.; Mirjalili, S. Artificial hummingbird algorithm: Anew bio-inspired optimizer with its engineering applications. *Comput. Methods Appl. Mech. Eng.* **2022**, *388*, 114194. [CrossRef]

sustainability

MDPI

Article

Machine Learning Classification and Prediction of Wind Estimation Using Artificial Intelligence Techniques and Normal PDF

Hiba H. Darwish and Ayman Al-Quraan *

Electrical Power Engineering Department, Hijjawi Faculty for Engineering Technology, Yarmouk University, Irbid 21163, Jordan
* Correspondence: aymanqran@yu.edu.jo

Abstract: Estimating wind energy at a specific wind site depends on how well the real wind data in that area can be represented using an appropriate distribution function. In fact, wind sites differ in the extent to which their wind data can be represented from one region to another, despite the widespread use of the Weibull function in representing the wind speed in various wind locations in the world. In this study, a new probability distribution model (normal PDF) was tested to implement wind speed at several wind locations in Jordan. The results show high compatibility between this model and the wind resources in Jordan. Therefore, this model was used to estimate the values of the wind energy and the extracted energy of wind turbines compared to those obtained by the Weibull PDF. Several artificial intelligence techniques were used (GA, BFOA, SA, and a neuro-fuzzy method) to estimate and predict the parameters of both the normal and Weibull PDFs that were reflected in conjunction with the actual observed data of wind probabilities. Afterward, the goodness of fit was decided with the aid of two performance indicators (RMSE and MAE). Surprisingly, in this study, the normal probability distribution function (PDF) outstripped the Weibull PDF, and interestingly, BFOA and SA were the most accurate methods. In the last stage, machine learning was used to classify and predict the error level between the actual probability and the estimated probability based on the trained and tested data of the PDF parameters. The proposed novel methodology aims to predict the most accurate parameters, as the subsequent energy calculation phases of wind depend on the proper selection of these parameters. Hence, 24 classifier algorithms were used in this study. The medium tree classifier shows the best performance from the accuracy and training time points of view, while the ensemble-boosted trees classifier shows poor performance regarding providing correct predictions.

Keywords: wind estimation; normal PDF; Weibull PDF; optimization algorithms; machine learning; prediction; classification; accuracy

check for updates

Citation: Darwish, H.H.; Al-Quraan, A. Machine Learning Classification and Prediction of Wind Estimation Using Artificial Intelligence Techniques and Normal PDF. *Sustainability* **2023**, *15*, 3270. https://doi.org/10.3390/su15043270

Academic Editor: Byungik Chang

Received: 11 January 2023
Revised: 6 February 2023
Accepted: 8 February 2023
Published: 10 February 2023

1. Introduction

Long ago, it was understood that the continuous usage of conventional energy sources (fossil fuel) jeopardizes and threatens the stability of life. As a result, humanity has tried to find other inexhaustible energy resources to tackle the issues of the undesired impacts of the dominant energy sources (fossil fuel). Renewable energy sources were the best alternative, which became grist, an integral part, and the interesting core of the energy sector due to their immense valuable features [1]. Furthermore, the lack of conventional energy resources boosts the harnessing of clean energy sources [2]. Inasmuch, the development of lifestyle is associated with energy demand. As such, the larger the energy demand in a certain area, the most sophisticated the area [3–7].

Wide choices of renewable energy are available, such as solar, the internal heat of the earth, wind, tidal, and biomass energy [8]. Wind energy has played a prominent, astounding, and marvelous role in contributing to the depreciation of carbon dioxide [9], which has encouraged some countries to invest in wind energy [10]. The key to wind energy is its kinetic energy; the energy that can be harvested by wind turbines fundamentally depends on the average wind speed. The most effective areas to install a wind farm are those located beside coasts, on the edge of water bodies, and in open terrain [11]. Figure 1 shows the worldwide distribution of wind energy [12].

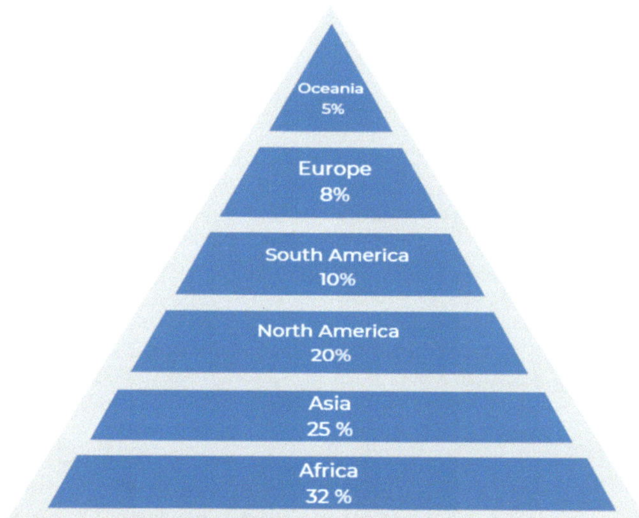

Figure 1. The worldwide distribution of wind energy.

Wind energy is defined as an inherently unfixed energy source, which varies rapidly over time [13]. It is dramatically growing and ubiquitous since this type of renewable energy has several strong points which make it outstrip fossil fuel; for example, it can meet the massive demand for energy and minimize the pollution resulting from fossil fuel usage up to a certain limit. Consequently, wind energy is deemed a green energy technology [14]. In addition, wind energy projects contribute to enhancing the situation of the environment, economy, and society [15].

Wind turbines can be installed on ranches or farms, which improves the economic situation, as mentioned before, chiefly in rural regions where the best sites for wind are found. These turbines do not generate any atmospheric emissions that are responsible for greenhouse gases and acid rain, which makes wind energy eco-friendly, as mentioned before [16,17].

Among the types of renewable energy, the consumption of wind energy is the largest category in most countries [18]. For illustration, it constitutes about more than 20% of the total renewable energy, and this percentage is increasing continuously [19,20]. The global capacity of wind was about 336.327 GW in June 2014. However, 17.613 GW of this was installed in the first half of the same year [21]. In particular, Jordan is considered a country where the wind is available in abundance. The attention to wind energy began in Jordan in 1979, and these days, Jordan has decided to provide 20% of the required energy from wind and solar energy [22]. Several wind projects exist, such as the Tafilah wind project, which delivers almost 132 MW of electricity to the national grid; meanwhile, some other wind projects deliver about 25 MW of electricity [23–28]. Moreover, Jordan has been deemed as one of the Arab countries which contributes to the spread of the culture of renewable energy exploitation; notably, by the end of 2021, the installed renewable energy projects will contribute to electricity generation by a percentage of 20.1% generally and 15.3% by the wind, as clarified in Figure 2. Accordingly, the awareness of wind energy has increased abruptly and attracted attention, which is the reason for installing two new wind farms in

2021 with a capacity of 51.75MW each [29]. Figure 3 represents the MW production from wind during the period from 2010 to 2021 in Jordan [30].

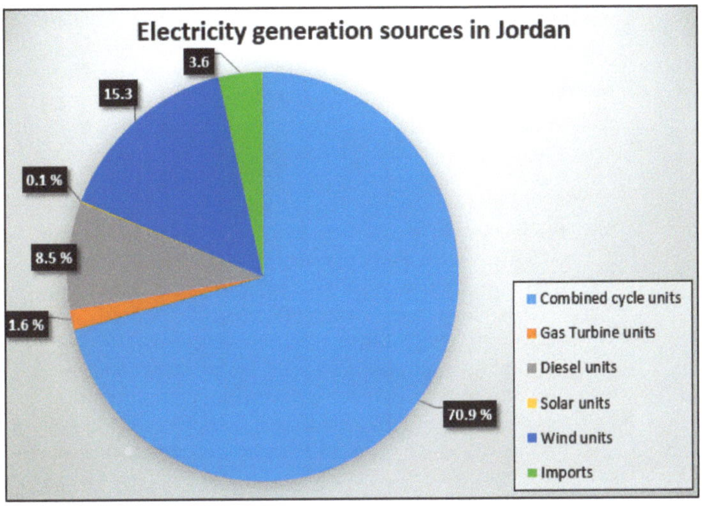

Figure 2. Electricity generation sources in Jordan.

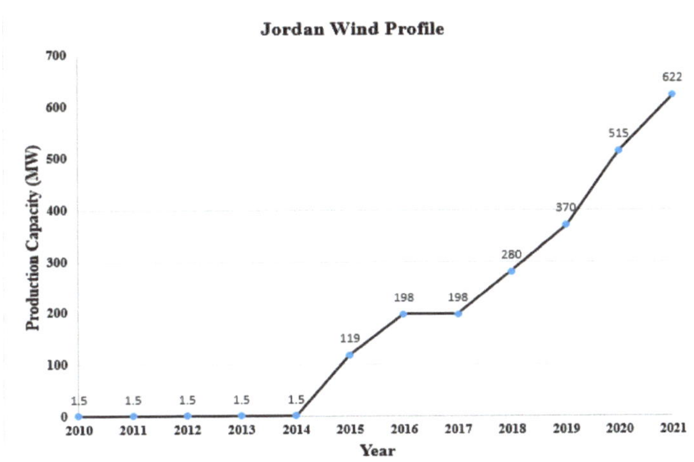

Figure 3. Wind Profile of Jordan.

The output of wind power depends mainly on the speed of the wind. Hence, it is irrefutable that evaluating the distribution of wind speed is considered the starting point for wind energy potential assessment purposes [31]. Usually, the distribution of wind speed is estimated and described by several probability distribution functions (PDFs) [32], especially by the Weibull PDF, since the estimated outcomes are close to the actual observed wind speed. In addition, the Rayleigh PDF, which is a special case of the previously mentioned PDF (Weibull), is used in some studies, and sometimes, it provides a better fitting [33]. Therefore, it can be understood that no particular PDF can fit the distribution of wind speed in all sites meticulously, bearing in mind that the estimation of wind speed is not simple because of the stochastic nature of the wind source and frictional and roughness effects [34–36].

Several recent studies stated that estimating the wind turbines' P-V curve is required in the preliminary assessment of the wind turbines' energy yield [37]. However, different methods have been proposed to make an initial assessment and estimation of wind speed with an uneven degree of resolution and accuracy [38]. Kevin et al. proposed a technique in [39] that endorsed a study conducted by Al-Mhairat et al. in [1], showed that gamma PDF outperforms the other PDFs in wind assessment. This study was performed in Kenya and aimed to specify the optimal parameters of the selected distribution functions, which were Weibull, log-normal, and gamma. The process of this study was conducted by estimating the PDFs' parameters by using a numerical approach, that is, the maximum likelihood method (MLM).

Consequently, Rejhana [40] decided to use two methods to estimate the parameters of the Weibull PDF and the wind power density, which are the MLM and the energy pattern factor methods. The outcomes showed that the wind energy that is available in Sarajevo is not enough to meet the required energy for that region.

Similarly, Boro et al. in [41] used the MLM in order to compare the accuracy of various PDFs, including the inverse Gaussian, gamma, Rayleigh, hybrid Weibull, and Weibull PDFs. Some statistical tools were used as indicators, such as the coefficient of determination (R^2) and root-mean-square error (RMSE). The results stated that there is not only one PDF that fits the whole region worldwide, such that in some sites, such as Ouahigouya, Dédougou, and Ouaga, the Weibull PDF was the most suitable one, while in other sites, such as Gaoua, Dori, and Boromo, it was found that the inverse Gaussian PDF was the most suitable one.

Saeed et al. in [42] aimed to improve the performance of the Weibull PDF by using artificial intelligence optimization techniques (AIOP) to obtain the highest possible precision from the Weibull PDF. This study was conducted in thirteen different sites in Pakistan and tried to provide an alternative method for the estimation of parameters for the Weibull PDF. Further, the convergence was enhanced in this study by three AIOTs. The results showed that the proposed method for estimating the parameters of the Weibull PDF outperforms the common Weibull PDF.

However, some studies used both Weibull and Rayleigh PDFs in wind speed estimation. The reason for being the two most common PDFs is their accuracy in predicting and describing wind speed. For instance, Bidaoui et al. in [43] evaluated the potential of wind energy by using stochastic models of Rayleigh and Weibull PDFs of five locations in Northern Morocco. Some indicators were utilized, such as the mean bias error (MBE), RMSE, Chi-square error (χ^2), and R^2. The outcomes of this study indicated that the accuracy of the Weibull PDF is higher than the Rayleigh PDF.

Abeysirigunawardena et al. in [44] claimed in their study that the maximum likelihood estimation (MLE) approach is the most commonly used in wind estimation. This method can mix various information with the parameters of the model. Moreover, the results showed that approximate standard errors for the estimated parameters may be shaped automatically.

Baloch et al. stated in [45] that a sensitivity analysis is usually conducted in order to test the effectiveness of varying the parameters. Several research studies were conducted to study the effect of distribution function parameter variations on the energy of both wind regimes and wind turbines [46–48]. However, the goal of proposing new approaches is to be able to select accurate parameters for each PDF such that the estimated probability becomes very close to the actual one. Accordingly, the subsequent applications based on these parameters will become more reliable. The previously mentioned recent studies are summarized in Table 1.

Table 1. Summary for the previous studies in the same field of research.

Study	Year	Proposed Region	Used PDF	Used Method	Objective Function	Data Resolution	Data Period
[39]	2020	Kenya	Weibull, log-normal, and gamma	Numerical approach (MLM)	NA	Hourly	From 2016 to 2018
[40]	2021	Sarajevo, Bosnia and Herzegovina	Weibull	MLM and energy pattern factor	NA	The rate of the recorded data was 48 per day in 30 min time intervals	1 January 2019 to 31 December 2019
[41]	2020	Dori, Ouahigouya, Ouagadougou, Fada N'goura, Gaoua, PO, Dédougou, Ouaga, Bobo Dioulasso, Bogandé, and Boromo	Inverse Gaussian, gamma, Rayleigh, hybrid Weibull, and Weibull	MLM	To identify the best PDF for the proposed sites	Every three hours	From January 2006 to December 2016
[42]	2021	Pakistan	Weibull	AI	Minimize cost Min $LEC = \frac{\sum_{n=0}^{N} C_{net}/(1+i)^n}{\sum_{n=0}^{N} E_{o,n}/(1+i)^n}$	NA	April 2015–January 2018
[43]	2019	Northern Morocco (Tangier, Tetuan, Al-Hoceima, Nador, Larach)	Weibull and Rayleigh	NA	Minimize the error between the estimated theoretical and the actual wind speed	Monthly	One year
[44]	2009	South coast of British Columbia, Canada	Generalized Pareto distribution	MLM	Test the validity	NA	NA
Our Study	2022	Jordan	Normal and Weibull	AI and machine learning (ML)	$Max\ E_{Total} = (1-\xi) \times (E_{ir} + E_{ro})$	Daily	From 1 October 2021 to 30 September 2022

The importance of the PDF is summarized by being able to make a description and prediction for the probability of a certain event. Each PDF carries its own parameters, and the right selection of these parameters will be reflected in a proper application. The normal PDF depicts one of the most common PDFs that is commonly used in estimating probability [49]. This PDF is represented by two main parameters, μ, which is the mean value that describes the central tendency, and σ, which is the standard deviation that describes how the probability values are dispersed around the central point. [50]. Moreover, the Weibull PDF is also considered an accurate PDF that is used in wind variation estimation. This PDF is represented by two main parameters, K, which is the shape factor, and C, which is the scaling factor [51].

The main contribution of this study compared to other studies in the same field of research can be summarized by the following:

- This study estimates the wind energy and extracted energy of a wind turbine using a new distribution function (normal) that has not regularly been used in the literature.
- The estimation method of this study is performed using several artificial intelligence methods with the aid of machine learning classifiers that have not previously been used in other studies.

However, wind speed can be measured by several apparatuses, such as the cup anemometer, which is represented in Figure 4a [52]; this technology responds promptly to wind movement, but this device cannot stop immediately once the wind stops [53]. Moreover, another technology that is used in wind speed measuring is the propeller anemometer, which is represented in Figure 4b [52], while sonic anemometers are used for both wind speed and direction measurement, which is represented in Figure 4c [54].

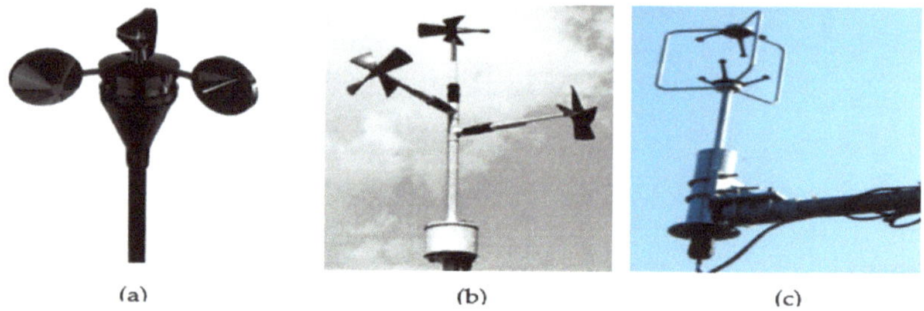

Figure 4. Technologies that are used in wind speed measurement. (**a**) Cup anemometer; (**b**) propeller anemometers; and (**c**) sonic anemometer.

2. Methodology

Innumerable choices for studying the extractable energy from wind are available nowadays; the chosen method strongly depends on statistical science. The general methodology adopted to accomplish this study is clarified and depicted in Figure 5.

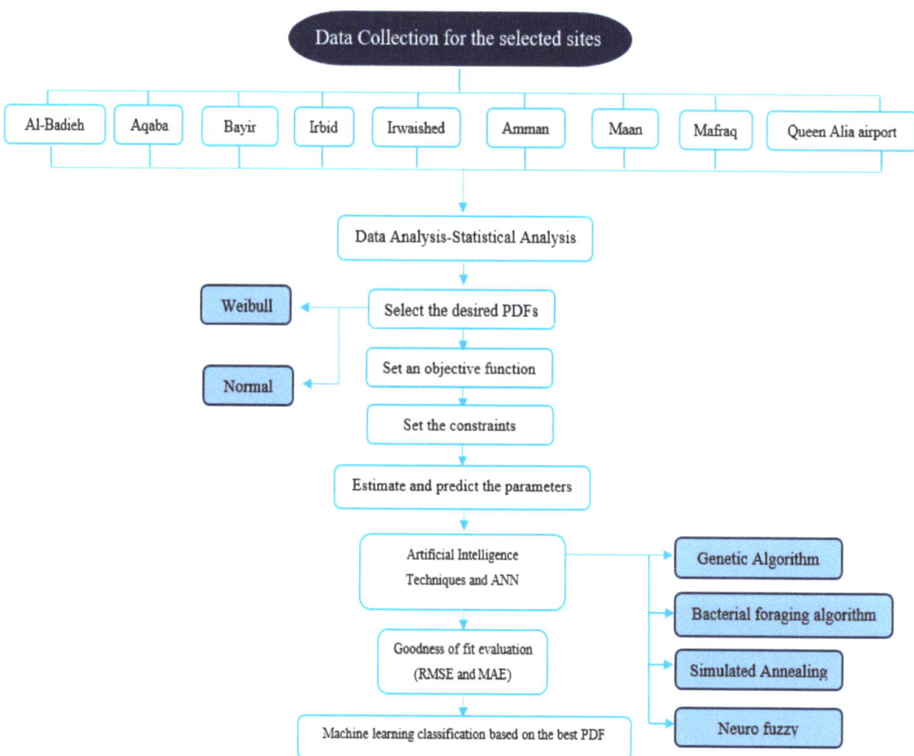

Figure 5. Flowchart steps of the intended study.

2.1. Data Collection for the Whole Site

An abundant number of marvelous research were conducted to achieve the goal of developing some reliable and sophisticated methods for wind forecasting. Each method has a tolerance percentage and does not match the actual wind measurements strictly.

In other words, the only essence disparity that differs vividly from one strategy to another is the uneven level of rigor. However, these methods are clarified in Figure 6 [55].

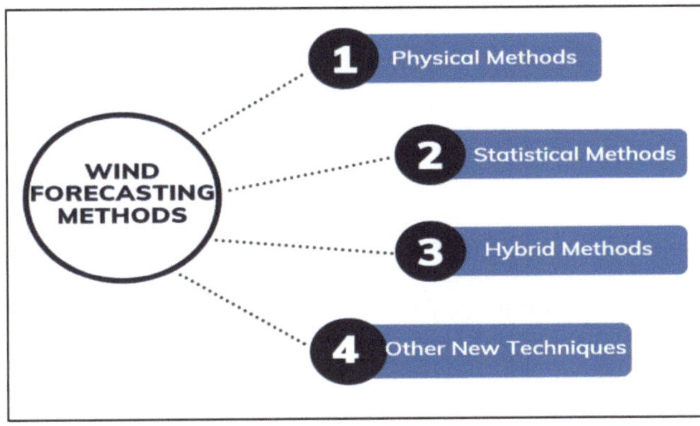

Figure 6. Forecasting methods for wind measurements.

The physical approaches use several types of physical information, including, to name a few, wind conditions, especially at the turbine hub's height, the power curve of the wind turbine, data from the meteorological departments, and the weather conditions, by and large. Regardless, these physical data are borne in mind to estimate wind readings.

In juxtaposition, statistical methods can forecast either wind speed power probability or wind speed/power value. Both are commonly obtained based on statistical analyses of time series, which depend on past observed wind data.

The third astounding method is the hybrid approach, which is a combination of several methods, such as physical methods in conjunction with statistical methods. Finally, the fourth method is other new techniques such as entropy-based training, ensemble predictions, wavelet transform, fuzzy logic, and spatial correlation.

In this paper, the daily wind speed data have been collected from RETScreen software for nine sites in Jordan, which are clarified in Table 2 with their corresponding details. The raw data period was for one complete year, starting from 1 October 2021 and going to 30 September 2022. All of these data were recorded at a 10 m height.

Table 2. The latitude, longitude, and elevation of the proposed sites [56].

Site	Latitude North (°)	Longitude East (°)	Elevation (m)
Al-Badieh	31.88	36.90	658
Amman	31.59	35.59	767
Aqaba	29.52	35.00	6
Bayir	30.76	36.68	831
Irbid	32.33	35.51	618
Irwaished	32.30	38.12	686
Ma'an	30.10	35.47	1069
Mafraq	32.22	36.15	686
Queen Alia Airport	31.43	35.59	722

2.2. Data Analysis and Statistical Analysis

Currently, modern turbines have a hub height of around 100 m, while the wind data were measured at 10 m, as mentioned afore. Therefore, the first step in wind data analysis is carried out by making a conversion for the obtained wind speed at 10 m into corresponding data at 100 m. The reflected wind speed at 100 m is obtained by applying Equation (1) [57].

$$V_2 = V_1\left(\frac{h_2}{h_1}\right)^{\alpha} \quad (1)$$

where V_2 represents the desired wind speed at the extrapolated height, V_1 represents the wind speed at the reference level, h_2 represents the desired height, and h_1 represents the reference height. Finally, the symbol α represents the wind shear exponent (WSE) that discerns and describes the terrain situation of the site. Table 3 clarifies several scenarios for the factor of the proposed sites. The following subsections clarify how the corrected wind speed was analyzed statistically.

Table 3. Details regarding the WSE [58].

Type of Terrain	The Corresponding Value of α
Large cities with tall buildings	0.4
Small towns with trees and shrubs	0.3
Wooded countryside, many trees	0.25
High crops, hedges, and shrubs	60.20
Tall grass on level ground	0.15
Smooth, hard ground, calm water	0.1

2.2.1. Selecting the Candidate PDFs

The current trend is to use the probability density function (PDF) approach to assess wind energy resources. The Weibull approach is the most ubiquitous PDF that has been used in most recent studies in wind assessment. This research sheds light on another PDF, the well-known normal PDF, to assess wind in several sites in Jordan. The mathematical representations for the proposed PDFs are clarified in the following points [59].

(A). Weibull

As stated before, the wind speed probability for a certain region is often expressed and represented by the Weibull PDF. Based on the following mathematical representation, it can be observed that this PDF has two main parameters, which are the shape and the scale factor.

$$f_W(v) = \frac{k}{c}\left(\frac{v}{c}\right)^{k-1} e^{-\left(\frac{v}{c}\right)^k} \tag{2}$$

(B). Normal

The second proposed PDF in this paper is the normal PDF, which is also known as the Gaussian distribution. This distribution function is called the Gaussian distribution by physicists and the bell curve by social scientists [60]. The following mathematical formula shows the PDF for the normal function:

$$f_N(v) = \frac{1}{\sqrt{2\pi\sigma^2}} \times e^{\frac{-(v-\mu)^2}{2\sigma^2}} \tag{3}$$

where σ and μ are the standard deviation and the mean wind speed.

Analyses of wind speed can be made by the PDFs. Thus, the more precise the selected parameters, the more accurate and more satisfactory the outcomes. In other words, the estimation of the parameters is the springboard, and it is considered a critical phase that plays an essential role in achieving the desired and accurate outcomes. The performance of the PDFs relies on several factors, such as the number of data, the evaluation criteria, and the period of data measuring.

2.2.2. Set an Objective Function

A power curve of a wind turbine is defined as a visual representation of the generated electrical output power for each corresponding wind speed. The lofty goal of capturing and exploiting the wind is to produce electrical power. Therefore, it is necessary to link these two parameters (electrical power and intermittent wind speed) in order to comprehend how they affect each other. The power curve of wind turbines that depicts the relationship between the electrical power output and the wind speed is known as the P-V curve, and each wind turbine model has its own P-V curve. This curve is essential for many purposes, chiefly for project conducting and planning, monitoring the turbines, and detecting the likelihood of the maloperation of turbines [61].

Producing electrical power from wind energy at the sites under investigation relies on immense factors, including the mean wind speed and wind turbine speed characteristics, which involve the cut-in speed, rated speed, and cut-out speed. The available energy in the wind varies with the variation in wind speed. Hence, understanding the P-V characteristics of wind is essential in wind assessment. Figure 7 clarifies a typical representation of the ideal P-V curve.

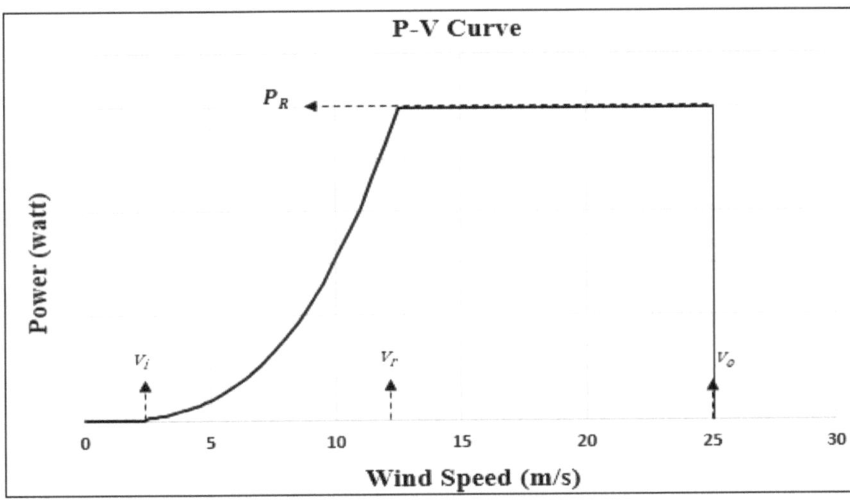

Figure 7. P-V curve of an ideal wind turbine.

It is apparent in Figure 7 that, in the region before the cut-in speed, the output power from the wind turbine is zero since the cut-in speed is quite low and cannot produce enough power to overbear the friction of the wind turbine. However, even if the friction of the wind turbine has been overcome and a rotation of the generator is observed, the corresponding generated electrical power may be slight and not sufficient to offset the required power by the generator field windings. Per contra, once the wind speed increases above the cut-in speed, the resultant output power rapidly increases until it reaches a critical point where the output power flattens out. At this point, the turbine is reaching its upper limit of generation, which is known as the rated output power. After a certain threshold value, which is around 25 m/s, the next wind speed is known as the cut-out speed, where the turbine initiates shut-down mode for protection purposes since the blades are at risk because of the large applied force. Hence, based on the previous clarification, the symbols v_i, v_r, v_o, P_R refer to cut-in speed, rated wind speed, cut-out speed, and rated power, respectively. However, the most important parameter of this curve is $Q\ (v)$, which describes the nonlinear region.

The enclave region between the cut-in speed and the rated wind speed (nonlinear region) can be represented using several mathematical formulas. Therefore, the output power from a wind turbine can be determined depending on the interval of wind speed, as illustrated below [62]:

$$P(v) = \begin{cases} 0, & v > v_o \ or \ v < v_i \\ Q(v), & v_i < v < v_r \ (enclave\ region) \\ P_R, & v_r < v < v_o \end{cases} \tag{4}$$

In this research, four representations of $Q(v)$ have been examined, which are [63]:

$$Q_1(v) = P_R\left(\frac{v^2 - v_i^2}{v_r^2 - v_i^2}\right) \tag{5}$$

$$Q_2(v) = P_R\left(1 - e^{-\left(\frac{v}{\beta}\right)^5}\right), \beta = 0.70335986v_r - 0.00049995 \tag{6}$$

$$Q_3(v) = \frac{P_R}{1 + e^{-(bv-7.5)}}, \quad b = 5.822e^{-0.3398vr} + 1.79e^{-0.0548vr} \tag{7}$$

$$Q_4(v) = P_R\left(\frac{v^3}{v_r^3}\right) \tag{8}$$

In this paper, the objective function is to maximize the energy captured by wind by the proper selection of the proposed PDFs' parameters as illustrated in the following equations [64]:

$$Maximize: \; E_{Total} = (1 - \xi) \times (E_{ir} + E_{ro}) \tag{9}$$

Subject to:

$$Minimise\ Error = \sqrt{(P_O - P_D)^2} \tag{10}$$

where:

E_{Total}: Total energy that can be generated by the wind turbine (KWh/m^2);

ξ: The overall loss percentage of the turbine;

T: Time period in an hour;

E_{ir}: Generated energy by the wind turbine in the region between the cut-in speed and rated speed in $\mathrm{KWh/m^2}$.

$$E_{ir} = T\int_{v_i}^{v_r} Q(v) \times f(v)dv \tag{11}$$

E_{ro}: Generated energy by the wind turbine in the region from the rated speed to the cut-out speed in $\mathrm{KWh/m^2}$.

$$E_{ro} = TP_R\int_{v_r}^{v_o} f(v)dv \tag{12}$$

P_O: The probability based on the observations (real data from RETScreen);

P_D: The probability based on the proposed PDFs.

In general, the energy that is available from wind resources can be determined based on the following expression:

$$E_D = \int_0^\infty WPD \times f(v)dv = \int_0^\infty \frac{1}{2} \times \rho_a \times V^3 \times f(v)\, dv \tag{13}$$

where:

E_D : The available energy in the regime;

WPD : The wind power density;

$f(v)$: The used PDF;

A : The effective area of the disk;

ρ_a : The air density;

V : The velocity of the wind.

The attention in this research goes to tracking the maximum energy by varying the two parameters of the normal and Weibull PDFs such that the estimated probabilities must be close to the observed probabilities for each wind speed class.

2.2.3. Estimating and Predicting the Parameters by Artificial Intelligence Techniques in Conjunction with Neural Fuzzy Methods

In the design stage of any project implementation, optimization is an essential tool by which the performance of the overall system can be enhanced effectively. Furthermore, this tool is used in the whole field and is not limited to only one field due to the diversity of the dilemma nowadays. The intriguing ideas of the optimization algorithms were put forth depending on the behavior of things around us.

The concentration of this research is oriented toward estimating the parameters of the various distribution functions by optimization algorithms. Nowadays, artificial intelligence (AI) has become a trend due to its features. This technique programs the machine to be capable of performing complex tasks. Further, it works in various fields, which makes it popular these days [65].

In this study, three distinct algorithms are employed to estimate the parameters of the proposed PDFs. The first one is the genetic algorithm (GA), the second one is the bacterial foraging optimization algorithm (BFOA), and the third one is the simulated annealing (SA) algorithm. It is no wonder that artificial intelligence (AI) techniques are more recommended than numerical approaches due to their high level of accuracy and flexibility.

In this paper, the initial population of each parameter started from 0 to 100 for each PDF, with a step of 0.05 for each iteration. The goal of the AI was to find the best parameters that made the estimated wind speed as close as possible to the actual wind speed. Thus, the stopping criteria were based on finding the most accurate parameters by evaluating the holistic possibilities of the initial population.

(A). Genetic Algorithm

The GA mimics the biological evolution natural process by selecting fit individuals for reproduction. It works with specific population sizes (individuals) that are evolving with time. The principle of this algorithm is inspired by human beings based on the three main operators: selection, crossover, and mutation. Figure 7 shows the implementation steps of GA based on [66].

Based on Figure 8, the principle of this algorithm can be simply explained as follows: the population of a certain number of chromosomes is generated randomly. The next step is to find the corresponding fitness value of each chromosome. Afterward, a single-point crossover is applied for the two chromosomes (two inputs of the optimization problem) to generate offspring. The applied crossover in this thesis is the partially matched crossover (PMX) method since it is the most commonly used crossover. The next step is to apply the mutation operation to the obtained offspring to generate a new population. Thereafter, the previous process (selection, crossover, and mutation) is applied again until a new population is obtained [67].

(B). Bacterial Foraging Optimization Algorithm

BFOA is considered an optimization approach that was developed depending on the base of Escherichia Coli (E. Coil) bacteria's foraging strategy. These bacteria live inside the human gut. The term "foraging" refers to the animals' behavior for ingesting, handling, or locating their food. In general, E. Coil bacteria have flagella, which enable the bacteria to rotate or move in a locomotion manner. For clarification, with the aid of the flagella, the bacteria may move in the same direction or change its orientation. The ultimate two goals of this bacteria are to find a place with a high level of nutrients and mitigate the noxious areas by moving in a certain motion. Hence, as a summary, when the bacteria reach a place with a higher level of nutrients compared with the previous place, the movement is described as "swimming" or "running". Otherwise, it will tumble. Figure 9a,b clarifies the previous discussion. In Figure 9a, the bacteria move from location L_1 into location L_2 since the nutrient level in L_2 is higher than in L_1. It can be observed that the bacteria move forward within the same path, "swim". Similarly, the bacteria move from L_2 into L_3, which indicates that L_3 has more nutrient levels compared with L_2. This process is repeated until the phase-out of the bacteria's life.

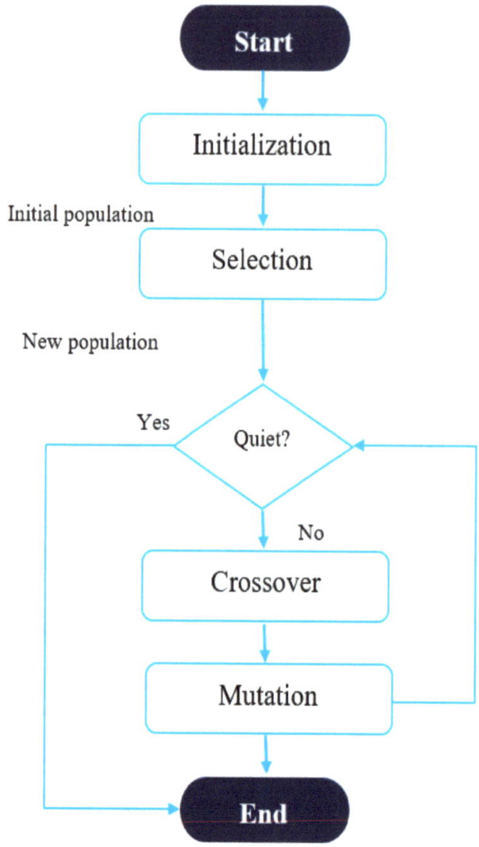

Figure 8. Flowchart of GA.

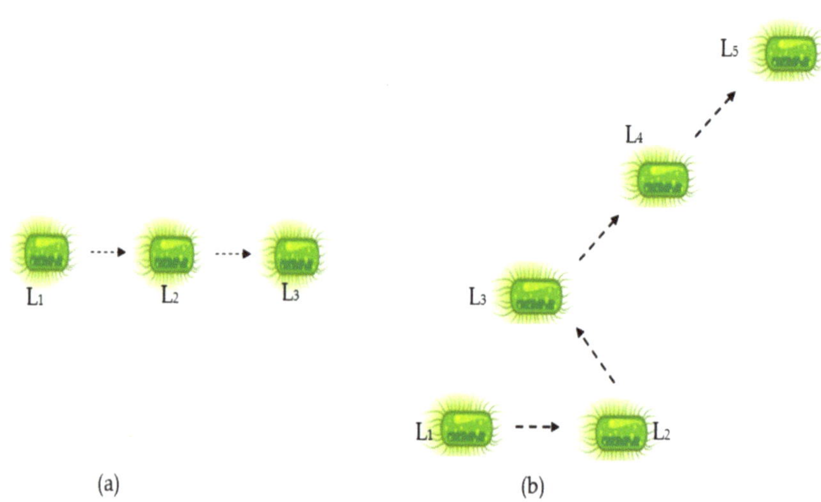

Figure 9. Bacteria movement based on the nutrient level in each location where (**a**) represents "swimming" case, and (**b**) represents "swimming and tumbling" case.

The second mentioned case is clarified in Figure 9b. This case may include swimming beside tumbling. In this case, the initial location is L_1; the bacteria try to find another location, and the nutrient levels of the two locations are compared. If the nutrient level of the new location is higher, the bacteria will go toward (swimming) this area; otherwise, it will search for another location with a higher nutrient level by changing its trajectory (tumbling). Since the nutrient level at L_2 is lower than that of L_1, the bacteria will not continue moving with the same path but will move in another path until reaching L_3, where the nutrient level is higher in comparison with the previous location. Another test for nutrient level will be conducted; if at L_3 the nutrient level is lower than that of L_2, the bacteria will tumble and move into another location, L_4. If the nutrient level at L_4 is higher than that of L_3, the bacteria will swim in the same path of the previous movement until a new location, L_5, is reached [68]. The flowchart of this algorithm is clarified in Figure 10, where N_d refers to the number of elimination-dispersal events, N_{re} refers to the number of reproduction steps, and N_c refers to the number of chemotactic steps [69].

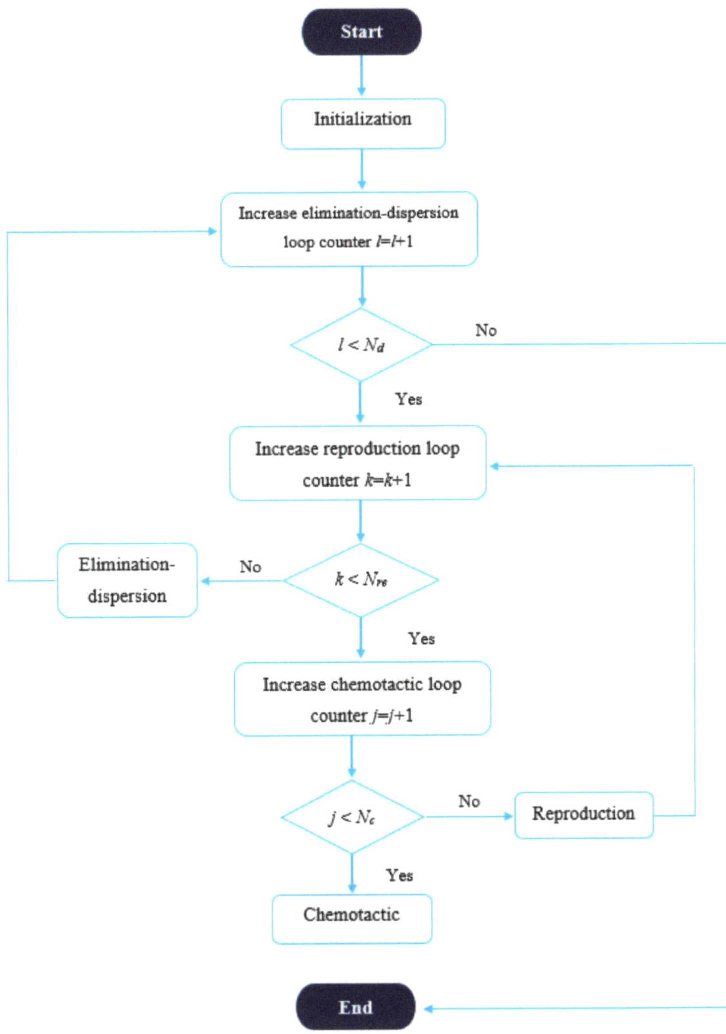

Figure 10. Flowchart of BFOA.

(C). Simulated Annealing Algorithm

The idea of the simulated annealing (SA) algorithm mimics the metal reshaping process, where a heated metal is reshaped from one structure to another after being cold. An initial solution is set randomly and then based on this procedure, which is clarified in Figure 11 [70]. The control parameter in this algorithm is the temperature, which controls the number of iterations of the process [71]. In the end, the best fit which meets the objective function will be selected. Based on Figure 11, it can be noticed that SA has a sequence of moving from the initial solution toward the next solution, where in some cases, the worst solution may be accepted based on a probability factor.

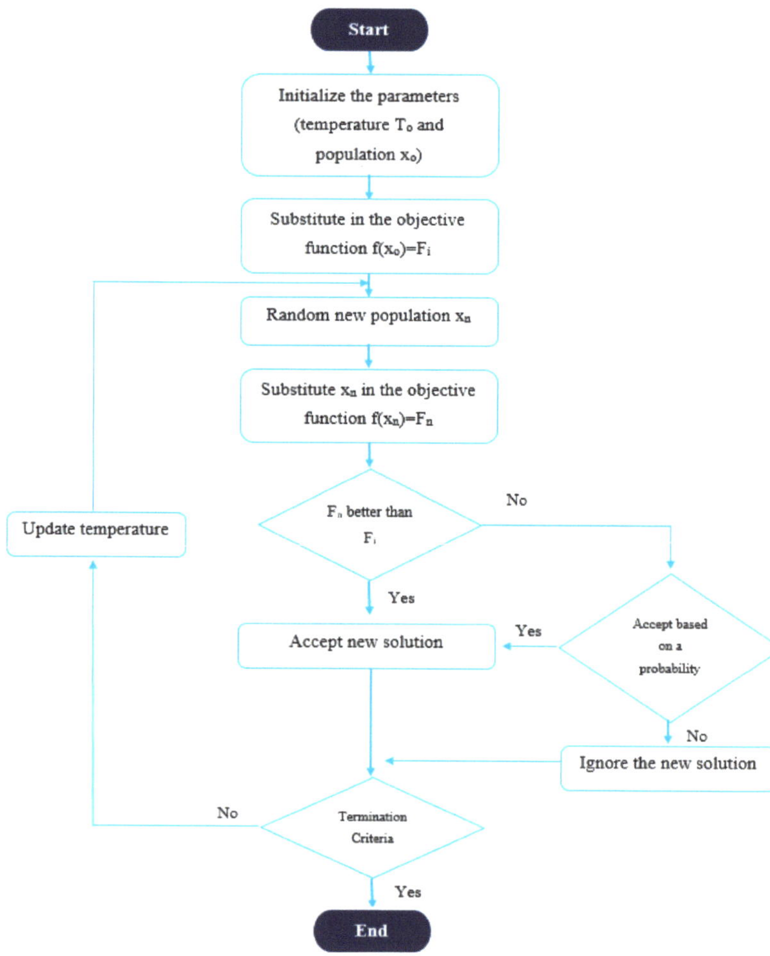

Figure 11. Flowchart of SA.

(D). Adaptive Neural Fuzzy Inference System (ANFIS)

Fuzzy logic is a worthwhile tool in conducting complex tasks, especially when it is difficult to obtain a mathematical model. In addition, fuzzy logic may be used in prediction problems for quality amelioration purposes. In contrast, neural networks (NNs) are networks that can link the input with the output data in a certain manner, where each input is assigned a certain value of weight. Subsequently, the output is determined based on the assigned weights [72].

However, in this paper, a combination of these two techniques is used, where ANFIS is classified as a hybrid AI model that coalesces and combines the intrinsic features of fuzzy logic in the parallel processing of a NN [72]. The architecture appears from the tool itself for the proposed problem, which has two inputs and a single output. The flowchart of ANFIS is presented in Figures 12 and 13 [73].

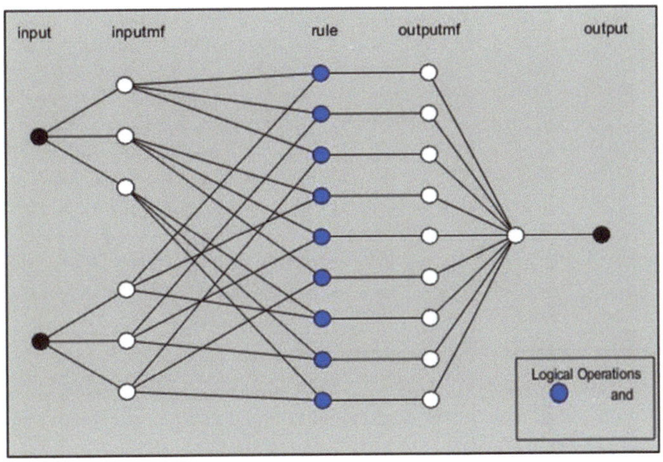

Figure 12. Architecture of ANFIS.

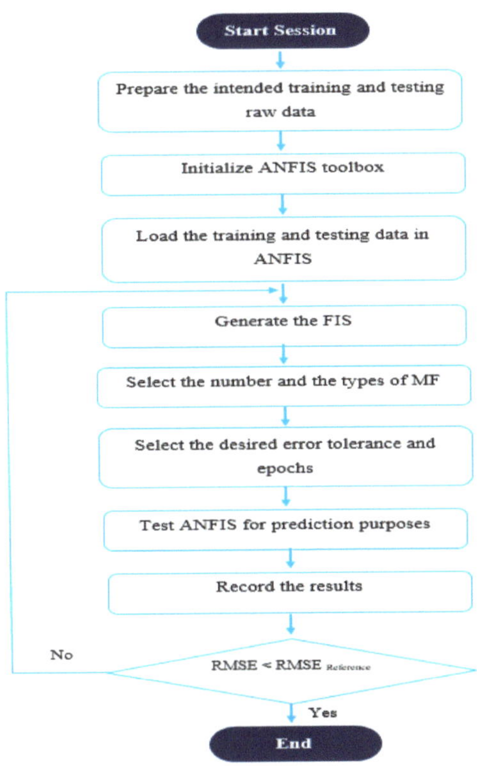

Figure 13. Flowchart of ANFIS.

ANFIS became a technique that can be used for predicting purposes, as it can make an input-output mapping and accordingly design an input-output prediction by the hybrid learning process. In general, ANFIS is commonly used to simulate nonlinear systems providing intrinsic and reliable outcomes [74].

2.2.4. Selecting the Most Accurate Approach Based on the Indicators (RMSE and MAE)

Estimating the parameters of the proposed PDFs is a non-trivial task, and the most challenging task is to opt for and decide which PDF model is the best. Based on the previously mentioned two methods of selecting the parameters (either by artificial intelligence techniques or by prediction), once the corresponding parameters of each PDF are selected, the next step is assigned to choose the most precise PDF. This can be accomplished with the aid of some goodness-of-fit (GOF) indicators, such as RMSE and MAE. Incontrovertibly, the lower the fitness value, the better the proposed model fit. Each proposed PDF is a candidate to be accepted if its parameters achieve a fitness value that is relatively small.

Several options for GOF tests are offered and available nowadays; in this study, RMSE and MAE are the proposed indicators since these two indicators are the most frequently employed in evaluating accuracy. The mathematical formulas for these indicators are illustrated in the following points [75].

(A). RMSE:

The root-mean-square error (*RMSE*) between *X* datasets (the observed probability) and *Y* datasets (the estimated probability) is defined as a measurement of the difference between their values and can be expressed as:

$$RMSE = \sqrt{\frac{1}{k}\sum_{i=1}^{k}(X_i - Y_i)^2} \tag{14}$$

This indicator is commonly used to assess how good the predicted probability is over the observed probability, where the smaller value of *RMSE* is, the better the accuracy of the proposed model, indicating that the selected parameters achieve the best results.

(B). Mean Absolute Error (MAE)

This test determines the absolute error between the observed value and the corresponding estimated value. This performance indicator looks similar to the RMSE since the lower the *MAE*, the better the outcomes obtained are. Its mathematical formula is represented below:

$$MAE = \frac{1}{n}\sum_{i=1}^{n}|y_i - x_i| \tag{15}$$

2.2.5. Machine Learning Classification and Prediction Based on the Best PDF

Machine learning (ML) is a description of a computer that has been programmed and trained based on a certain data pattern, thus becoming able to predict the situation for newly inserted data. This term (ML) is divided into two major categories: classification (supervised learning) and clustering (unsupervised learning). In classification, two phases are required: the training phase and the testing phase. In the training phase, the inserted data should be divided into a certain number of categories, where each category carries a particular percentage from the overall data that have the same classification. Commonly, the first columns in the intended trained data represent the input(s), while the last column represents how each input(s) has been classified. Within the same phase and based on several built-in classifier algorithms, the machine will be able to understand and learn the data. Accordingly, by generating a learning model while in the data testing phase, the data is classified based on the generated trained model in addition to the classifier model.

On the other hand, in clustering, the input data are grouped based on their similarities without any trained model. The purpose of using ML in this study is to be able to predict the difference (error) between the actual observed wind speed data and the estimated wind speed for the best PDF, which will be decided based on the performance indicators, as mentioned before. Here, a small error indicates that the selected parameters are good, while a medium error indicates that the selected parameters are not good enough, and finally, a Large error indicates that the selected parameters achieve a large difference between the observed and the estimated value.

In this paper, 24 classification algorithms have been used to run the data. These algorithms have been summarized in the last section of this paper. The significance of using ML in this study came from the necessity of the more precise prediction of the parameters since the remaining phases of energy estimation depend on the selected parameters.

3. Results and Discussion

The two parameters of the normal PDF, μ and σ, were selected intelligently by GA, BFOA, SA, and the neuro-fuzzy method. The corresponding energy regime in Kwh/year was determined accordingly based on each method for all sites, as clarified in Table 4. These parameters are reflected concurrently with the observed actual data of wind speed for all sites, as represented in Figure 14.

Table 4. Outcomes for the normal PDF parameters by GA, BFOA, SA, and ANFIS along with the energy regime per year.

	GA		BFOA		SA		Fuzzy Neural		Energy Regime kWh/m^2 (Year)			
Site	μ	σ	μ	σ	μ	σ	μ	σ	**GA**	**BFOA**	**SA**	**ANFIS**
Al-Badieh	4.5	1.44	4.31	1.41	4.35	1.41	4.4	1.4	641.207	570.84	581.838	595.31
Amman	4.05	1.07	4.02	1.09	3.95	1.14	4.1	1.1	431.509	425.21	413.435	449.1
Aqaba	3.75	1.09	3.81	1.09	3.85	1.1	3.75	1.1	355.470	370.41	380.653	355.47
Bayir	4.45	1.59	4.45	1.66	4.45	1.66	4.4	1.61	655.642	669.977	669.977	641.37
Irbid	2.7	0.591	2.36	0.62	2.35	0.59	2.5	0.63	120.85	85.217	82.893	99.958
Irwaished	4.45	1.56	3.66	1.18	3.65	1.18	4	1.18	648.476	345.16	343.249	433.64
Maan	4	1.09	3.91	1.06	3.9	1.05	4.1	1.12	420.761	390.83	387.428	452.4
Mafraq	3.65	1.02	3.75	1.02	3.75	1.02	3.7	1	322.68	346.414	346.414	331.42
Queen Alia airport	4.5	1.51	4.16	1.48	4.2	1.47	4.3	1.48	655.699	533.69	543.012	579.02

Similarly, the parameters of the Weibull PDF, K and C, were assigned by the same approaches of GA, BFOA, SA, and the neuro-fuzzy method. The corresponding energy regime in Kwh/year has been determined accordingly based on each method for all sites, as clarified in Table 5. These parameters are reflected in conjunction with the observed actual data of wind speed for all sites, as represented in Figure 15.

Table 5. Outcomes for the Weibull PDF parameters by GA, BFOA, SA, and ANFIS along with the energy regime per year.

	GA		BFOA		SA		Fuzzy Neural		Energy Regime kWh/m^2 (Year)			
Site	K	C	K	C	K	C	K	C	**GA**	**BFOA**	**SA**	**ANFIS**
Al-Badieh	3.65	4.6	3.5	4.8	3.5	4.8	3.7	4.5	489.56	562.41	562.41	456.81
Amman	3.25	4.8	3.2599	4.6326	3.2	4.6	3	4.9	575.51	516.84	509.29	631.25
Aqaba	3.2	4.35	3.9	4.25	3.9	4.25	3.3	4.1	430.68	380.41	380.41	356.82
Bayir	2.95	5.55	3.0571	4.993	3.1	5	2.9	5.3	923.93	662.69	661.83	810.84
Irbid	4.55	2.35	4.4	2.6	4.4	2.6	4.2	2.4	62.778	85.375	85.375	67.603
Irwaished	3.3	4.1	3.5064	4.031	3.55	4.05	3.1	4.2	356.82	332.93	336.51	392.27
Maan	4.3	4.1	4.1651	4.1869	4.2	4.2	4.4	4	335.85	359.41	362.31	310.88
Mafraq	4.05	3.95	4.25	4.1	4.25	4.1	4	3.85	303.23	336.43	336.43	281.41
Queen Alia airport	3.55	4.6	3.25	4.7	3.25	4.7	3.6	4.6	493.07	540.28	540.28	491.26

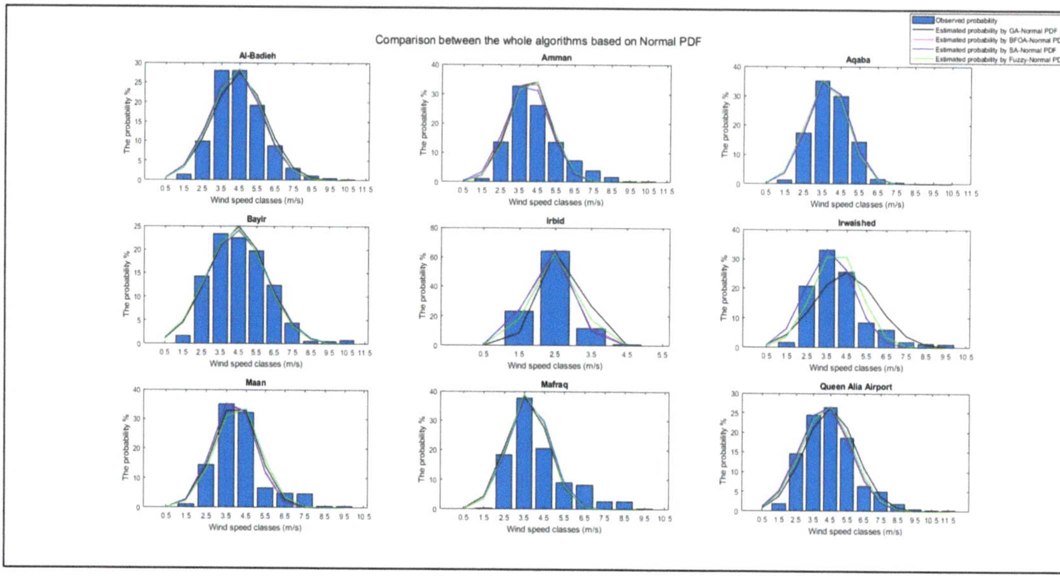

Figure 14. Outcomes based on Normal PDF.

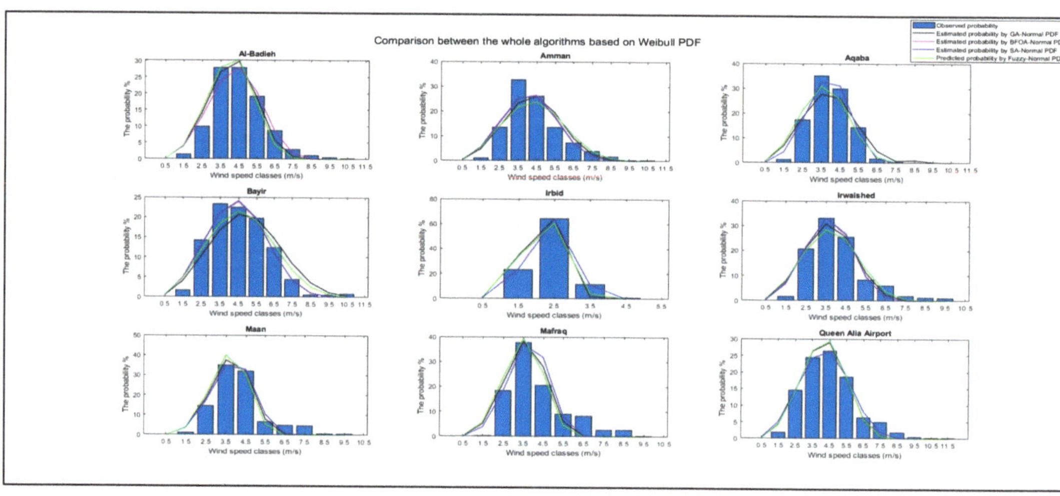

Figure 15. Outcomes based on Weibull PDF.

In the end, the extractable energy from wind has been determined based on the four listed P-V models for both the normal and Weibull PDFs. The outcomes are summarized in Tables 6 and 7, respectively, assuming the same wind turbine brand parameters for a fair comparison.

Table 6. Extractable energy (MWh per year) based on Normal PDF.

	Extractable Energy Based on Q_1 Model				Extractable Energy Based on Q_2 Model				Extractable Energy Based on Q_3 Model				Extractable Energy Based on Q_4 Model			
Site	**GA**	**BFOA**	**SA**	**ANFIS**	**GA**	**BFOA**	**SA**	**ANFIS**	**GA**	**BFOA**	**SA**	**ANFIS**	**GA**	**BFOA**	**SA**	**ANFIS**
Al-Badieh	2299	1938	1993	2056	2453	2044	2105	2169	3087	2572	2648	2728	2053	1779	1822	1875
Amman	1126	1106	1074	1222	1162	1141	1111	1256	1459	1434	1396	1577	1209	1185	1141	1283
Aqaba	795	856	904	795	834	894	942	834	1051	1126	1186	1051	902	961	1003	902
Bayir	2403	2486	2486	2337	2618	2733	2733	2546	3290	3437	3437	3203	2100	2162	2162	2053
Irbid	6	1	1	4	13	3	2	8	17	5	3	11	20	5	3	12
Irwaished	2361	794	787	1188	2550	831	823	1226	3220	1046	1037	1541	2080	877	870	1227
Maan	1087	934	911	1248	1120	972	950	1283	1410	1223	1196	1611	1160	1040	1024	1298
Mafraq	624	721	721	646	668	721	721	691	845	763	763	874	758	853	853	787
Queen Alia Airport	2386	1777	1819	1999	2573	1883	1926	2129	2573	2370	2423	2679	3238	1638	1673	1812

Table 7. Extractable energy (MWh per year) based on Weibull PDF.

	Extractable Energy Based on Q_1 Model				Extractable Energy based on Q_2 Model				Extractable Energy Based on Q_3 Model				Extractable Energy Based on Q_4 Model			
Site	**GA**	**BFOA**	**SA**	**ANFIS**	**GA**	**BFOA**	**SA**	**ANFIS**	**GA**	**BFOA**	**SA**	**ANFIS**	**GA**	**BFOA**	**SA**	**ANFIS**
Al-Badieh	1503	1899	1899	1328	1542	1980	1980	1358	1935	2490	2490	1703	1463	1749	1749	1331
Amman	1988	1686	1654	2292	2109	1766	1736	2504	2655	2222	2185	3153	1798	1571	1542	2009
Aqaba	1257	914	914	879	1304	943	943	910	1639	1186	1186	1145	1234	1016	1016	937
Bayir	3731	2446	2439	3184	4355	2682	2666	3648	5434	3377	3357	4568	3103	2129	2126	2682
Irbid	0.002	0.3	0.3	0.04	0.01	1	1	0.1	0.01	1	1	0.2	0.01	1	1	0.2
Irwaished	879	736	746	1081	910	768	779	1123	1145	967	980	1412	937	832	845	1085
Maan	651	780	790	525	697	816	826	0579	881	1029	1041	737	816	921	933	704
Mafraq	530	659	659	442	576	703	703	490	731	890	890	624	685	820	820	595
Queen Alia Airport	1532	1808	1808	1518	1578	1903	1903	1559	1981	2395	2395	1957	1477	1662	1662	1470

3.1. Assessment of the Proposed Approaches

An assessment of the proposed approaches was carried out based on the GOF by two performance indicators (RMSE, MAE), as clarified in Table 8, to ascertain the accuracy of the selected parameters. The goal of this step was to rank the selectivity level of each approach depending on the minimal difference (error) between the observed data and the foreshadowed data. Hence, based on Table 8, it can be perceived that BFOA and SA are the most accurate and predominant approaches in selecting the parameters in both the normal and Weibull PDFs, with a slight ignorable difference between them compared with their counterparts. In addition, based on the same table, it can be noticed that the normal PDF gives more accurate and precise estimated outcomes compared with the Weibull PDF. Therefore, the classification of classes was conducted based on the normal PDF. Accordingly, based on the outcomes that were obtained from this study, the worthiest regions in Jordan that are rich in wind resources, headed by Bayir and wrapped up by Irbid, are arranged in Figure 16 based on the BFOA and SA algorithms.

Table 8. Comparison between normal and Weibull PDFs based on several performance indicators for the four methods of parameter selection.

	RMSE								MAE							
	GA		BFOA		SA		ANFIS		GA		BFOA		SA		ANFIA	
Sites	GA Normal	GA Weibull	BFOA Normal	BFOA Weibull	SA Normal	SA Weibull	ANFIS Normal	ANFIS Weibull	GA Normal	GA Weibull	BFOA Normal	BFOA Weibull	SA Normal	SA Weibull	ANFIS Normal	ANFIS Weibull
Al-Badieh	2.2154	2.0480	1.6860	1.8402	1.7127	1.8402	1.7438	2.5164	1.4284	1.6278	1.1883	1.2964	1.1829	1.2964	1.2329	1.9336
Amman	3.1027	3.6164	2.8906	3.1293	2.5751	3.1179	3.1258	4.0767	1.9644	2.2439	1.8973	2.1273	1.9080	2.1301	2.1746	5.6719
Aqaba	1.9567	3.7269	1.4537	1.5499	1.3942	1.5499	2.7570	3.5467	1.4130	2.8002	0.9072	1.1287	0.9028	1.1287	1.4130	2.9174
Bayir	1.6636	3.0264	1.6757	1.4336	1.6757	1.4242	1.6422	2.2053	1.3706	2.3625	1.3043	1.0807	1.3043	1.0758	1.3435	1.7186
Irbid	9.4640	6.7671	0.8988	0.5492	1.0263	0.5492	3.6928	5.9630	6.1133	4.5869	0.7881	0.5072	0.9501	0.5072	2.6720	4.6240
Irwaished	6.3403	2.6331	2.2236	2.4685	2.2426	2.5946	3.5435	3.0769	4.4746	2.0076	1.6651	1.7918	1.6686	1.7943	2.8742	2.3617
Maan	3.0444	2.6829	2.9661	2.6439	2.5240	2.6300	3.4508	3.2724	2.0540	2.0683	1.7651	1.9797	1.7627	1.9842	2.5257	2.5927
Mafraq	3.7953	4.2286	4.0225	4.6509	4.0225	4.6509	3.9485	4.3644	2.9061	3.3184	2.5749	2.9943	2.5749	2.9943	2.8589	3.6309
Queen Alia Airport	2.2031	1.7485	1.4870	1.4602	1.4295	1.4602	1.5079	1.8163	1.6565	1.2139	1.0363	0.9552	1.0471	0.9552	1.2436	1.2702

Figure 16. The ranked sites in Jordan based on the availability of wind.

The bottom line of Table 8 is to emphasize that the Weibull PDF is not always the most accurate PDF but rather the opposite in this study, as it was found that the normal PDF overwhelmed the Weibull PDF.

3.2. Machine Learning Classification Outcomes

Several studies used ML in several fields, for instance, [76], in the diagnosis of the crime rate against women by using k-fold cross-validation. In [77], ML was used to conduct the sensitivity analysis of k-fold cross-validation, especially in error prediction and estimation. Another study conducted by [78] stated that ML is an effective and powerful tool, notably when massive amounts of data are collected.

In this study, a novel approach is proposed to classify and predict wind estimation by a MATLAB environment-classification learner application based on the datasets that were gathered from RETScreen, analyzed by the SPSS environment, and finally tested by AI codes. Hence, a huge dataset has been investigated to evaluate the performance of the normal PDF, where the nominated approaches try to find the best parameters of μ and σ that attain the least difference (error) between the observed and estimated probabilities for each site. The sample size comprises 3000 species for each candidate site; these 3000 were divided into three main categories. Thus, three distinct classifications were assigned, low error, medium error, and large error, based on each case of μ and σ and the resultant error. Each classifier has its accuracy percentage, cost misclassification, and training time. Therefore, the trade-off between them was based on the accuracy percentage in the first level, then on the training time if several classifiers gave the same accuracy percentage.

The validation process of the inserted data can be made by three options, cross-validation (x-validation), holdout validation, and no validation. Each approach has its features, where the cross-validation approach mitigates the situation of overfitting by dividing the whole raw data into a certain number of folds, while holdout validation is recommended to be used in large data sets. Finally, with no validation option, there is no protection against overfitting. In this study, the k-fold cross-validation method was used to validate the behavior of the generated learned model, where a certain number of folds equal to 5 was set. In other words, the inserted data was divided into five groups; one group was used for testing in the testing phase, while the remaining four groups were used for learning purposes in the training phase, as clarified in Figure 17. Generally, the k-fold cross-validation method has five main steps to be implemented, which are summarized in Figure 18. The principle of k-fold cross-validation can be explained by splitting the data into k groups, where each group carries an equal data sample weight; afterward, each group will be used as a test group for one time and as a training group for k-1 times. This validation approach is very common since it is easy to understand [76].

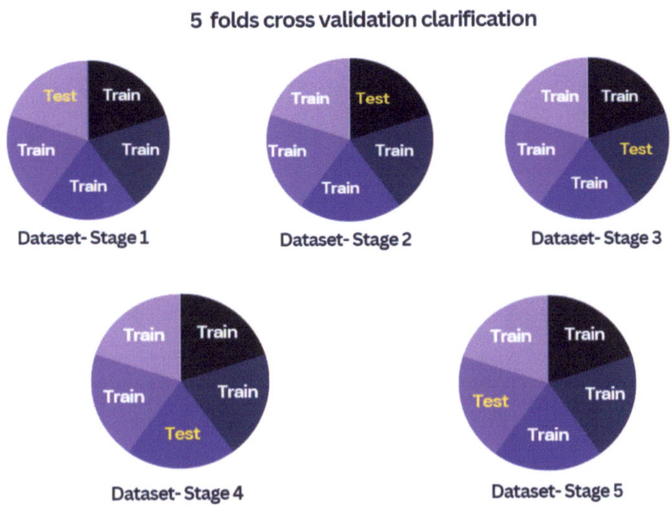

Figure 17. Five-fold cross-validation demonstration.

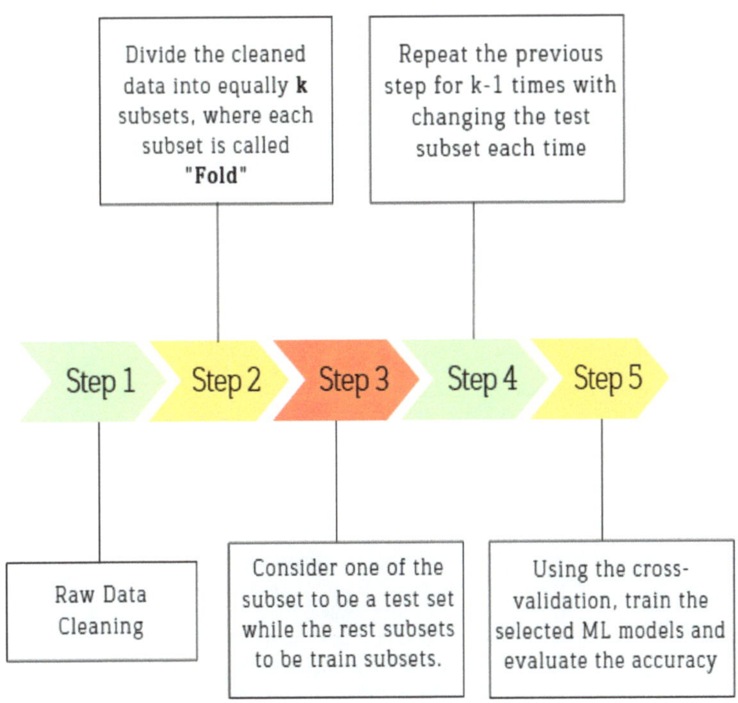

Figure 18. Steps of applying K-fold validation.

In general, the corresponding performance of any classifier is represented by the confusion matrix from an accuracy point of view, as clarified in Figure 19 in terms of true-positive rate (TPR) and false-positive rate (FPR). In a scatter plot, the performance of the classifier is represented in another manner, where the dot sign represents the correct prediction, and the cross sign indicates an incorrect prediction, as shown in Figure 20.

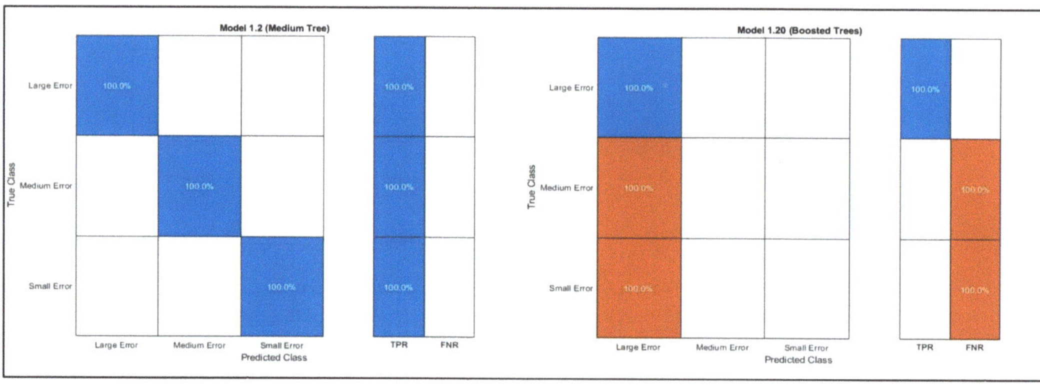

Figure 19. Obtained confusion matrices for all sites based on the best and the worst classifier.

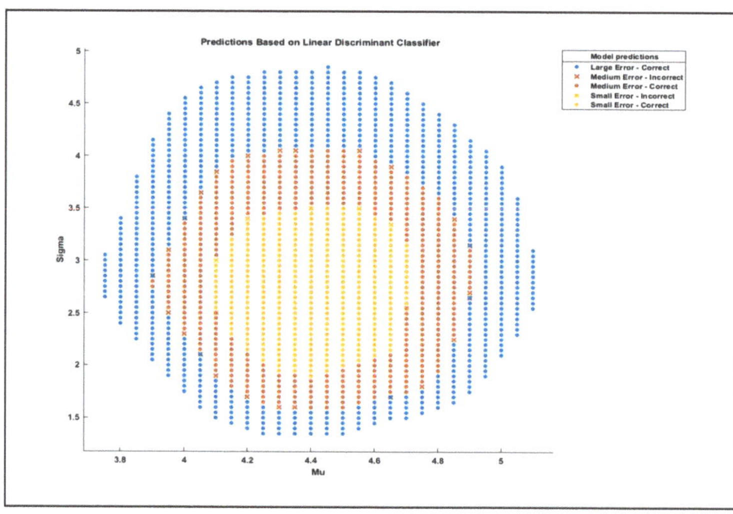

Figure 20. Scatter plot shows that the uncorrected prediction occurs on the edge of two classes when the accuracy is high (90%).

A confusion matrix is a square matrix that is divided into n × n, where n is the number of classes. In this paper, low error, medium error, and large error were the three classes. Thus, the resulting confusion matrix was 3 × 3. However, it can be observed that the percentage of prediction for each class in this matrix is contained inside a square, where the reflection of this square on the x-axis represents how this classifier classifies this class, while the reflection on the y-axis represents the true classification. The diagonal line of each matrix represents the accurate predictions, and the out-of-diagonal squares are the incorrect predictions. Regardless, some other classifier algorithms showed a good performance, where the accuracy was close to 100%. Interestingly, it was observed that for the classifiers with 90% accuracy, the misclassified points were those which were located at the boundary (edge) between two classes, as represented in Figure 20. Surprisingly, another observation that is in dire need to be mentioned is that in ML classification, the data sample for each class affects the accuracy. For instance, if a certain class contains merely one row of data (one case), while the other classes have a relatively higher sample of data, the prediction of this class will not be detected easily or maybe at all. For example, there is the best selection of μ and σ that achieves the best fit for each site. Hence, when a new class was created and named "Best", this class was not predicted since it contained merely one case, while the other classes contained around 500 cases or more.

Figure 20 shows that there is a certain range of μ and σ that commences and achieves a small error between the estimated and observed data. Once the values of μ and σ exceed this range, another region representing a medium error will be entered. Finally, when the range of μ and σ exceeds the range of medium error, the last region will be entered, which represents a large error.

$$Accuracy = \left(100 - round\frac{total\ misclassification\ cost}{Overall\ number\ of\ inserted\ data}\right)\% \tag{16}$$

The overall number of inserted data refers to the number of observations that were inserted to be trained and tested, or in other words, the same as the sample size. However, the previous table can be summarized by two main figures, as represented in Figures 21 and 22. Based on Table 9, the performance of 24 classifier algorithms was evaluated. Hence, from the accuracy and training time points of view, it was deduced that the medium tree classifier showed the most accurate and swiftest results for all sites, with an accuracy percentage of 100% and with minimal training time compared with the other classifier algorithms. It can predict the error level based on the given values of μ and σ. Therefore, the medium tree classifier is a second-to-none trustworthy classifier since it can effectively predict the error. On the other hand, the ensemble-boosted trees classifier shows poor and awful performance in predicting the error level based on the values of μ and σ of each site. Figure 19 shows the confusion matrices based on these two classifier algorithms, where for all sites, the same matrices were obtained for these two classifiers. These classifier algorithms were trained and tested automatically by the classification learner application in MATLAB software, and the prediction process was carried out by exporting the trained model into a workspace.

In the end, the usage of ML in this paper is justified since it was concluded based on Equations (11) and (12), which clarify how to calculate wind energy, that the energy calculation of wind depends on the turbine specifications, which are constant, in addition to the parameters of the proposed PDF. Hence, the precise selection of the parameter gives actual values for the wind energy regime and for the extractable energy from wind turbines. Otherwise, if the selection of the parameters is not accurate enough, all corresponding calculations cannot be trusted. Accordingly, the selection of these parameters is a critical phase and must be carried out wisely.

Hence, it was necessary to predict and double-check if the chosen parameters attained low error between the actual and the estimated wind speed values, and that was the role of the ML in this paper based on the most accurate classifier, which was the medium tree classifier.

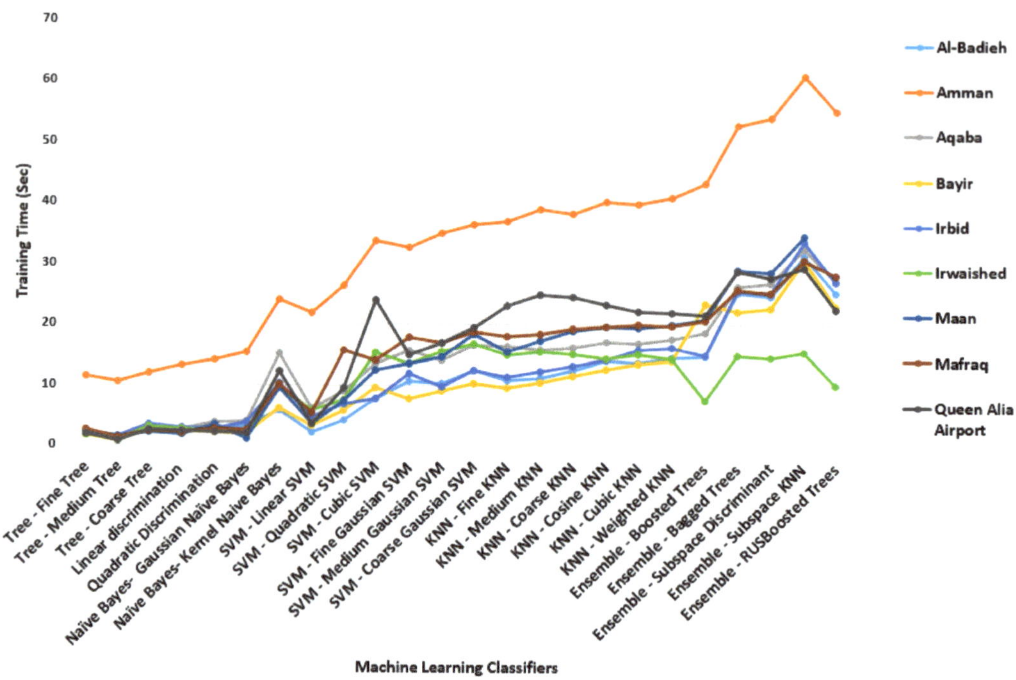

Figure 21. Comparison between all 24 classifiers for all locations from the training time point of view.

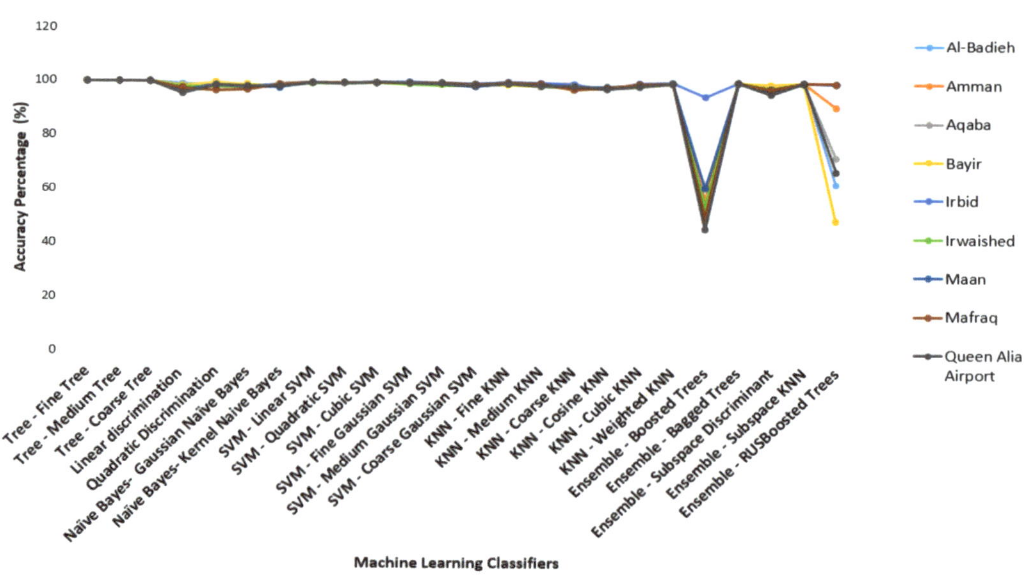

Figure 22. Comparison between all 24 classifiers for all locations from an accuracy point of view.

Table 9. The tested 24 classifiers for all candidate sites.

	Al-Badieh		Amman		Aqaba		Bayir		Irbid		Irwaished		Maan		Mafraq		Queen Alia Airport	
	Accuracy	Training Time	Accuracy	Training Time	Accuracy	Training Time	Accuracy	Training Time	Accuracy	Training Time	Accuracy	Training Time	Accuracy	Training Time	Accuracy	Training Time	Accuracy	Training Time
Fine Tree	100	1.91	100	11.44	100	1.79	100	1.65	100	1.92	100	1.78	100	2.22	100	2.67	100	1.91
Medium Tree *	**100**	**1.36**	**100**	**10.55**	**100**	**1.30**	**100**	**0.82**	**100**	**1.52**	**100**	**0.87**	**100**	**1.45**	**100**	**1.41**	**100**	**0.84**
Coarse Tree	100	2.78	100	11.87	100	2.44	100	2.62	100	3.44	100	3.10	100	2.26	100	2.47	100	2.58
Linear discrimination	98.9	2.59	97.7	13.19	98.3	2.68	98.3	2.45	97	2.97	98	2.85	96.6	1.94	97.3	2.14	95.3	2.39
Quadratic Discrimination	98.5	2.32	98.1	14.04	98.6	3.84	99.3	2.26	98.5	2.73	98	2.45	98.5	3.57	96.5	2.83	98.5	2.18
Gaussian Naïve Bayes	98.6	3.31	98.2	15.41	98.1	4.03	98.7	2.05	98	3.79	97.9	2.16	97.7	1.18	96.7	2.59	98.2	2.00
Kernel Naive Bayes	98.4	5.78	98.2	24	98	15.19	98.3	6.12	98.9	10.42	98.1	9.87	97.8	9.63	98.8	10.09	98.3	12.38
Linear SVM	99.5	2.24	99.5	21.83	99.6	6.15	99.5	3.30	99.6	4.79	99.1	5.84	99.5	3.65	99.5	5.39	99.4	3.65
Quadratic SVM	99.3	4.20	99.6	26.30	99.5	9.15	99.4	5.86	99.6	6.84	99.3	7.56	99.5	7.37	99.6	15.70	99.3	9.70
Cubic SVM	99.6	7.78	99.6	33.85	99.8	13.62	99.5	9.63	99.6	7.74	99.6	15.38	99.5	12.50	99.7	14.20	99.8	23.96
Fine Gaussian SVM	99.3	10.69	99.6	32.76	99.2	15.81	99.4	7.89	99.9	12.03	99	13.51	99.6	13.59	99.5	17.94	99.4	15.01
Medium Gaussian SVM	99.1	10.41	99.3	34.95	99.1	14.05	99	9.04	99.7	9.80	98.7	15.57	99.3	14.71	99.5	16.92	99.3	17.02
Coarse Gaussian SVM	98.6	12.42	98.8	36.39	98.6	16.65	99.1	10.30	99.1	12.53	98.6	16.89	98.2	18.52	98.8	18.79	98.6	19.51
Fine KNN	99.3	10.84	99.3	37.10	99.1	16.43	98.9	9.53	99.9	11.39	99.5	14.98	99.5	15.48	99.6	18.14	99.2	23.16
Medium KNN	98.5	11.22	98.6	38.93	98.3	15.87	98.6	10.48	99.5	12.35	98.7	15.62	99.1	17.41	99.1	18.41	98.5	24.95
Coarse KNN	98.3	12.42	97.9	38.29	98.4	16.24	98.5	11.63	99.1	13.16	98	15.28	98.1	18.93	97	19.37	98.5	24.52
Cosine KNN	97.8	14.12	97.5	40.21	97.6	17.19	97.7	12.76	97.3	14.36	98.2	14.46	98.2	19.71	98	19.78	97.4	23.29
Cubic KNN	98.4	13.69	98.6	39.95	98.3	16.97	98.7	13.57	99.5	15.99	98.7	15.15	99	19.48	99.1	20.15	98.6	22.28
Weighted KNN	99.6	14.62	99.7	40.98	99.5	17.76	99.4	14.11	99.9	16.34	99.7	14.46	99.7	20.15	99.6	19.95	99.5	21.96
Ensemble-Boosted Trees *	**53.1**	**14.88**	**56.3**	**43.39**	**54**	**18.75**	**48.9**	**23.45**	**94.7**	**14.96**	**53.9**	**7.62**	**60.8**	**20.90**	**49.9**	**20.80**	**45.7**	**21.63**
Ensemble-Bagged Trees	100	25.22	100	52.83	100	26.43	100	22.32	100	25.58	100	15.01	100	29.04	100	25.89	100	28.99
Subspace Discriminant	98.4	24.81	98.7	54.21	97.2	26.86	99.1	22.83	97.9	25.31	97.7	14.73	96.9	28.75	97.6	25.35	95.8	27.75
Subspace KNN	99.8	31.35	100	60.89	99.9	32.65	99.8	30.77	100	33.72	99.8	15.60	99.9	34.61	100	30.67	99.9	29.51
RUSBoosted Trees	62.3	25.23	90.9	55.29	72.2	27.19	48.9	23.2	99.6	27.12	99.8	10.23	99.8	28.13	99.8	28.10	67.3	22.64

* These rows indicate the best and worst classifiers respectively.

4. Conclusions

This paper shed light on a PDF that is not regularly used in wind estimation, the normal PDF, which overwhelmed the most commonly used PDF in wind estimation, the Weibull PDF. The decision was made based on two performance indicators (RMSE and MAE). The goal of using these PDFs was to estimate the extractable energy from wind in nine sites in Jordan by the proper selection of the parameters of each PDF. The outcomes showed that Bayir is the richest wind source site. Finally, this paper used machine learning with 24 classifier algorithms for the purpose of predicting suitable parameters for each site based on previously trained data based on the k-fold cross-validation method. It was noticed that several classifier algorithms achieve an accuracy of 100%, which justifies the comparison between them based on the training time point of view. The medium tree classifier was the most accurate and swiftest classifier for all sites. On the contrary, the ensemble-boosted trees classifier was the worst one, with the lowest accuracy for the nine sites. Finally, it was observed that for the classifiers with accuracy in the range of 90%, the misclassified points were those which are located at the boundary (edge) between two classes.

Author Contributions: Methodology, H.H.D.; Formal analysis, H.H.D.; Investigation, A.A.-Q.; Writing—draft, H.H.D.; Writing—review & editing, A.A.-Q.; Supervision, A.A.-Q.; Project administration, A.A.-Q.; Funding acquisition, A.A.-Q. All authors have read and agreed to the published version of the manuscript.

Funding: This research received no external funding.

Institutional Review Board Statement: Not applicable.

Informed Consent Statement: Not applicable.

Data Availability Statement: Data available on request.

Acknowledgments: The authors would like to acknowledge Yarmouk University for their support in this research study.

Conflicts of Interest: The authors declare no conflict of interest.

References

1. Al-Mhairat, B.; Al-Quraan, A. Assessment of Wind Energy Resources in Jordan Using Different Optimization Techniques. *Processes* **2022**, *10*, 105. [CrossRef]
2. Al-Quraan, A.; Al-Qaisi, M. Modelling, Design and Control of a Standalone Hybrid PV-Wind Micro-Grid System. *Energies* **2021**, *14*, 4849. [CrossRef]
3. Alrwashdeh, S.S. Energy sources assessment in Jordan. *Results Eng.* **2022**, *13*, 100329. [CrossRef]
4. Stephan, A.; Stephan, L. Achieving net zero life cycle primary energy and greenhouse gas emissions apartment buildings in a Mediterranean climate. *Applied Energy* **2020**, *280*, 115932. [CrossRef]
5. Thormark, C. A low energy building in a life cycle—Its embodied energy, energy need for operation and recycling potential. *Build. Environ.* **2002**, *37*, 429–435. [CrossRef]
6. Arvesen, A.; Hertwich, E.G. More caution is needed when using life cycle assessment to determine energy return on investment (EROI). *Energy Policy* **2015**, *76*, 1–6. [CrossRef]
7. Sahu, O. Sustainable and clean treatment of industrial wastewater with microbial fuel cell. *Results Eng.* **2019**, *4*, 100053. [CrossRef]
8. Bull, S.R. Renewable energy today and tomorrow. *Proc. IEEE* **2001**, *89*, 1216–1226. [CrossRef]
9. Neupane, D.; Kafle, S.; Karki, K.R.; Kim, D.H.; Pradhan, P. Solar and wind energy potential assessment at provincial level in Nepal: Geospatial and economic analysis. *Renew. Energy* **2022**, *181*, 278–291. [CrossRef]
10. Emblemsvåg, J. Wind energy is not sustainable when balanced by fossil energy. *Appl. Energy* **2022**, *305*, 117748. [CrossRef]
11. Wind Energy Project Analysis Clean Energy Project Analysis: Retscreen ®Engineering & Cases Textbook. Available online: https://unfccc.int/resource/cd_roms/na1/mitigation/Module_5/Module_5_1/b_tools/RETScreen/Manuals/Wind.pdf (accessed on 11 January 2023).
12. Imdadullah; Alamri, B.; Hossain, M.A.; Asghar, M.S.J. Electric Power Network Interconnection: A Review on Current Status, Future Prospects and Research Direction. *Electronics* **2021**, *10*, 2179. [CrossRef]
13. Bitar, E.Y.; Rajagopal, R.; Khargonekar, P.P.; Poolla, K.; Varaiya, P. Bringing Wind Energy to Market. *IEEE Trans. Power Syst.* **2012**, *27*, 1225–1235. [CrossRef]

14. Saidur, R.; Rahim, N.A.; Islam, M.R.; Solangi, K.H. Environmental impact of wind energy. *Renew. Sustain. Energy Rev.* **2011**, *15*, 2423–2430. [CrossRef]
15. Billinton, R.; Gao, Y. Multistate Wind Energy Conversion System Models for Adequacy Assessment of Generating Systems Incorporating Wind Energy. *IEEE Trans. Energy Convers.* **2008**, *23*, 163–170. [CrossRef]
16. Varun; Prakash, R.; Bhat, I.K. Energy, economics and environmental impacts of renewable energy systems. *Renew. Sustain. Energy Rev.* **2009**, *13*, 2716–2721. [CrossRef]
17. Kikuchi, R. Adverse impacts of wind power generation on collision behaviour of birds and anti-predator behaviour of squirrels. *J. Nat. Conserv.* **2008**, *16*, 44–55. [CrossRef]
18. Sadorsky, P. Wind energy for sustainable development: Driving factors and future outlook. *J. Clean. Prod.* **2021**, *289*, 125779. [CrossRef]
19. Renewable Capacity Statistics 2019. Irena.org. 2019. Available online: https://www.irena.org/publications/2019/Mar/Renewable-Capacity-Statistics-2019 (accessed on 11 January 2023).
20. Statistics Time Series. Available online: https://www.irena.org/Statistics/View-Data-by-Topic/Capacity-and-Generation/Statistics-Time-Series (accessed on 11 January 2023).
21. Siddique, S.; Wazir, R. A review of the wind power developments in Pakistan. *Renew. Sustain. Energy Rev.* **2016**, *57*, 351–361. [CrossRef]
22. Alrwashdeh, S.S. Map of Jordan governorates wind distribution and mean power density. *Int. J. Eng. Technol.* **2018**, *7*, 1495. [CrossRef]
23. Alsaad, M.A. Wind energy potential in selected areas in Jordan. *Energy Convers. Manag.* **2013**, *65*, 704–708. [CrossRef]
24. Dalabeeh, A.S.K. Techno-economic analysis of wind power generation for selected locations in Jordan. *Renew. Energy* **2017**, *101*, 1369–1378. [CrossRef]
25. Al-omary, M.; Kaltschmitt, M.; Becker, C. Electricity system in Jordan: Status & prospects. *Renew. Sustain. Energy Rev.* **2018**, *81*, 2398–2409. [CrossRef]
26. Ammari, H.D.; Al-Rwashdeh, S.S.; Al-Najideen, M.I. Evaluation of wind energy potential and electricity generation at five locations in Jordan. *Sustain. Cities Soc.* **2015**, *15*, 135–143. [CrossRef]
27. Bataineh, K.M.; Dalalah, D. Assessment of wind energy potential for selected areas in Jordan. *Renew. Energy* **2013**, *59*, 75–81. [CrossRef]
28. Feilat, E.A.; Azzam, S.; Al-Salaymeh, A. Impact of large PV and wind power plants on voltage and frequency stability of Jordan's national grid. *Sustain. Cities Soc.* **2018**, *36*, 257–271. [CrossRef]
29. *National Electric Power Company (NEPCO), Annual Report*; NEPCO: Amman, Jordan, 2021.
30. Online Store and Quote Request—The Wind Power—Wind Energy Market Intelligence. Available online: https://www.thewindpower.net/store_en.php (accessed on 11 January 2023).
31. Filom, S.; Radfar, S.; Panahi, R. A Comparative Study of Different Wind Speed Distribution Models for Accurate Evaluation of Onshore Wind Energy Potential: A Case Study on the Southern Coasts of Iran. *Energy Fuel Technol.* **2020**. [CrossRef]
32. Mazzeo, D.; Oliveti, G.; Labonia, E. Estimation of wind speed probability density function using a mixture of two truncated normal distributions. *Renew. Energy* **2018**, *115*, 1260–1280. [CrossRef]
33. Li, M.; Li, X. MEP-type distribution function: A better alternative to Weibull function for wind speed distributions. *Renew. Energy* **2005**, *30*, 1221–1240. [CrossRef]
34. Stathopoulos, T.; Alrawashdeh, H.; Al-Quraan, A.; Blocken, B.; Dilimulati, A.; Paraschivoiu, M.; Pilay, P. Urban wind energy: Some views on potential and challenges. *J. Wind. Eng. Ind. Aerodyn.* **2018**, *179*, 146–157. [CrossRef]
35. Al-Masri, H.M.K.; Al-Quraan, A.; AbuElrub, A.; Ehsani, M. Optimal Coordination of Wind Power and Pumped Hydro Energy Storage. *Energies* **2019**, *12*, 4387. [CrossRef]
36. Usta, I. An innovative estimation method regarding Weibull parameters for wind energy applications. *Energy* **2016**, *106*, 301–314. [CrossRef]
37. Al-Quraan, A.; Al-Mahmodi, M.; Radaideh, A.; Al-Masri, H.M.K. Comparative study between measured and estimated wind energy yield. *Turk. J. Electr. Eng. Comput. Sci.* **2020**, *28*, 2926–2939. [CrossRef]
38. Al-Quraan, A.; Stathopoulos, T.; Pillay, P. Comparison of wind tunnel and on site measurements for urban wind energy estimation of potential yield. *J. Wind. Eng. Ind. Aerodyn.* **2016**, *158*, 1–10. [CrossRef]
39. Kevin, O.O.; Otumba, E.; David, A.A.; Matuya, J. Fitting Wind Speed to a Two Parameter Distribution Model Using Maximum Likelihood Estimation Method. *Int. J. Stat. Distrib. Appl.* **2020**, *6*, 57. [CrossRef]
40. Blažević, R. Statistical Analysis and Assessment of Wind Energy Potential in Sarajevo, Bosnia and Herzegovina. *Teh. Vjesn.-Tech. Gaz.* **2020**, *28*, 71–83.
41. Boro, D.; Thierry, K.; Kieno, F.P.; Bathiebo, J. Assessing the Best Fit Probability Distribution Model for Wind Speed Data for Different Sites of Burkina Faso. *Curr. J. Appl. Sci. Technol.* **2020**, *39*, 71–83. [CrossRef]
42. Saeed, M.A.; Ahmed, Z.; Zhang, W. Optimal approach for wind resource assessment using Kolmogorov–Smirnov statistic: A case study for large-scale wind farm in Pakistan. *Renew. Energy* **2021**, *168*, 1229–1248. [CrossRef]
43. Bidaoui, H.; Abbassi, I.E.; Bouardi, A.E.; Darcherif, A. Wind Speed Data Analysis Using Weibull and Rayleigh Distribution Functions, Case Study: Five Cities Northern Morocco. *Procedia Manuf.* **2019**, *32*, 786–793. [CrossRef]

44. Abeysirigunawardena, D.S.; Gilleland, E.; Bronaugh, D.; Wong, P. Extreme wind regime responses to climate variability and change in the inner south coast of British Columbia, Canada. *Atmosphere-Ocean* **2009**, *47*, 41–62. [CrossRef]
45. Baloch, Z.A.; Tan, Q.; Kamran, H.W.; Nawaz, M.A.; Albashar, G.; Hameed, J. A multi-perspective assessment approach of renewable energy production: Policy perspective analysis. *Environ. Dev. Sustain.* **2021**, *24*, 2164–2192. [CrossRef]
46. Sumair, M.; Aized, T.; Gardezi, S.A.R.; Bhutta, M.M.A.; Rehman, S.M.S.; ur Rehman, S.U. Comparison of three probability distributions and techno-economic analysis of wind energy production along the coastal belt of Pakistan. *Energy Explor. Exploit.* **2020**, *39*, 2191–2213. [CrossRef]
47. Hemalatha, S.; Selwyn, T.S. Computation of mechanical reliability for Sub- assemblies of 250 kW wind turbine through sensitivity analysis. *Mater. Today Proc.* **2021**, *46*, 3180–3186. [CrossRef]
48. Trevisi, F.; McWilliam, M.; Gaunaa, M. Configuration optimization and global sensitivity analysis of Ground-Gen and Fly-Gen Airborne Wind Energy Systems. *Renew. Energy* **2021**, *178*, 385–402. [CrossRef]
49. Gupta, A.; Mishra, P.; Pandey, C.; Singh, U.; Sahu, C.; Keshri, A. Descriptive Statistics and Normality Tests for Statistical Data. *Ann. Card. Anaesth.* **2019**, *22*, 67. [CrossRef]
50. Viti, A.; Terzi, A.; Bertolaccini, L. A practical overview on probability distributions. *J. Thorac. Dis.* **2015**, *7*, E7–E10. [CrossRef]
51. Serban, A.; Paraschiv, L.S.; Paraschiv, S. Assessment of wind energy potential based on Weibull and Rayleigh distribution models. *Energy Rep.* **2020**, *6*, 250–267. [CrossRef]
52. Honrubia, A.; Vigueras, A.; Gomez, E.; Mejıas, M.; Lainez, I. Comparative analysis between lidar technologies and common wind speed meters. In Proceedings of the World Wind Energy Conference, Istanbul, Turkey, 27 June 2010.
53. Manwell, J.F.; Mcgowan, J.G.; Rogers, A.L. *Wind Energy Explained: Theory, Design and Application;* John Wiley & Sons, Ltd.: Chichester, UK, 2011; Available online: https://www.wiley.com/en-us/Wind+Energy+Explained%3A+Theory%2C+Design+and+Application%2C+2nd+Edition-p-9780470015001 (accessed on 11 January 2023).
54. An, Y.; Quan, Y.; Gu, M. Field Measurement of Wind Characteristics of Typhoon Muifa on the Shanghai World Financial Center. *Int. J. Distrib. Sens. Netw.* **2012**, *8*, 893739. [CrossRef]
55. Bagiorgas, H.S.; Mihalakakou, G.; Rehman, S.; Al-Hadhrami, L.M. Wind power potential assessment for three buoys data collection stations in the Ionian Sea using Weibull distribution function. *Int. J. Green Energy* **2015**, *13*, 703–714. [CrossRef]
56. GPS Coordinates of Jordan Latitude Longitude Elevation—CountryCoordinate.com. Available online: https://www.countrycoordinate.com/country-jordan/ (accessed on 11 January 2023).
57. Çakmakçı, B.A.; Hüner, E. Evaluation of wind energy potential: A case study. *Energy Sources Part A Recovery Util. Environ. Eff.* **2020**, *44*, 834–852. [CrossRef]
58. Masters, G.M. *Renewable and Efficient Electric Power Systems;* Wiley-Blackwell: Chichester, UK, 2013.
59. Forbes, C.; Evans, M.G.A.; Hastings, N.A.; Peacock, B. *Statistical Distributions;* Wiley: Hoboken, NJ, USA, 2011.
60. Weisstein, E.W. Normal Distribution. Available online: https://mathworld.wolfram.com/NormalDistribution.html (accessed on 11 January 2023).
61. Milan, P.; Wächter, M.; Barth, S.; Peinke, J. Power curves for wind turbines. *Wind. Power Gener. Wind. Turbine Des.* **2010**, *44*, 595–612. [CrossRef]
62. Al-Quraan, A.; Al-Mhairat, B. Intelligent Optimized Wind Turbine Cost Analysis for Different Wind Sites in Jordan. *Sustainability* **2022**, *14*, 3075. [CrossRef]
63. Eminoglu, U.; Turksoy, O. Power curve modeling for wind turbine systems: A comparison study. *Int. J. Ambient. Energy* **2019**, *42*, 1912–1921. [CrossRef]
64. Mölders, N.; Khordakova, D.; Dlugi, R.; Kramm, G. Sustainability of Wind Energy under Changing Wind Regimes—A Case Study. *Atmos. Clim. Sci.* **2016**, *6*, 158–173. [CrossRef]
65. Mellit, A.; Kalogirou, S.A. Artificial intelligence techniques for photovoltaic applications: A review. *Prog. Energy Combust. Sci.* **2008**, *34*, 574–632. [CrossRef]
66. Singh, S.; Patel, B.; Upadhyay, R.K.; Singh, N.K. Improvement of process performance of powder mixed electrical discharge machining by optimisation—A Review. *Adv. Mater. Process. Technol.* **2021**, *8*, 3074–3104. [CrossRef]
67. Katoch, S.; Chauhan, S.S.; Kumar, V. A review on genetic algorithm: Past, present, and future. *Tools Appl.* **2021**, *80*, 8091–8126. [CrossRef]
68. Liu, Y.; Passino, K.M. Biomimicry of Social Foraging Bacteria for Distributed Optimization: Models, Principles, and Emergent Behaviors. *J. Optim. Theory Appl.* **2002**, *115*, 603–628. [CrossRef]
69. Li, Z.; Wang, Z. Bacterial Foraging Algorithm With Potential Field Guidance Mechanism. *Research Square* **2021**. [CrossRef]
70. Odziemczyk, W. Application of simulated annealing algorithm for 3D coordinate transformation problem solution. *Open Geosci.* **2020**, *12*, 491–502. [CrossRef]
71. Venkateswaran, C.; Ramachandran, M.; Ramu, K.; Prasanth, V.; Mathivanan, G. Application of Simulated Annealing in Various Field. *Mater. Its Charact.* **2022**, *1*, 100299. [CrossRef]
72. Mewada, K.M.; Sinhal, A.; Verma, B. Adaptive neuro-fuzzy inference system (ANFIS) based software evaluation. *Int. J. Comput. Sci. Issues (IJCSI)* **2013**, *10*, 244.
73. Onu, C.E.; Nweke, C.N.; Nwabanne, J.T. Modeling of thermo-chemical pretreatment of yam peel substrate for biogas energy production: RSM, ANN, and ANFIS comparative approach. *Appl. Surf. Sci. Adv.* **2022**, *11*, 100299. [CrossRef]

74. Zeinalnezhad, M.; Chofreh, A.G.; Goni, F.A.; Klemeš, J.J. Air pollution prediction using semi-experimental regression model and Adaptive Neuro-Fuzzy Inference System. *J. Clean. Prod.* **2020**, *261*, 121218. [CrossRef]
75. Hodson, T.O. Root mean square error (RMSE) or mean absolute error (MAE): When to use them or not. *Geosci. Model Dev.* **2022**, *15*, 5481–5487. [CrossRef]
76. Tamilarasi, P.; Rani, R. Diagnosis of Crime Rate against Women using k-fold Cross Validation through Machine Learning. In Proceedings of the 2020 Fourth International Conference on Computing Methodologies and Communication (ICCMC), Erode, India, 11–13 March 2020. [CrossRef]
77. Rodriguez, J.D.; Perez, A.; Lozano, J.A. Sensitivity Analysis of k-Fold Cross Validation in Prediction Error Estimation. *IEEE Trans. Pattern Anal. Mach. Intell.* **2010**, *32*, 569–575. [CrossRef]
78. Breck, E.; Polyzotis, N.; Roy, S.; Whang, S.; Zinkevich, M. Data Validation for Machine Learning. In Proceedings of the 2nd SysML Conference, Palo Alto, CA, USA, 31 March–2 April 2019.

sustainability

MDPI

Article

Mechanisms for Choosing PV Locations That Allow for the Most Sustainable Usage of Solar Energy

Syed Hammad Mian [1,*], Khaja Moiduddin [1], Hisham Alkhalefah [1], Mustufa Haider Abidi [1], Faraz Ahmed [2] and Faraz Hussain Hashmi [2]

[1] Department of Industrial Engineering, College of Engineering, King Saud University, P.O. Box 800, Riyadh 11421, Saudi Arabia
[2] Department of Mechanical Engineering, College of Engineering, King Saud University, P.O. Box 800, Riyadh 11421, Saudi Arabia
* Correspondence: smien@ksu.edu.sa

Abstract: The electrical power need in the Kingdom of Saudi Arabia (KSA) has been escalating at a rapid rate of about 7.5% annually. It has the third highest usage rate in the world as stated by World Energy Council statistics. The rising energy demand has a significant impact on the country's economy since oil is considered to be its mainstay. Additionally, conventional energy production using fossil fuels is a leading contributor to ecological degradation and adversely influences human health. As a result, Saudi Arabia has taken significant steps to shift from its current status of total reliance on oil to new frontiers of exploration of other kinds of renewable energies. Photovoltaic (PV) solar energy is the most preferred renewable energy to be harnessed in Saudi Arabia. In accordance with Vision 2030, the KSA intends to generate at least 9.5 GW of electricity from green sources, a significant portion of which will come from solar PV power. Since the site peculiarities have a huge influence on the project's technical and economic dimensions, the scaled-up deployment of solar projects calls for a judicious selection of PV sites. Undoubtedly, performing a thorough solar site survey is the foremost step to establishing a financially viable and successful solar project. Multiple criterion decision-making (MCDM) strategies can be very helpful in making judgments, given that a number of criteria might influence PV site selection. The objective of this research is to provide valuable information on various MCDM approaches that can be utilized to select optimal locations for PV solar plants. A number of variables, including topography, air temperature, dust storms, solar radiation, etc., are considered in this analysis. This study has combined various MCDM techniques in order for the strengths of each method to outweigh the weaknesses of the others. It has been deduced from this analysis that the most crucial factors in choosing PV sites are solar radiation and sunshine hours. It has also been concluded that of the surveyed cities, Tabuk is the optimum location for the construction of a solar power plant due to its high GHI value of 5992 W/m^2/day and abundant sunshine hours of 12.16 h/day. Additionally, the FAHP-VIKOR method is noted as being the most rigorous, whereas Entropy-GRA is the simplest method.

Keywords: solar energy; renewable energy; multi criteria decision-making approaches; clean energy; site selection

Citation: Mian, S.H.; Moiduddin, K.; Alkhalefah, H.; Abidi, M.H.; Ahmed, F.; Hashmi, F.H. Mechanisms for Choosing PV Locations That Allow for the Most Sustainable Usage of Solar Energy. *Sustainability* **2023**, *15*, 3284. https://doi.org/10.3390/su15043284

Academic Editors: Ayman Al-Quraan and Ahmad M.A. Malkawi

Received: 10 January 2023
Revised: 2 February 2023
Accepted: 6 February 2023
Published: 10 February 2023

1. Introduction

The electricity usage in the Kingdom of Saudi Arabia (KSA) is expanding every year, and it has the third-highest utilization rate worldwide as per World Energy Council statistics [1]. In the last ten years or so, KSA's demand for energy has been growing at a rapid rate of about 7.5% annually [1]. One of the primary issues the power companies in Saudi Arabia are dealing with is the rising energy demand, which leads to the burning of more barrels of carbon-based fuel, which has an impact on the country's economy since oil is considered to be its mainstay. Furthermore, it is well-known that the use of fossil fuels in traditional energy generation contributes significantly to environmental deterioration and has a detrimental effect on human health. Saudi Arabia has thus made tremendous efforts to move away from its current state of complete reliance on oil and move toward new horizons of exploration of other types of renewable energies.

Solar energy is one of the most prominent sustainable energy sources, which is rising in popularity due to its many benefits, including its ability to reduce reliance on energy sources, e.g., oil and gas [2,3]. In addition, solar energy is a dependable, unending, clean alternative that is safe for the environment. Solar photovoltaic (PV) electricity is the most popular green energy source in Saudi Arabia. Its location in the sun belt and favorable spatial conditions make it one of the world's top solar energy producers [4]. The KSA is fortunate to have access to a wealth of solar energy resources. There are vast undeveloped land areas that might be used to house solar projects, and Saudi Arabia's average daily global horizontal irradiance (GHI) receives between 5.7 and 6.7 kWh/m^2 on average [5]. In accordance with Vision 2030, the KSA intends to generate 9.5 GW of electricity from renewable sources, a significant portion of which will come from solar PV power [6]. Additionally, the cost of PV modules has decreased by an average of 70% over the past ten years, mostly due to improvements in their efficiency [7]. The scaled-up deployment of solar projects necessitates a precise evaluation and analysis of PV site selection. A thorough solar site survey is the first step in ensuring a solar project is both affordable and efficient.

The location of a solar PV power plant is vital since the site's attributes directly affect the project's technical and financial elements, and consequently, its viability [8]. The selection of the ideal solar energy site, which is crucial to their installation, is influenced by a variety of variables. These factors should be addressed to obtain more energy while also lowering startup and ongoing operating expenses [9]. These variables must be considered in the first stages of solar energy installation to appropriately locate the plant. Considering that a number of criteria might drive the choice of location, implementing multiple criteria decision-making (MCDM) methodologies can greatly help in making judgments regarding site selection for PV solar energy systems by taking important elements into consideration [10]. Numerous investigations have been conducted in the literature in order to locate solar power plants [11–14]. MCDM provides a potent decision-making method that may be used in numerous applications, including the site selection for solar power projects [15–17]. Seven MCDM techniques were used by Villacreses et al. [18] to identify suitable locations for PV solar farm installation. The analysis used nine variables, four constraints, and the Analytic Hierarchy Process (AHP) approach to weigh the factors. The Pearson correlation coefficient was used to analyze seven MCDM outcomes. Wang et al. [14] consolidated three techniques, such as the fuzzy analytical hierarchy process (FAHP), data envelopment analysis (DEA), and the technique for order of preference by similarity to ideal solution (TOPSIS), to discover the best location for a solar power facility combining both quantitative and qualitative metrics. Similarly, Mirzaei [19] coupled MCDM and fuzzy logic to assess possible Turkish cities for the construction of newer solar power stations. Stepwise Weight Assessment Ratio Analysis (SWARA) was used in the suggested procedure to measure weights, and the Pythagorean Fuzzy Form of TOPSIS was used to grade the locations. A two-step methodology premised on FAHP and DEA was presented by Lee et al. [20] for assessing the viability of potential locations for sustainable energy plant sites. The FAHP was applied as the initial step to establish the assurance region (AR) of the input variables, and the AR was integrated into DEA to evaluate the efficacy of potential plant sites. The best-worst method (BWM) was employed to quantify the criterion and sub-criteria in a hybrid MCDM approach [13]. Grey relational analysis (GRA) and Vlsekriterijumska Optimizacija I Kompromisno Resenje (VIKOR) were used to rank the prospective locations. Additionally, a novel Monte Carlo simulation-based (MCSB) methodology was used to examine the sensitivity of GRA and VIKOR. Similarly, to identify the ideal PV system for Saudi Arabia, Al-Shammari et al. [21] utilized AHP and TOPSIS effectively. An MCDM strategy for the site location of PV charging stations was also used by Dang et al. [22] to choose the optimal alternative in China. The fuzzy VIKOR approach was used to rank the solutions, and the fuzzy measure technique was utilized to calculate the weight of the criteria. AHP, the entropy weight method, the fuzzy measure method, and VIKOR were all successfully used in the hybrid fuzzy approach. Seker and Kahraman [23] established a thorough two-stage MCDM model including the AHP and Multiplicative Multi-Objective Ratio Analysis (MULTIMOORA) approaches to select the best PV panel maker for solar power plants. The robustness and verification of the proposed strategy for the solar power industry were also demonstrated by the sensitivity and comparison analyses. Similarly, Wang et al. [24] established a fuzzy MCDM approach by integrating FAHP with DEA to identify a solar panel supplier for a PV system design. They also used a number of DEA models for rating possible suppliers in the last phases.

The preceding works cited in this research serve as proof of the extensive literature that is available in various domains of solar energy applications employing MCDM. The prior studies provide examples of different MCDM methodology behavior when several conflicting variables are considered. It is true that earlier studies have used MCDM to determine the ideal location for PV installation. None of them have highlighted the basis for choosing an appropriate MCDM approach, rather, they have all emphasized choosing a suitable location utilizing any MCDM based on their convenience. The key challenge for the decision-makers has always been deciding which MCDM method to use. Consequently, this study is an effort in which several MCDM methods have been employed and evaluated. Thus, the objective of this research is to provide valuable information on various MCDM combinations that can be utilized to select optimal locations for PV solar plants. This work explores MCDM methods for evaluating possible solar PV sites by considering a variety of factors, such as solar radiation, air temperature, dust storms, topology, etc. The suggested methodology is based on a combination of different MCDM techniques, such as the group eigenvalue method (GEM), FAHP, entropy method, VIKOR, and GRA. The application of various MCDM techniques, which are coupled and employed as an adaptive approach for the PV site selection problem in Saudi Arabia, is where this research makes a scientific contribution. First, the weights of the criteria are determined using GEM, FAHP, and entropy approaches. Next, the PV sites are ranked and prioritized using VIKOR, and GRA. To demonstrate the reliability of the evaluation methodologies, comparison, and sensitivity analyses are carried out. The investigation of a specific instance in Saudi Arabia serves to support the applicability of the proposed methodology. This study, which is still in its early phases of development in terms of solar energy, is expected to help the authorities gain a better knowledge of the potential investment in solar energy.

2. MCDM Methods

This work intends to put into practice the combination of different MCDM techniques, including GEM, FAHP, Entropy, VIKOR, and GRA approaches. The combinations, including GEM-VIKOR, FAHP-VIKOR, Entropy-VIKOR, GEM–GRA, FAHP–GRA, and Entropy-GRA have been used to obtain the optimal PV site for the generation of solar energy. The following is a succinct summary of various MCDM techniques.

2.1. Determination of Weights

2.1.1. Group Eigen Value Method

GEM [25] is deployed to give attribute weights or assess their significance by developing an expert judgment matrix. Since one expert only has a limited amount of information and experience, a judgment matrix derived from a group of various specialists should be employed. This approach aims to locate the ideal expert, whose evaluation is most reliable, accurate, and repeatable with the opinions of fellow experts in the team. Following are the phases and equations (Equations (1)–(6)) for the GEM technique [25].

Take into account the assessment of n variables $N_1, N_2, N_3, \ldots, N_n$ by a team of m analysts $M_1, M_2, M_3, \ldots, M_m$. Let x_{ij} represent the analyst's opinion of the j-th variable by *i*-th analyst. Suppose $x = (x_{ij})_{m \times n}$ be the $m \times n$ order matrix (Equation (1)) assessed by the specialists.

$$x = (x_{ij})_{m\times n} = \begin{bmatrix} x_{11} & x_{12} & \dots & x_{1n} \\ x_{21} & x_{22} & \dots & x_{2n} \\ \dots & \dots & \dots & \dots \\ x_{m1} & x_{m2} & \dots & x_{mn} \end{bmatrix} \quad (1)$$

x_* as shown in Equation (2) can be used to express the evaluation vector of a specialist with the greatest judgment quality and exact assessment.

$$x_* = (x_{*1}, x_{*2}, x_{*3}, \dots, x_{*n})^T \quad (2)$$

The summation of angles at which the preferable expert judgment vector and the assessment vectors of other specialists' overlap should be as small as possible. It indicates that x_* can be calculated once the function $f = \sum_{i=1}^{m} (b^T x_i)^2$ reaches its highest value. Consequently, it is possible to determine an idealized expert's assessment vector, x_* using Equation (3).

$$\max_{||b||_2=1} \sum_{i=1}^{m} \left(b^T x_i\right)^2 = \sum_{i=1}^{m} \left(x_*^T x_i\right)^2 \quad (3)$$

The expression $\forall b = (b_1, b_2, \dots, b_n)^T$ is satisfied by the parameter, b^T, which is an eigenvector of F. x_* designates the positive eigenvector equivalent to ρ_{max}, which is the highest possible positive eigenvalue of the matrix $F = x^T.x$. After the eigenvector is normalized to match the highest eigenvalue, the standardized weight vector for every specialist can be found. The procedure listed below should be considered to estimate the criterion weights.

1. Analysts designate assessment scores to specified criteria.
2. Transpose of the assessment matrix and then multiply it by the transposed one as shown in Equation (4).

$$F = x^T.x \quad (4)$$

3. The power method, as employed by Qiu in 1997 [25], can be utilized to derive the eigenvector x_*.
   - Suppose $k = 0$, $y_0 = (1/n, 1/n, \dots ..,1/n)^T$

$$y_1 = Fy_0;\ z_1 = \frac{y_1}{||y_1||_2} \quad (5)$$

   - For $k = 1, 2, 3, \dots$, $y_{k+1} = Fz_k$, and $z_{k+1} = \frac{y_{k+1}}{||y_{k+1}||_2}$
   - Verify if $|z_{k\to k+1}| \le \varepsilon$, if it does, then, z_{k+1} equates to x_*, else return to the prior step (with ε representing the precision and $|z_{k\to k+1}|$ is the maximum absolute value of the difference between z_k and z_{k+1})
4. Normalization of the derived eigenvector using Equation (6).

$$w_j = x_{*j} / \sum_{j=1}^{n} x_{*j} \quad (6)$$

where, $j = 1, 2, 3, \dots, n$, so that $\sum_{j=1}^{n} x_{*j} = 1$ and x_{*j} is the weight obtained for criteria utilizing GEM.

2.1.2. Fuzzy Analytic Hierarchy Process

The fuzzy set principle proposed by Zadeh [26] is used to deal with the ambiguity and vagueness of specialist views as well as information gathering. The FAHP can methodically oversee design choices since it is founded on the fundamentals of fuzzy set theory and hierarchical structural analysis. The FAHP, however, is an expansion of a conventional AHP technique using fuzzy numbers in a fuzzy environment [27]. The initial conversion of the typical AHP into FAHP was posited by Van Laarhoven and Pedrycz [28], Buckley [29], and

Chang [30]. They used fuzzy numbers with triangular membership functions to indicate the relative importance of the Saaty. In this work, the crisp values are modeled by applying a trapezoidal membership function because of its greater effectiveness over the triangular membership function [31]. The following are the steps used to estimate fuzzy weights.

Step 1: translating crisp values into fuzzy numbers.

The decision matrices generated by many experts are adapted into trapezoidal fuzzy numbers in this stage. The membership function of the F-designated trapezoidal fuzzy number with the characteristics (l, m, n, u) can be depicted using Equation (7) [32].

Membership function,

$$\mu_F(x) = \begin{cases} 0 & x \leq l \\ \frac{x-l}{m-l} & l \leq x \leq m \\ 1 & m \leq x \leq n \\ \frac{u-x}{u-n} & n \leq x \leq u \\ 0 & x \geq u \end{cases} \tag{7}$$

The Saaty scale-based crisp values are initially transformed into triangular fuzzy numbers and subsequently into trapezoidal fuzzy numbers to accomplish this conversion.

The triangle fuzzy numbers are converted into trapezoid numbers by consistently retaining the upper and bottom limits of the fuzzy numbers and slightly increasing the center [33]. As an illustration, consider the transformation of the crisp value of x into a triangular fuzzy number (a, b, c) where $a = x - 1$; $b = x$ and $c = x + 1$. Following this transformation, the triangular fuzzy numbers have been changed into trapezoidal fuzzy numbers (l, m, n, u), where $l = a$; $u = c$; $m = l + 0.5$ and $n = u - 0.5$.

Step 2: weights computation.

The Chang extent analysis is adopted in this work to derive the weights of the performance indicators [30]. For computing priori weights, the extent analysis is outlined as follows [34,35].

Suppose $X = \{x_1, x_2, x_3, \ldots, x_n\}$ symbolize the array of objects and $G = \{g_1, g_2, g_3, \ldots, g_n\}$ signify the list of objectives. symbolize the collection of objects. Every object must experience extent analysis for each objective of the problem, according to the extent analysis framework. As a result, m extent analysis values for each object can be obtained as $M^1{}_{gi}, M^2{}_{gi}, M^3{}_{gi}, M^4{}_{gi}, \ldots, M^m{}_{gi}$, $i = 1, 2, 3, 4, \ldots, n$. Here, $M^j{}_{gi}$ ($j = 1, 2, 3, 4, \ldots, m$) is an illustration of a fuzzy trapezoidal number. The succeeding steps (Equations (8)–(15)) can be undertaken to complete Chang's extent analysis [30].

The synthetic fuzzy values for i-th object can be derived using Equation (8).

$$S_i = \sum_{j=1}^{m} M_{gi}^{j} \otimes \left[\sum_{i=1}^{n}\sum_{j=1}^{m} M_{gi}^{j}\right]^{-1} \tag{8}$$

Because FAHP evaluates the relevant attribute using a trapezoidal fuzzy number with four values. As a result, the fuzzy addition action for a specific matrix can be accomplished by employing the expression in Equation (9) to estimate $\sum_{j=1}^{m} M_{gi}^{j}$.

$$\sum_{j=1}^{m} M_{gi}^{j} = \left(\sum_{j=1}^{m} l_j, \sum_{j=1}^{m} m_j, \sum_{j=1}^{m} n_j, \sum_{j=1}^{m} u_j\right) \tag{9}$$

The value of $\left[\sum_{i=1}^{n}\sum_{j=1}^{m} M_{gi}^{j}\right]^{-1}$ can be calculated by the fuzzy addition process of $M^j{}_{gi}$ ($j = 1, 2, 3, 4, \ldots, m$) values in the following way by applying Equation (10).

$$\sum_{i=1}^{n}\sum_{j=1}^{m} M_{gi}^{j} = \left(\sum_{i=1}^{n} l_i, \sum_{i=1}^{n} m_i, \sum_{i=1}^{n} n_i, \sum_{i=1}^{n} u_i\right) \tag{10}$$

Additionally, the vector's inverse can be defined using Equation (11).

$$\left[\sum_{i=1}^{n}\sum_{j=1}^{m} M_{gi}^{j}\right]^{-1} = \left(\frac{1}{\sum_{i=1}^{n} u_i}, \frac{1}{\sum_{i=1}^{n} n_i}, \frac{1}{\sum_{i=1}^{n} m_i}, \frac{1}{\sum_{i=1}^{n} l_i}\right) \tag{11}$$

Suppose $M_1 = (l_1, m_1, n_1, u_1)$ and $M_2 = (l_2, m_2, n_2, u_2)$ denotes two trapezoidal fuzzy numbers. The following condition in Equation (12) affects how likely it is that $M_2 = (l_2, m_2, n_2, u_2) \geq M_1 = (l_1, m_1, n_1, u_1)$.

$$V\,(M_2 \geq M_1) = \sup_{y \geq x} \lfloor \min(\mu_{M_1}(x), \mu_{M_2}(y)) \rfloor \tag{12}$$

where, $V\,(M_2 \geq M_1)$ can be expressed using the definition in Equation (13).

$$V\,(M_2 \geq M_1) = \text{hgt}(M_1 \cap M_2) = \mu_{M_2}(d) = \begin{cases} 1, & m_1 \geq m_2 \\ 0, & (m_2 - n_1) > (u_1 + l_2) \\ \frac{((n_1 - m_2) + (u_1 + l_2))}{(u_1 + l_2)}, & 0 < (m_2 - n_1) < (u_1 + l_2) \\ \frac{((m_2 - n_1) + (u_1 + l_2))}{(u_1 + l_2)}, & (m_2 - n_1) < (u_1 + l_2), \text{ where } m_2 < n_1 \text{ and } m_1 < m_2 \end{cases} \tag{13}$$

A fuzzy number with a probability greater than k can be represented as $V\,(M_2 \geq M_1, M_2, M_3, M_4, \ldots, M_k) = V\,[(M \geq M_1)$ and $V\,[(M \geq M_2)$ and $V\,[(M \geq M_3)$ and $V\,[(M \geq M_4)$ and $V\,[(M \geq M_k)] = \min V\,(M \geq M_i), i = 1, 2, 3, \ldots, k$.

Assume that $d\,(A_i) = \min V\,(S_i \geq S_k)$ for $k = 1, 2, 3, 4, \ldots, n; k \neq i$. The weight vector can be estimated as depicted in Equation (14).

$$w' = \left(d'(A_1),\ d'(A_2), d'(A_3), d'(A_4), \ldots, d'(A_n)\ \right)^T \tag{14}$$

where A_i ($i = 1, 2, 3, 4, \ldots, n$) indicates n elements. Normalization is needed to obtain the weight vector for the individual elements. The weight vector is normalized to get the vector normalized as in Equation (15).

$$w = (d(A_1),\ d(A_2), d(A_3), d(A_4), \ldots, d(A_n)\ )^T \tag{15}$$

where w is a vector of non-fuzzy numbers or crisp values.

2.1.3. Entropy Method

Entropy weight calculation is an empirical approach that uses the attribute's innate knowledge to compute its weights. Accordingly, noise in the resultant weights can be reduced, leading to the production of impartial outcomes [36]. The attribute's higher relevance is indicated by its greater entropy weight. The entropy weight approach makes use of the attribute's intrinsic information to make the resulting weights more realistic than subjective [37].

If there are m alternatives and n criteria, x_{ij} reflects the value for j-th criterion and i-th alternative. The entropy weight can be calculated as seen below using Equations ((16)–(21)) [36,37]:

Standardization of attribute values, r_{ij} using Equations (16)–(17).

$$\text{Benefit attribute} = \frac{x}{x_{max}}\ (i = 1,2,3,4\ldots,m; j = 1,2,3,\ldots,n) \tag{16}$$

$$\text{Cost attribute} = \frac{x_{min}}{x}(i = 1,2,3,4\ldots,m; j = 1,2,3,\ldots,n) \tag{17}$$

Quantification of entropy using Equation (19).

$$H_j = -\frac{\sum_{i=1}^{m} P_{ij} \ln P_{ij}}{\ln m}(i = 1,2,3,4\ldots,m; j = 1,2,3,\ldots,n) \tag{18}$$

$$P_{ij} = \frac{r_{ij}}{\sum_{i=1}^{m} r_{ij}}(i = 1,2,3,4\ldots,m; j = 1,2,3,\ldots,n) \tag{19}$$

Estimation of Entropy weight using Equation (20).

$$w_j = 1 - \frac{1 - H_j}{n - \sum_{j=1}^{n} H_j} \tag{20}$$

$$\sum_{j=1}^{n} w_j = 1 (j = 1, 2, 3, \ldots, n) \tag{21}$$

2.2. *Ranking Approaches*

2.2.1. VIKOR Method

The various procedures required to establish VIKOR-based MCDM [38] tools are covered in this section. Assume there are m possibilities S_i (i = 1, 2, ... , m) and n criteria A_j (j = 1, 2, ... , n). The VIKOR approach's major goal is to assess both the positive and negative ideal positions in the candidate solutions. The applied equations (Equations (22)–(25)) to implement the VIKOR approach are discussed in the following steps [38].

Step 1. The first step is to create the decision matrix, which is written as $X = (x_{ij})_{m \times n}$. Here, x_{ij} stands for real numbers that represent the values of the j-th criterion for the alternative i.

Step 2. Apply Equation (22) to determine the normalized decision matrix, $(r_{ij})_{m \times n}$.

$$(r_{ij})_{m \times n} = \frac{x_{ij}}{\sqrt{\sum (x_{ij})^2}} \tag{22}$$

Step 3. Specify the upper and lower bounds for the normalized decision matrix. For the benefit criteria, $v_j^+ = \max_j r_{ij}$ and $v_j^- = \min_j r_{ij}$. For the non-benefit condition, $v_j^+ = \min_j r_{ij}$ and $v_j^- = \max_j r_{ij}$.

Step 4. Using Equations (23) and (24), determine how far away each alternative is from the best choice, which is represented by the utility (S_i) and regret (R_i) measures.

$$S_i = \sum_{j=1}^{n} w_j \frac{v_j^+ - v_{ij}}{v_j^+ - v_j^-} \tag{23}$$

$$R_i = \max_j (w_j \frac{v_j^+ - v_{ij}}{v_j^+ - v_j^-}) \tag{24}$$

Step 5. Estimate the values for Q_i (rank indexes) by applying Equation (25).

$$Q_i = \left(\varphi * \frac{S_i - S^+}{S^- - S^+} + (1 - \varphi) * \frac{R_i - R^+}{R^- - R^+} \right) \tag{25}$$

where, $S^+ = \min_i S_i$, $S^- = \max_i S_i$, $R^+ = \min_i R_i$, $R^- = \max_i S_i$

The weight for the "majority criteria" (also known as "the maximum group utility") technique is represented by $\varnothing \in [0,1]$. The lowest scores are based on R_i values, while the rate of quality is dependent on S_i values.

Step 6. In accordance with the Q_i values, arrange the best options in ascending order.

The VIKOR technique suggests a compromise option for the alternative, that is the best rated by Q (minimum) when the underlying two requirements are met [38,39].

C1. Allowable advantage

$Q_2 - Q_1 \geq DQ$, where i = 2 is the second-best choice using Q and $DQ = 1/(m - 1)$.

C2. Allowable stability

Additionally, R and/or S must also rank the option with i = 1 as the highest. It gives a number of potential options in the event that one of the requirements is not met.

- Alternatives i = 1 and i = 2 if only condition C_2 is not achieved or

- Alternatives $i = 1, 2, \ldots, m$, if the condition C_1 is not attained, where m is specified by the $Q_M - Q_1 < DQ$ relationship, for maximum i.

2.2.2. Grey Relation Analysis

Prof. Deng first established the concept of grey theory from the grey set in conjunction with control theory and space theory [40]. Grey theory is being used with the intention of exploiting its capacity to account for the complexity and ambiguity related to user preferences and data collection. The following method and Equations (26)–(32) can be used to implement GRA [41,42].

If $x_i^*(k)$ denotes the sequence following the data processing, $x_i^{(O)}(k)$ indicates the original sequence of responses, (where $i = 1, 2, \ldots, m$ and $k = 1, 2, \ldots, n$), max $x_i^{(O)}(k)$ and min $x_i^{(O)}(k)$ denote the highest and lowest values of $x_i^{(O)}(k)$, respectively. The following is how the data is standardized (see Equations (26)–(28)).

$$x_i^*(k) = \frac{x_i^{(O)}(k) - min\ x_i^{(O)}(k)}{\max x_i^{(O)}(k) - \min x_i^{(O)}(k)} \text{Larger-the-better} \tag{26}$$

$$x_i^*(k) = \frac{max\ x_i^{(O)}(k) - x_i^{(O)}(k)}{\max x_i^{(O)}(k) - \min x_i^{(O)}(k)} \text{Smaller-the-better} \tag{27}$$

$$x_i^*(k) = \frac{\left|x_i^{(O)}(k) - IV\right|}{\max\left\{max\ x_i^{(O)}(k) - IV,\ IV - \min x_i^{(O)}(k)\right\}} \text{Nominal-the-better (define the response as intended value (IV))} \tag{28}$$

A reference sequence is established employing the comparability sequences. Upon data processing, the preprocessed sequences are used to estimate the grey relational coefficient (*GRC*). Equation (29) is used to calculate the *GRC*.

$$GRC\ (x_O^*(k),\ x_i^*(k)) = \frac{\Delta_{min} + \varepsilon\ \Delta_{max}}{\Delta_{oi}(k) + \epsilon\ \Delta_{max}} \tag{29}$$

where $\Delta_{oi}\ (k)$ is the deviation sequence of the reference sequence $x_O^*(k)$ and the comparability $x_i^*(k)$, i.e., $\Delta_{oi}\ (k) = |x_O^*(k) - x_i^*(k)|$ is the absolute magnitude of the difference between $x_O^*(k)$ and $x_i^*(k)$.

$$\Delta_{min} = \min_{\forall i} \min_{\forall k} \Delta_{oi}(k) \tag{30}$$

$$\Delta_{max} = \max_{\forall i} \max_{\forall k} \Delta_{oi}(k) \tag{31}$$

where $i = 1, 2, \ldots, m$ and $k = 1, 2, \ldots, n$ ϵ: distinguishing coefficient, ϵ [0,1]. The value of ϵ is set as 0.5. The grey relational grade (GRG), which is calculated using Equations (30) and (31), can be described as a weighting sum of the GRC.

$$GRG\ (x_o, x_i^*) = \sum_{k=1}^{n} \beta_k \gamma\ (x_o * (k), x_i * (k)) \tag{32}$$

where β_k represented the weighting value of the kth performance characteristic, and $\sum_{k=1}^{n} \beta_k = 1$.

The rating is calculated for each user and also for an imaginary user whose needs are the sum of each of the four unique users.

2.3. Sensitivity Analysis

The consistency and dependability of the ranking produced using the suggested approach are assessed using sensitivity analysis. It can be described as a technique for understanding how changes in input values affect a model's results [43]. Undoubtedly, it is important to consider the changes in the proposed model's output parameters brought on by changes in the values of the input variable [44,45]. The model is reliable and trustworthy if the result is not highly sensitive to changes in input. The most popular sensitive analysis method is changing the performance criterion weights and examining how the outcomes evolve [46]. Previous studies have also shown that the ranks of the alternatives strongly depend on the weight coefficients of the criterion [47]. Sensitivity analysis based on variations in the weight coefficients is used in this study to confirm the model's validity and check the accuracy of the findings. The first stage in sensitivity analysis is the random assignment of a criterion. The chosen criterion's weight is then changed by a certain percentage (increase or decrease). As a result, Equation (33) is used to calculate the weights for the remaining criteria [48].

$$w_n^* = \frac{w_n\left(1 - w_i^*\right)}{\left(1 - w_i\right)} \tag{33}$$

where,

w_i: original weights for attribute i

w_i^*: weight derived after adjusting the original weight by 10% for attribute i

w_n: original weight for attribute n

w_n^*: recalculated weight for attribute n

To assess the robustness of the proposed method and to examine the similarity of ranks obtained by varying the weights, Kendall's coefficient (Z) of concordance is used [49,50]. Kendall's coefficient, whose values range from 0 to 1, illustrates analogies in the ranking of sorted quantities. For instance, a value of 1 indicates that all of the different ranking orders perfectly match each other. It implies that the robustness will be better the closer the value of Z is to 1. Equations (34)–(36) can be used to calculate the value of Z [49,50].

$$Z = \frac{12R}{m^2\left(k^3 - k\right)} \tag{34}$$

$$R_i = \sum\nolimits_{j=1}^{m} r_{ij} \tag{35}$$

$$R = \sum\nolimits_{i=1}^{k} \left(R_i - \overline{R}\right)^2 \tag{36}$$

where m is the number of scenarios, k is the number of options, and r_{ij} is the score that scenario i gives to alternative j.

As shown in Table 1, six different scenarios based on weight percentage are devised. These scenarios involve a percentage in weight for a chosen criterion and subsequently changing the weights of the remaining criteria using Equation (36).

Table 1. Scenarios explored in the sensitivity analysis.

Scenario	Description
1	20% increase in weight for average temperature
2	40% increase in weight for the population
3	20% decrease in weight for topography
4	30% increase in weight for global horizontal irradiance
5	40% decrease in weight for dust storm
6	30% decrease in weight for wind speed

3. Data Collection

3.1. Study Area

The KSA is located in the farthest region of southwestern Asia, surrounded by the Red Sea to the west and the Arabian Gulf to the east. The KSA, with a total area of around 2,000,000 square kilometers, makes up approximately four-fifths of the Arab Peninsula [51]. Saudi Arabia has a desert climate, with summers that are exceedingly hot and dry, with temperatures ranging from 27 °C to 43 °C inland and 27 °C to 38 °C by the shore [52].

3.2. Criteria for MCDM

Factors from technological, economical, ecological, social, and risk considerations are examined in order to discover the ideal location for a PV solar power plant to produce electricity [21,53–56]. The following is a brief description of these criteria.

1. Solar radiation (C1). The yearly solar radiation is a meteorological factor that is applied to quantify the intensity of sunlight for a prospective site. The uninterrupted functioning of a PV plant relies upon solar irradiance. The opportunity for producing energy in a location increases with the amount of solar radiation available. In this research, the amount of solar radiation received by a surface is measured using the Global Horizontal Irradiance (GHI) [57]. The unit of measurement for GHI is in Watts per square meter per day (W/m^2/day).
2. Average air temperature (C2). The efficiency of power generation in PV systems is significantly influenced by ambient temperature [58]. The efficiency of power conversion in solar cells decreases with increasing ambient temperature, resulting in a reduction in the amount of power produced. The air temperature is measured in °C.
3. Wind speed (C3). The efficacy of a PV system's energy production is influenced by wind speed. The rate at which solar systems cool down increases with wind speed, which is accompanied by an increase in power production. It is measured in kilometers per hour (km/h).
4. Sunshine hours (C4). The territory with more sun hours has the ability to generate more power when taking into account that various regions receive the same quantity of solar radiation. It is measured in hours (h).
5. Sand and dust storm (C5). A suitable parameter for solar PV systems can be sand and dust storm [59]. Among the globe's places where sand and dust storm existence are particularly intense is the Arabian Peninsula. The extent of radiation reaching the surface of PV panels is decreased with a greater incidence of storm phenomena. As a result, the amount of power generated in locations prone to higher sand storms is also decreased. It is calculated using the average yearly number of storms.
6. Topography (C6). Minimal elevation fluctuations aid in lowering the high construction costs and flat topography is generally preferred for the placement of PV plants. Due to low economic viability, more change in topography or terrain is not acceptable. A maximum elevation change in a specific location can be approximated in feet.
7. Population (C7). A high population in a region causes both a larger energy demand as well as a higher need for employment. Consequently, job generation increases with an increase in population. It implies that having a higher population in a specific area makes it a good location for a PV installation. As a result, the site of a solar system must be preferable to one where there is enough consumption and trained personnel to run and manage the PV system.

Table 2 displays the values of the chosen criteria that have been acquired from various sources [21,52,54,60–65]. Since the information has been gathered from a number of sources, the authors cannot guarantee that it is updated and highly accurate. The main focus of the authors' research is the implementation and viability of MCDM techniques in the problem of solar site selection, even though they have made every effort to acquire data that is as accurate as possible.

We have solely taken technical factors into account due to our consideration of cities. It is presumed that economic considerations, such as distance to power lines and distance to urban centers, are not very significant and can be assumed to be the same for all cities. They have not been found to be important as past literature has claimed [21,53,56].

Table 2. Criteria and their values.

Cities	Criteria						
	C1	C2	C3	C4	C5	C6	C7
Arar	5392	22	14	12.16	0.30	174	148,540
Al-Jouf	5016	21.87	13.5	12.16	0.57	82	102,903
Tabuk	5992	22.98	10	12.16	0.16	125	455,450
Hail	5359	23.3	12	12.15	0.28	387	267,005
Dhahran	5205	26.4	15	12.16	0.15	525	99,540
Al-Ahsa	5278	26.9	12	12.15	0.91	157	293,179
Taif	5671	22.2	14	12.15	0.08	459	530,848
Makkah	5303	28.6	3	12.15	0.02	1106	1,323,624
Jeddah	5525	28.1	13	12.13	0.17	112	2,867,446
Yanbu	5765	27.7	11	12.14	0.48	190	200,161
Medina	5376	27.2	11	12.14	0.09	276	1,300,000
Riyadh	6040	27	11	12.15	0.16	200	4,205,961
Abha	5674	19.2	11	12.15	0.05	1942	210,886
Jizan	4663	28.55	11	12.13	0.60	190	105,198
Najran	6684	23.9	11	12.15	0.26	1486	258,573

4. Implementation

The effectiveness and efficiency of solar power generation are influenced by the site location. A lower benefit-to-cost ratio may result from the site's random selection. A variety of MCDM techniques are combined to find the optimal site. The decision-making tools used in this research have advantages and limitations of their own. The various strategies used to determine the ideal location for solar power generation are as follows. Additionally, the estimation of the weight or importance of various criteria is the most significant step in the use of any MCDM method. Numerous techniques including GEM, FAHP, and entropy approaches have been used in this work. The development of the decision matrix comes before the estimate of weights using these techniques. These decision tables are created with the assistance of professionals with extensive experience in the solar energy sector. Three specialists (E1, E2, and E3), including academics, researchers, etc., are taken into consideration. These experts are asked to rate the chosen criteria on a scale of 1 to 10, with 1 representing the lowest priority and 10 the most important. Table 3 displays the ranking outcomes of the experts.

Table 3. Assessment of criteria by experts.

Criteria	E1	E2	E3
C1	9	10	9
C2	8	7	8
C3	7	8	6
C4	9	9	8
C5	7	8	6
C6	6	5	5
C7	4	6	4

The decision matrix obtained from the experts must be analyzed to see whether their subjective assessment is valid. Thus, the consistency ratio-based approach established by Prof. Saaty is adopted in the paper [66]. In this method, the ratio between the consistency index (*CI*) and random consistency index (*RI*) yields a consistency ratio (*CR*). According to this method, if the *CR* is 10% or less, the inconsistency is acceptable; otherwise, the subjective judgment needs to be changed. An earlier study described in [67] provides an illustration of *CR* estimation. For experts E1, E2, and E3, the calculated *CR* is, respectively, 3.91%, 3.97%, and 4.02%, which is substantially less than 10%. This suggests that the expert's subjective judgment is reasonable.

4.1. GEM-VIKOR

The encoding of ratings by the evaluation matrix x, as illustrated below, is the preliminary step in the use of GEM-VIKOR.

$$x = \begin{pmatrix} 9 & 8 & 7 & 9 & 7 & 6 & 4 \\ 10 & 7 & 8 & 9 & 8 & 5 & 6 \\ 9 & 8 & 6 & 8 & 6 & 5 & 4 \end{pmatrix};\ x^T = \begin{pmatrix} 9 & 10 & 9 \\ 8 & 7 & 8 \\ 7 & 8 & 6 \\ 9 & 9 & 8 \\ 7 & 8 & 6 \\ 6 & 5 & 5 \\ 4 & 6 & 4 \end{pmatrix}$$

$$F = \begin{pmatrix} 262 & 214 & 197 & 243 & 197 & 149 & 132 \\ 214 & 177 & 160 & 199 & 160 & 123 & 106 \\ 197 & 160 & 149 & 183 & 149 & 112 & 100 \\ 243 & 199 & 183 & 226 & 183 & 139 & 122 \\ 197 & 160 & 149 & 183 & 149 & 112 & 100 \\ 149 & 123 & 112 & 139 & 112 & 86 & 74 \\ 132 & 106 & 100 & 122 & 100 & 74 & 68 \end{pmatrix}$$

The ideal evaluation vector x^* has been derived in Table 4 by applying the power approach and assuming precision, = 0.0005.

Table 4. Outcomes from the power method.

k	0	1	2	3
y_k^T	(0.14, 0.14, 0.14, 0.14, 0.14, 0.14, 0.14)	(199.14, 162.71, 150, 185, 150, 113.57, 100.29)	(539.20, 440.77, 406.01, 500.91, 406.01, 307.58, 271.36)	(539.20, 440.77, 406.01, 500.91, 406.01, 307.58, 271.36)
$\|\|y_k\|\|_2$	-	410.262	1110.79	1110.79
z_k^T	-	(0.4854, 0.3966, 0.3656, 0.4509, 0.3656, 0.2768, 0.2444)	(0.4854, 0.3968, 0.3655, 0.4509, 0.3655, 0.2769, 0.2443)	(0.4854, 0.3968, 0.3655, 0.4509, 0.3655, 0.2769, 0.2443)

Ideal evaluation vector, $x_* = (0.4854, 0.3968, 0.3655, 0.4509, 0.3655, 0.2769, 0.2443)^T$

The GHI and sunshine hours, according to experts, are the most important criteria to consider when choosing a location for a solar power plant, whereas population and topology are regarded to be less important. The weights acquired for various criteria after normalization can be rewritten as

w = (0.1877, 0.1535, 0.1414, 0.1744, 0.1414, 0.1071, 0.0945)

The establishment of the decision matrix, as demonstrated in Table 2, is the first step in the VIKOR implementation process. The weights generated from the GEM technique are then used to construct the weighted normalized decision matrix (Table 5).

To compute v_j^+ and v_j^- as shown in Table 6, the different attributes have been categorized as benefit and cost attributes.

Set of benefit criteria = C1, C3, C4, and C7

Set of cost criteria = C2, C5, and C6

The values of S_i, R_i, and Q_i are computed in the final steps of the VIKOR technique as shown in Table 7. The options are then ranked in order of decreasing Q values, with the best choice having the lowest Q value.

Table 5. Weighted normalized decision matrix.

Cities	Criteria						
	C1	C2	C3	C4	C5	C6	C7
Arar	0.0471	0.0346	0.0433	0.0451	0.0293	0.0065	0.0026
Al-Jouf	0.0438	0.0344	0.0417	0.0451	0.0557	0.0031	0.0018
Tabuk	0.0523	0.0361	0.0309	0.0451	0.0158	0.0047	0.0078
Hail	0.0468	0.0366	0.0371	0.0450	0.0276	0.0145	0.0046
Dhahran	0.0455	0.0415	0.0464	0.0451	0.0146	0.0197	0.0017
Al-Ahsa	0.0461	0.0423	0.0371	0.0450	0.0886	0.0059	0.0050
Taif	0.0495	0.0349	0.0433	0.0450	0.0076	0.0173	0.0091
Makkah	0.0463	0.0449	0.0093	0.0450	0.0020	0.0416	0.0228
Jeddah	0.0483	0.0441	0.0402	0.0450	0.0166	0.0042	0.0493
Yanbu	0.0504	0.0435	0.0340	0.0450	0.0466	0.0071	0.0034
Medina	0.0470	0.0427	0.0340	0.0450	0.0089	0.0104	0.0224
Riyadh	0.0528	0.0424	0.0340	0.0450	0.0159	0.0075	0.0723
Abha	0.0496	0.0302	0.0340	0.0450	0.0046	0.0730	0.0036
Jizan	0.0407	0.0448	0.0340	0.0450	0.0588	0.0071	0.0018
Najran	0.0584	0.0375	0.0340	0.0450	0.0249	0.0559	0.0044

Table 6. Estimation of v_j^+ and v_j^-.

	Criteria						
	C1	C2	C3	C4	C5	C6	C7
v_j^+	0.0584	0.0302	0.0464	0.0451	0.0020	0.0031	0.0723
v_j^-	0.0407	0.0449	0.0093	0.0450	0.0886	0.0730	0.0017

Table 7. Computation of S_i, R_i, and Q_i.

Cities	Criteria							S_i	R_i	Q_i
	C1	C2	C3	C4	C5	C6	C7			
Arar	0.0471	0.0346	0.0433	0.0451	0.0293	0.0065	0.0026	0.3206	0.1200	0.1929
Al-Jouf	0.0438	0.0344	0.0417	0.0451	0.0557	0.0031	0.0018	0.3983	0.1550	0.4497
Tabuk	0.0523	0.0361	0.0309	0.0451	0.0158	0.0047	0.0078	0.2961	0.0863	0.0000
Hail	0.0468	0.0366	0.0371	0.0450	0.0276	0.0145	0.0046	0.4335	0.1231	0.3309
Dhahran	0.0455	0.0415	0.0464	0.0451	0.0146	0.0197	0.0017	0.3955	0.1374	0.3601
Al-Ahsa	0.0461	0.0423	0.0371	0.0450	0.0886	0.0059	0.0050	0.5856	0.1414	0.5866
Taif	0.0495	0.0349	0.0433	0.0450	0.0076	0.0173	0.0091	0.3284	0.0941	0.0737
Makkah	0.0463	0.0449	0.0093	0.0450	0.0020	0.0416	0.0228	0.6066	0.1535	0.6692
Jeddah	0.0483	0.0441	0.0402	0.0450	0.0166	0.0042	0.0493	0.5073	0.1744	0.6644
Yanbu	0.0504	0.0435	0.0340	0.0450	0.0466	0.0071	0.0034	0.5586	0.1388	0.5446
Medina	0.0470	0.0427	0.0340	0.0450	0.0089	0.0104	0.0224	0.5049	0.1306	0.4458
Riyadh	0.0528	0.0424	0.0340	0.0450	0.0159	0.0075	0.0723	0.3219	0.1274	0.2305
Abha	0.0496	0.0302	0.0340	0.0450	0.0046	0.0730	0.0036	0.4024	0.1071	0.2182
Jizan	0.0407	0.0448	0.0340	0.0450	0.0588	0.0071	0.0018	0.7553	0.1877	1.0000
Najran	0.0584	0.0375	0.0340	0.0450	0.0249	0.0559	0.0044	0.3910	0.0908	0.1256

Tabuk is ranked first in the VIKOR ranking scheme, followed by Taif and Najran. In VIKOR, a compromise solution is put into practice in order to verify the two conditions mentioned in Section 2.2.2 above. The condition 1 (acceptable advantage) is satisfied since Q_2–Q_1 (0.0737) $\geq$ DQ (0.07143). As a result, it is feasible to differentiate the best one between

Tabuk and Taif. Additionally, the alternative that received the highest Q value also received the highest R and S rankings. Thus, it meets both criteria for acceptability. It implies that condition 2 is also satisfied. Consequently, based on the GEM-VIKOR approach, Tabuk is suggested as the ideal location for a solar power plant.

4.2. FAHP-GRA

In this method, the alternatives are ranked using the GRA technique, and weights are estimated using FAHP. Below are the steps for putting this strategy into practice. First, trapezoidal fuzzy numbers are generated from the crisp values collected from each expert in the form of a decision matrix. Table 8 shows the trapezoidal fuzzy numbers for different levels of significance.

Table 8. Conversion of crisp values to trapezoidal fuzzy numbers.

Importance	Explanation	Trapezoidal Fuzzy Number	Importance	Trapezoidal Fuzzy Number
1	Equal importance	(1,1,1,1)	1	(1,1,1,1)
3	Moderate importance	(2, 2.5, 3.5, 4)	0.3333	(0.25, 0.286, 0.4, 0.5)
5	Strong importance	(4, 4.5, 5.5, 6)	0.2	(0.167, 0.182, 0.222, 0.25)
7	Very strong importance	(6, 6.5, 7.5, 8)	0.1429	(0.125, 0.133, 0.154, 0.167)
9	Extreme importance	(9, 9, 9, 9)	0.1111	(0.111, 0.111, 0.111, 0.111)
2	Intermediate values	(1, 1.5, 2.5, 3)	0.5	(0.333, 0.4, 0.667, 1)
4		(3, 3.5, 4.5, 5)	0.25	(0.2, 0.222, 0.286, 0.333)
6		(5, 5.5, 6.5, 7)	0.1667	(0.143, 0.154, 0.182, 0.2)
8		(7, 7.5, 8.5, 9)	0.125	(0.111, 0.118, 0.133, 0.143)

After transformation into trapezoidal fuzzy numbers, Chang's extent analysis is applied to estimate the weights for various criteria for a specific expert. In this calculation, expert 1′s judgment matrix for several criteria is used as an illustration. The first step is to create a fuzzy pairwise matrix as shown in Table 9.

Table 9. Fuzzy pairwise comparison matrix.

Criteria	C1	C2	C3	C4	C5	C6	C7
C1	(1,1,1,1)	(1, 1.5, 2.5, 3)	(2, 2.5, 3.5, 4)	(1,1,1,1)	(2, 2.5, 3.5, 4)	(3, 3.5, 4.5, 5)	(5, 5.5, 6.5, 7)
C2	(0.333, 0.4, 0.667, 1)	(1,1,1,1)	(1, 1.5, 2.5, 3)	(0.333, 0.4, 0.667, 1)	(1, 1.5, 2.5, 3)	(2, 2.5, 3.5, 4)	(4, 4.5, 5.5, 6)
C3	(0.25, 0.286, 0.4, 0.5)	(0.333, 0.4, 0.667, 1)	(1,1,1,1)	(0.25, 0.286, 0.4, 0.5)	(1,1,1,1)	(1, 1.5, 2.5, 3)	(3, 3.5, 4.5, 5)
C4	(1,1,1,1)	(1, 1.5, 2.5, 3)	(2, 2.5, 3.5, 4)	(1,1,1,1)	(2, 2.5, 3.5, 4)	(3, 3.5, 4.5, 5)	(5, 5.5, 6.5, 7)
C5	(0.25, 0.286, 0.4, 0.5)	(0.333, 0.4, 0.667, 1)	(1,1,1,1)	(0.25, 0.286, 0.4, 0.5)	(1,1,1,1)	(1, 1.5, 2.5, 3)	(3, 3.5, 4.5, 5)
C6	(0.2, 0.222, 0.286, 0.333)	(0.25, 0.286, 0.4, 0.5)	(0.333, 0.4, 0.667, 1)	(0.2, 0.222, 0.286, 0.333)	(0.333, 0.4, 0.667, 1)	(1,1,1,1)	(2, 2.5, 3.5, 4)
C7	(0.143, 0.154, 0.182, 0.2)	(0.167, 0.182, 0.222, 0.25)	(0.2, 0.222, 0.286, 0.333)	(0.143, 0.154, 0.182, 0.2)	(0.2, 0.222, 0.286, 0.333)	(0.25, 0.286, 0.4, 0.5)	(1,1,1,1)

Subsequently, the synthetic fuzzy values are measured utilizing Equations (7)–(14).

$S_1 = (15, 17.5, 22.5, 25) \otimes (1/103.982, 1/91.632, 1/69.994, 1/59.751)$.

$S_2 = (9.666, 11.8, 16.334, 19) \otimes (1/103.982, 1/91.632, 1/69.994, 1/59.751)$.

$S_3 = (6.833, 7.972, 10.467, 12) \otimes (1/103.982, 1/91.632, 1/69.994, 1/59.751)$.

$S_4 = (15, 17.5, 22.5, 25) \otimes (1/103.982, 1/91.632, 1/69.994, 1/59.751)$.

$S_5 = (6.833, 7.972, 10.467, 12) \otimes (1/103.982, 1/91.632, 1/69.994, 1/59.751)$.

$S_6 = (4.316, 5.03, 6.806, 8.166) \otimes (1/103.982, 1/91.632, 1/69.994, 1/59.751)$.

$S_7 = (2.103, 2.22, 2.558, 2.816) \otimes (1/103.982, 1/91.632, 1/69.994, 1/59.751)$.

These fuzzy values are then compared using Equation (12).

V ($S_1 \geq S_2$) = 1, V ($S_1 \geq S_3$) = 1, V ($S_1 \geq S_4$) = 1, V ($S_1 \geq S_5$) = 1, V ($S_1 \geq S_6$) = 1, V ($S_1 \geq Sc_7$) = 1.

V ($S_2 \geq S_1$) = 1, V ($S_2 \geq S_3$) = 1, V ($S_2 \geq S_4$) = 1, V ($S_2 \geq S_5$) = 1, V ($S_2 \geq S_6$) = 1, V ($S_1 \geq S7$) = 1.

V ($S_3 \geq S_1$) = 0.8799, V ($S_3 \geq S_2$) = 1, V ($S_3 \geq S_4$) = 0.8799, V ($S_3 \geq S_5$) = 1, V ($S_3 \geq S_6$) = 1, V ($S_3 \geq S_7$) = 1.

V ($S_4 \geq S_1$) = 1, V ($S_4 \geq S_2$) = 1, V ($S_4 \geq S_3$) = 1, V ($S_4 \geq S_5$) = 1, V ($S_4 \geq S_6$) = 1, V ($S_4 \geq S_7$) = 1.

V ($S_5 \geq S_1$) = 0.8799, V ($S_5 \geq S_2$) = 1, V ($S_5 \geq S_3$) = 0.8799, V ($S_5 \geq S_4$) = 1, V ($S_5 \geq S_6$) = 1, V ($S_5 \geq S_7$) = 1.

V ($S_6 \geq S_1$) = 0.6663, V ($S_6 \geq S_2$) = 0.8627, V ($S_6 \geq S_3$) = 1, V ($S_6 \geq S_4$) = 0.6663, V ($S_6 \geq S_5$) = 1, V ($S_6 \geq S_7$) = 1.

V ($S_7 \geq S_1$) = 0.1931, V ($S_7 \geq S_2$) = 0.3416, V ($S_7 \geq S_3$) = 0.5529, V ($S_7 \geq S_4$) = 0.1931, V ($S_7 \geq S_5$) = 0.5529, V ($S_7 \geq S_6$) = 0.7930.

The weights for various criteria can be estimated by using Equation (13).

$d'(C1) = \min(1, 1, 1, 1, 1, 1) = 1$.

$d'(C2) = \min(1, 1, 1, 1, 1, 1) = 1$.

$d'(C3) = \min(0.8799, 1, 0.8799, 1, 1, 1) = 0.8799$.

$d'(C4) = \min(1, 1, 1, 1, 1, 1) = 1$.

$d'(C5) = \min(0.8799, 1, 0.8799, 1, 1, 1) = 0.8799$.

$d'(C6) = \min(0.6663, 0.8627, 1, 0.6663, 1, 1) = 0.6663$.

$d'(C7) = \min(0.1931, 0.3416, 0.5529, 0.1931, 0.5529, 0.7930) = 0.1931$.

The weight vector can now be expressed as below.

$w' = (1, 1, 0.8799, 1, 0.8799, 0.6663, 0.1931)$.

Upon normalization, the weight vector can be inferred as

$w = (0.1780, 0.1780, 0.1566, 0.1780, 0.1566, 0.1186, 0.0344)$.

The remaining weights for the other decision matrices are calculated in a similar manner. The individual weights are also aggregated by applying the geometric mean (GM) method and normalized as indicated in Table 10.

Table 10. Weights computed using FAHP.

Criteria	E1	E2	E3	Aggregated	Normalized
C1	0.1780	0.1888	0.1835	0.1834	0.1862
C2	0.1780	0.1384	0.1835	0.1654	0.1679
C3	0.1566	0.1762	0.1476	0.1597	0.1622
C4	0.1780	0.1888	0.1835	0.1834	0.1862
C5	0.1566	0.1762	0.1476	0.1597	0.1622
C6	0.1186	0.0404	0.1042	0.0794	0.0806
C7	0.0344	0.0910	0.0500	0.0539	0.0547

After computing the weights, the best location for a solar power plant is determined using the GRA approach. Data normalization is the first step in the GRA implementation process to produce comparability sequences, followed by deviation sequences and GRC. The comparability sequences, deviation sequences, and GRC are shown in Table 11, Table 12, and Table 13, respectively.

The weights acquired through FAHP are then used to obtain the GRG and the corresponding ranking after the GRC has been estimated. Table 14 shows the GRG and the city's respective rankings.

The Arar and Tabuk as stated in Table 14 can be recommended for the construction of a solar power plant based on the study utilizing FAHP and GRA.

Table 11. Comparability sequences after data normalization.

C1	C2	C3	C4	C5	C6	C7
0.3607	0.7021	0.9167	1.0000	0.6857	0.9505	0.0119
0.1747	0.7160	0.8750	1.0000	0.3799	1.0000	0.0008
0.6576	0.5979	0.5833	1.0000	0.8415	0.9769	0.0867
0.3444	0.5638	0.7500	0.6667	0.7045	0.8360	0.0408
0.2682	0.2340	1.0000	1.0000	0.8546	0.7618	0.0000
0.3043	0.1809	0.7500	0.6667	0.0000	0.9597	0.0472
0.4988	0.6809	0.9167	0.6667	0.9353	0.7973	0.1050
0.3167	0.0000	0.0000	0.6667	1.0000	0.4495	0.2981
0.4265	0.0532	0.8333	0.0000	0.8321	0.9839	0.6740
0.5453	0.0957	0.6667	0.3333	0.4859	0.9419	0.0245
0.3528	0.1489	0.6667	0.3333	0.9203	0.8957	0.2923
0.6813	0.1702	0.6667	0.6667	0.8396	0.9366	1.0000
0.5002	1.0000	0.6667	0.6667	0.9700	0.0000	0.0271
0.0000	0.0053	0.6667	0.0000	0.3443	0.9419	0.0014
1.0000	0.5000	0.6667	0.6667	0.7364	0.2452	0.0387

Table 12. Deviation sequences after data pre-processing.

C1	C2	C3	C4	C5	C6	C7
0.6393	0.2979	0.0833	0.0000	0.3143	0.0495	0.9881
0.8253	0.2840	0.1250	0.0000	0.6201	0.0000	0.9992
0.3424	0.4021	0.4167	0.0000	0.1585	0.0231	0.9133
0.6556	0.4362	0.2500	0.3333	0.2955	0.1640	0.9592
0.7318	0.7660	0.0000	0.0000	0.1454	0.2382	1.0000
0.6957	0.8191	0.2500	0.3333	1.0000	0.0403	0.9528
0.5012	0.3191	0.0833	0.3333	0.0647	0.2027	0.8950
0.6833	1.0000	1.0000	0.3333	0.0000	0.5505	0.7019
0.5735	0.9468	0.1667	1.0000	0.1679	0.0161	0.3260
0.4547	0.9043	0.3333	0.6667	0.5141	0.0581	0.9755
0.6472	0.8511	0.3333	0.6667	0.0797	0.1043	0.7077
0.3187	0.8298	0.3333	0.3333	0.1604	0.0634	0.0000
0.4998	0.0000	0.3333	0.3333	0.0300	1.0000	0.9729
1.0000	0.9947	0.3333	1.0000	0.6557	0.0581	0.9986
0.0000	0.5000	0.3333	0.3333	0.2636	0.7548	0.9613

Table 13. GRC computed depending on the deviation sequences.

C1	C2	C3	C4	C5	C6	C7
0.4389	0.6267	0.8571	1.0000	0.6141	0.9100	0.3360
0.3773	0.6377	0.8000	1.0000	0.4464	1.0000	0.3335
0.5935	0.5542	0.5455	1.0000	0.7593	0.9558	0.3538
0.4327	0.5341	0.6667	0.6000	0.6285	0.7530	0.3426
0.4059	0.3950	1.0000	1.0000	0.7747	0.6773	0.3333
0.4182	0.3790	0.6667	0.6000	0.3333	0.9254	0.3442
0.4994	0.6104	0.8571	0.6000	0.8854	0.7116	0.3584
0.4225	0.3333	0.3333	0.6000	1.0000	0.4759	0.4160
0.4658	0.3456	0.7500	0.3333	0.7486	0.9688	0.6054
0.5237	0.3561	0.6000	0.4286	0.4931	0.8960	0.3389
0.4358	0.3701	0.6000	0.4286	0.8625	0.8274	0.4140
0.6108	0.3760	0.6000	0.6000	0.7571	0.8874	1.0000
0.5001	1.0000	0.6000	0.6000	0.9434	0.3333	0.3395
0.3333	0.3345	0.6000	0.3333	0.4326	0.8960	0.3336
1.0000	0.5000	0.6000	0.6000	0.6548	0.3985	0.3422

Table 14. GRG and ranking of the cities for installation of the solar power plant.

Cities	C1	C2	C3	C4	C5	C6	C7	Sum	Rank
Arar	0.0817	0.1052	0.1390	0.1862	0.0996	0.0733	0.0184	0.7035	1
Al-Jouf	0.0703	0.1071	0.1297	0.1862	0.0724	0.0806	0.0183	0.6645	6
Tabuk	0.1105	0.0931	0.0884	0.1862	0.1231	0.0770	0.0194	0.6978	2
Hail	0.0806	0.0897	0.1081	0.1117	0.1019	0.0607	0.0188	0.5715	9
Dhahran	0.0756	0.0663	0.1622	0.1862	0.1256	0.0546	0.0182	0.6887	3
Al-Ahsa	0.0779	0.0636	0.1081	0.1117	0.0541	0.0746	0.0188	0.5088	13
Taif	0.0930	0.1025	0.1390	0.1117	0.1436	0.0573	0.0196	0.6668	5
Makkah	0.0787	0.0560	0.0541	0.1117	0.1622	0.0384	0.0228	0.5237	12
Jeddah	0.0867	0.0580	0.1216	0.0621	0.1214	0.0781	0.0331	0.5611	10
Yanbu	0.0975	0.0598	0.0973	0.0798	0.0800	0.0722	0.0185	0.5051	14
Medina	0.0812	0.0621	0.0973	0.0798	0.1399	0.0667	0.0227	0.5496	11
Riyadh	0.1137	0.0631	0.0973	0.1117	0.1228	0.0715	0.0547	0.6349	8
Abha	0.0931	0.1679	0.0973	0.1117	0.1530	0.0269	0.0186	0.6685	4
Jizan	0.0621	0.0562	0.0973	0.0621	0.0702	0.0722	0.0183	0.4382	15
Najran	0.1862	0.0840	0.0973	0.1117	0.1062	0.0321	0.0187	0.6362	7

Table 15. Normalized decision matrix.

Cities	Criteria						
	C1	C2	C3	C4	C5	C6	C7
Arar	0.8067	1.1458	0.9333	1.0000	14.4000	2.1220	0.0353
Al-Jouf	0.7504	1.1391	0.9000	1.0000	27.4400	1.0000	0.0245
Tabuk	0.8965	1.1969	0.6667	1.0000	7.7600	1.5244	0.1083
Hail	0.8018	1.2135	0.8000	0.9992	13.6000	4.7195	0.0635
Dhahran	0.7787	1.3750	1.0000	1.0000	7.2000	6.4024	0.0237
Al-Ahsa	0.7896	1.4010	0.8000	0.9992	43.6400	1.9146	0.0697
Taif	0.8484	1.1563	0.9333	0.9992	3.7600	5.5976	0.1262
Makkah	0.7934	1.4896	0.2000	0.9992	1.0000	13.4878	0.3147
Jeddah	0.8266	1.4635	0.8667	0.9975	8.1600	1.3659	0.6818
Yanbu	0.8625	1.4427	0.7333	0.9984	22.9200	2.3171	0.0476
Medina	0.8043	1.4167	0.7333	0.9984	4.4000	3.3659	0.3091
Riyadh	0.9037	1.4063	0.7333	0.9992	7.8400	2.4390	1.0000
Abha	0.8489	1.0000	0.7333	0.9992	2.2800	23.6829	0.0501
Jizan	0.6976	1.4870	0.7333	0.9975	28.9600	2.3171	0.0250
Najran	1.0000	1.2448	0.7333	0.9992	12.2400	18.1220	0.0615

4.3. *Entropy-VIKOR*

The criteria weights are computed in the consolidated Entropy-VIKOR scheme and VIKOR is used to rank the alternatives. The weighted normalized decision matrix is computed first, as depicted in Table 15, before implementing the entropy technique.

After standardization, the entropy H_j and weights are determined for different criteria as presented in Table 16.

After the entropy weights are determined, the choices are ranked using the VIKOR based on various criteria. The VIKOR approach indicates that Tabuk or Taif are both viable options because only condition 2 (Section 2.2.1) is met. Tabuk and Taif are therefore the best candidates for the location of a solar power plant according to the results of the Entropy-VIKOR technique.

Similarly, the remaining approaches are applied and the findings are listed in Table 17.

Table 16. Estimation of entropy and weight.

Criteria	C1	C2	C3	C4	C5	C6	C7
H_j	0.9987	0.9975	0.9873	1.0000	0.8778	0.8237	0.7382
m				15			
$\ln m$				2.7081			
$1 - H_j$	0.0013	0.0025	0.0127	0.0000	0.1222	0.1763	0.2618
$n - \sum H_j$				0.5767			
w_j	0.0022	0.0043	0.0219	0.0000	0.2118	0.3057	0.4540
$1 - w_j$	0.9978	0.9957	0.9781	1.0000	0.7882	0.6943	0.5460
Normalized w_j	0.1663	0.1659	0.1630	0.1667	0.1314	0.1157	0.0910

Table 17. Findings from several MCDM techniques.

MCDM Approach	Ideal Location
GEM-VIKOR	Tabuk
FAHP-VIKOR	Tabuk
Entropy-VIKOR	Tabuk or Taif
GEM-GRA	Tabuk
FAHP-GRA	Arar (Tabuk is second-ranked)
Entropy-GRA	Arar (Tabuk is second-ranked)

Kendall's coefficient of concordance, W (described in Section 2.3) is calculated to assess the degree of ranking agreement between various MCDM approaches. The W is calculated to be 0.8741 for this site selection problem for solar power plants. It shows a high degree of agreement amongst the various methodologies.

The sensitivity analysis is undertaken by varying the weights in accordance with the scenarios in Table 1 and calculating the rank using each approach. The ranking order for the six scenarios generated using each approach is examined for similarity in order to obtain the values of W. The greater values of W imply that the established ranking order is dependable and not overly sensitive to changes in weight. The outputs of the sensitivity study, which are presented in Table 18 show that the ranking order ascertained using various methodologies is reliable and stable. It makes sense given that each method's W value is more closely related to 1.

Table 18. Outcomes obtained in the sensitivity analysis.

MCDM Technique	GEM-VIKOR	FAHP-VIKOR	Entropy-VIKOR	GEM-GRA	FAHP-GRA	Entropy-GRA
W	0.9598	0.9759	0.9694	0.9752	0.9821	0.9841

5. Results and Discussion

The combinations of MCDM techniques used in this study qualify as logical, practical, and systematic approaches. They are effective in assisting to choose the best location for the generation of solar energy. Because there are so many MCDM strategies already in use, each with advantages and disadvantages, it is necessary to choose the right approach. The effort required and the outcomes produced by various MCDM strategies vary. As a result, the selection of a suitable MCDM approach has a big impact on the decision's quality as well as the work that must be conducted. As illustrated in Table 19, different MCDM techniques can rank identical alternatives in distinct sequences. The various approaches have varied computational needs and varying degrees of difficulty. Some strategies work well for large-sized problems, while others are better suited to small-sized problems.

Table 19. Comparative analysis of MCDM methods.

Technique	Combination		Ideal Alternative	Benefits	Limitations	Difficulty	Problem Size
	Weights	Ranking					
GEM-VIKOR	GEM	VIKOR	Tabuk	Higher decision reliability. Precise and consistent evaluation. Suitable for medium-scale problems.	Generation of the judgment matrix is relatively difficult, i.e., shortlisting experts and collecting precise information from them is a hectic task.	6	Medium-sized
FAHP-VIKOR	FAHP	VIKOR	Tabuk	Higher precision in small data problems. Checks inconsistency through the consistency index. Utilizes inherent information of criteria.	Large-sized problems can be demanding. Rank reversal problem—the final ranking can be reversed with the addition or elimination of an alternative. Geometric aggregation approach is used, so there is a possibility that some information may be lost.	8	Small-sized
Entropy-VIKOR	Entropy	VIKOR	Tabuk or Taif	Unbiased results. Capable of handling multiple inputs and outputs. Flexible to fit small and medium-sized problems.	Sensitive to inconsistent data.	5	Large -sized
GEM-GRA	GEM	GRA	Tabuk	Straightforward and uncomplicated method. Require precise information. Suitable for medium and small-sized problems.	Generation of judgement matrix is relatively difficult. Can be difficult for large-sized problems.	5	Medium-sized
FAHP-GRA	FAHP	GRA	Arar (Tabuk is second-ranked)	Takes into account uncertainty and vagueness. Checks inconsistency through consistency index. Full use of inherent information of criteria. Suitable for small-sized problems.	Complicated and computationally exhaustive even for medium-sized problems. Possibility of rank reversal.	6	Small-sized
Entropy-GRA	Entropy	GRA	Arar (Tabuk is second-ranked)	Suitable for large-sized problems. Simple and easy to use. Easy to understand. Number of steps remains the same regardless of the number of criteria.	Consistency is not controlled. Results may differ from other methods.	4	Large-sized

Large-sized > 15 criteria. Medium-sized = 10–15 criteria. Small-sized < 10 criteria.

The degree of difficulty of the various approaches has been graded on a scale from 1 to 10, with 10 denoting the greatest degree of difficulty. These scores may change depending on the user's comprehension and the execution method. The Entropy-GRA method is the easiest of all the techniques used, whilst the FAHP-VIKOR method is the most difficult. Keep in mind that some strategies work effectively with medium-sized problems while others perform better with larger-sized problems. Large-sized problems are those that involve more than 15 criteria, whereas small-sized problems only require less than ten criteria. Table 19 presents a detailed summary of the advantages and drawbacks of several MCDM techniques. The FAHP-VIKOR method provides a number of advantages, including improved precision, the use of inherent information in the criteria, and checking for inconsistency using the consistency index, despite the fact that it can be challenging, especially for large-scale situations. Entropy-GRA is different in that it is the simplest and best suited for large-scale problems, but consistency cannot be maintained. This shows that each MCDM strategy has its own unique advantages and constraints. Therefore, it is crucial to choose the right strategy based on the problem, the information that is available, the simplicity of use, consistency, etc.

It is evident that choosing the option that can be replicated by a number of MCDM procedures is the best option out of a range of alternatives. Tabuk can therefore be chosen as the best location for the production of solar energy. This outcome is comparable to that of past research. For instance, Al-Shammari et al. [21] ranked Tabuk city best for PV installation after analyzing 17 Saudi Arabian cities using an AHP-TOPSIS-based approach. Similarly, Tabuk was proven to be the optimal site for both PV and concentrated solar power in a study by Awan et al. [68] that compared three cities, namely Majmaah, Najran, and Tabuk. Tabuk station was recognized to be the preferred venue for a PV power plant with an energy power of 110,250 kWh in other investigations by Awan et al. [1]. Moreover, Tabuk was classified as one of the cities with the best prospects for renewable energy plants by Brumana et al. [69] due to the city's year-round adequate solar radiation and wind velocity. Mohammed et al. [70] also came to the conclusion that Tabuk city is the ideal place to deploy a PV system for residential structures.

The factors responsible for Tabuk's selection can be traced to its high GHI. However, it can be demonstrated that Riyadh likewise has a high GHI value but also has a greater ambient temperature, making it less suitable than Tabuk for the generation of solar energy. Similarly, Najran has a very high GHI value but is not chosen because it also has a higher ambient temperature, more dust storms, and more topographic variation. The Tabuk region typically features clear days with minimal clouds and low dust, which is another factor contributing to the higher average GHI value [5]. Furthermore, if a simple and a sophisticated method yield similar rankings, the simplest approach should be used because it can reduce computing time and effort. The strategy that has the fewest steps to estimate weight and rank is the simplest. Moreover, the stakeholders or users can easily comprehend the results using the simple and direct approaches, which are only marginally impacted by the problem's size. From an ecological standpoint, it has been discovered, according to Rehman et al. [71], that on average 8182 tons of greenhouse emissions can be inhibited from invading the local atmosphere every year. This highlights how important it is to use renewable energy sources and to increase their effectiveness and efficiency, especially by choosing an ideal location for the installation of their generation plants and so by employing proper MCDM methodologies.

The established ranking order is dependent and not unduly susceptible to variations in weight, as evidenced by the higher values of W for the various scenarios developed utilizing each approach. Additionally, the results are adequate and steady as indicated by the ranking order from several MCDM techniques, which is implied by a considerably larger value of W that is nearer to 1.

The work is distinctive since it takes into account a number of factors connected to the choice of location for a solar power facility. The selection procedure in this study is made more flexible, accurate, and useful by the employment of numerous MCDM approaches. When choosing locations for solar power plants or other similar problems, where the level of uncertainty is predicted to vary according to environmental factors, the utilized MCDM methodologies are very useful. However, as the number of criteria and alternative locations increases, some of these selection processes, may become repetitive, computationally difficult, and tiresome. Additionally, because the effectiveness of most of these decision-making methods depends on expert assessments, they frequently tend to exaggerate the ranking process. Because the basic problem of solar power plant site selection utilizing consolidated MCDM techniques has not been treated with proactive diligence due to its dynamic nature, this research intends to assist stakeholders in selecting the most appropriate location. The repository needs to be updated frequently with new criteria and reliable data due to the constantly changing solar energy industry and shifting environmental concerns. For this reason, new efforts should investigate more precise information and different sites in order to make a more trustworthy conclusion regarding the location of solar power plants.

6. Conclusions

Many MCDM approaches have been proposed to address the issue of selecting the appropriate site for solar power generation. A concise guideline for choosing an MCDM approach for solar power plant site selection is provided by the examination of several methods. Consequently, the following conclusions about the outcomes of this study can be drawn.

- The fact that different strategies might produce different outcomes when used to solve the same problem is a key critique of MCDM. A decision-maker need to select a course of action that comes the closest to the ideal. As a result, the best solution can be one that is repeated by numerous MCDM procedures.
- This paper emphasizes the significance of evaluating several decision-making strategies and choosing the most suitable methodology for the specific application, without implying that any one MCDM method is superior to other methods.
- The outcomes of the different techniques might not be the same. This is explained by various weights and their distributions, as well as various solution algorithms.
- All three weight calculation methods have established solar radiation and sunshine hours as the most important criteria. For instance, using GEM, solar radiation and sunlight hours are given weights of 0.1877 and 0.1744, respectively, thereby contributing 18.77% and 17.44% to the choice of the PV site. Similarly, solar radiation and sunshine hour receive the same weight value of 0.1862 from the FAHP. Solar radiation and sunlight hours have entropy weights of 0.1663 and 0.1667, respectively.
- The type of MCDM approach selected affects the decision's quality and the amount of work necessary. The various approaches display differing degrees of difficulty and demand varying degrees of computation.
- It makes sense to employ one of the simplest techniques. However, the use of multiple techniques is suggested in order to verify consistency and improve the trustworthiness of the results.
- It should be noted that some strategies work well for large-sized problems while others are better suited for small-scale problems.
- Of all the methods utilized in this study, the FAHP-VIKOR technique is the most exhaustive with a difficulty level of 8, while Entropy-GRA is the easiest with a complexity degree of 4.
- A decision among numerous options that is replicated by several MCDM approaches can be regarded as the best option. As a result, Tabuk is the ideal location for the construction of a solar power plant. High solar radiation (GHI value of 5992 $W/m^2/day$) and more sunshine hours (12.16 h/day) are the main factors that contribute to its selection.
- The ranking consistency among the various MCDM techniques employed in the study is reasonable, as indicated by Kendall's coefficient of concordance value of 0.8741, which is very close to 1.

- The likelihood of uncertainty in expert decision-making as well as the lack of precise data collection is the work's limitations. Future studies will find ways to circumvent these restrictions. Future versions of the work will also be expanded by integrating additional accurate data, expert opinions, different cities or locations, and improvised MCDM procedures.

Author Contributions: Conceptualization, S.H.M. and K.M.; methodology, S.H.M. and K.M.; software, S.H.M., K.M. and M.H.A.; validation, S.H.M. and M.H.A.; formal analysis, S.H.M., K.M. and H.A.; investigation, S.H.M., F.H.H. and F.A.; resources, H.A. and M.H.A.; data curation, F.H.H. and F.A.; writing—original draft preparation, S.H.M. and K.M.; writing—review and editing, S.H.M., K.M. and H.A.; visualization, S.H.M. and M.H.A.; supervision, H.A.; project administration, S.H.M. and H.A.; funding acquisition, H.A. All authors have read and agreed to the published version of the manuscript.

Funding: This research was funded through the Researchers Supporting Project number (RSP2023R499), King Saud University, Riyadh, Saudi Arabia.

Institutional Review Board Statement: Not Applicable.

Informed Consent Statement: Not Applicable.

Data Availability Statement: The data presented in this study are available in the article.

Acknowledgments: The authors extend their appreciation to King Saud University for funding this work through Researchers Supporting Project number (RSP2023R499), King Saud University, Riyadh, Saudi Arabia.

Conflicts of Interest: The authors declare no conflict of interest.

References

1. Awan, A.B.; Zubair, M.; Prasannakumari, P.R.; Abokhalil, A.G. Solar Energy Resource Analysis and Evaluation of Photovoltaic System Performance in Various Regions of Saudi Arabia. *Sustainability* **2018**, *10*, 1129. [CrossRef]
2. Akkas, O.P.; Erten, M.Y.; Cam, E.; Inanc, N. Optimal Site Selection for a Solar Power Plant in the Central Anatolian Region of Turkey. *Int. J. Photoenergy* **2017**, *2017*, e7452715. [CrossRef]
3. Bączkiewicz, A.; Kizielewicz, B.; Shekhovtsov, A.; Yelmikheiev, M.; Kozlov, V.; Sałabun, W. Comparative Analysis of Solar Panels with Determination of Local Significance Levels of Criteria Using the MCDM Methods Resistant to the Rank Reversal Phenomenon. *Energies* **2021**, *14*, 5727. [CrossRef]
4. Almasoud, A.H.; Gandayh, H.M. Future of solar energy in Saudi Arabia. *J. King Saud Univ. Eng. Sci.* **2015**, *27*, 153–157. [CrossRef]
5. Zell, E.; Gasim, S.; Wilcox, S.; Katamoura, S.; Stoffel, T.; Shibli, H.; Engel-Cox, J.; Subie, M.A. Assessment of solar radiation resources in Saudi Arabia. *Sol. Energy* **2015**, *119*, 422–438. [CrossRef]
6. Almulhim, T.; Al Yousif, M. An Analysis of Renewable Energy Investments in Saudi Arabia: A Hybrid Framework Based on Leontief and Fuzzy Group Decision Support Models. *SSRN* **2022**. [CrossRef]
7. National Renewable Energy Laboratory. *Documenting a Decade of Cost Declines for PV Systems*; National Renewable Energy Laboratory: Golden, CO, USA, 2021.
8. Ozdemir, S.; Sahin, G. Multi-criteria decision-making in the location selection for a solar PV power plant using AHP. *Measurement* **2018**, *129*, 218–226. [CrossRef]
9. Anandaraj, S.; Ayyasamy, M.; Marquez, F.P.G.; Athikesavan, M.M. Experimental studies of different operating parameters on the photovoltaic thermal system using a flattened geometrical structure. *Environ. Sci. Pollut. Res.* **2023**, *30*, 1116–1132. [CrossRef]
10. Kaya, İ.; Çolak, M.; Terzi, F. Use of MCDM techniques for energy policy and decision-making problems: A review. *Int. J. Energy Res.* **2018**, *42*, 2344–2372. [CrossRef]
11. Wang, C.-N.; Dang, T.-T.; Nguyen, N.-A.-T.; Wang, J.-W. A combined Data Envelopment Analysis (DEA) and Grey Based Multiple Criteria Decision Making (G-MCDM) for solar PV power plants site selection: A case study in Vietnam. *Energy Rep.* **2022**, *8*, 1124–1142. [CrossRef]
12. Saraswat, S.K.; Digalwar, A.K.; Yadav, S.S.; Kumar, G. MCDM and GIS based modelling technique for assessment of solar and wind farm locations in India. *Renew. Energy* **2021**, *169*, 865–884. [CrossRef]
13. Kannan, D.; Moazzeni, S.; Darmian, S.M.; Afrasiabi, A. A hybrid approach based on MCDM methods and Monte Carlo simulation for sustainable evaluation of potential solar sites in east of Iran. *J. Clean. Prod.* **2021**, *279*, 122368. [CrossRef]
14. Wang, C.-N.; Nguyen, V.T.; Thai, H.T.N.; Duong, D.H. Multi-Criteria Decision Making (MCDM) Approaches for Solar Power Plant Location Selection in Viet Nam. *Energies* **2018**, *11*, 1504. [CrossRef]

15. Malemnganbi, R.; Shimray, B.A. Solar Power Plant Site Selection: A Systematic Literature Review on MCDM Techniques Used. In Proceedings of the Electronic Systems and Intelligent Computing; Mallick, P.K., Meher, P., Majumder, A., Das, S.K., Eds.; Springer: Singapore, 2020; pp. 37–48.
16. Gnanasekaran, S.; Venkatachalam, N. A Review on Applications of Multi-Criteria Decision Making (MCDM) For Solar Panel Selection. *Int. J. Mech. Prod. Eng. Res. Dev.* **2019**, *9*, 11–20. [CrossRef]
17. Levosada, A.T.A.; Ogena, R.P.T.; Santos, J.R.V.; Danao, L.A.M. Mapping of Suitable Sites for Concentrated Solar Power Plants in the Philippines Using Geographic Information System and Analytic Hierarchy Process. *Sustainability* **2022**, *14*, 12260. [CrossRef]
18. Villacreses, G.; Martínez-Gómez, J.; Jijón, D.; Cordovez, M. Geolocation of photovoltaic farms using Geographic Information Systems (GIS) with Multiple-criteria decision-making (MCDM) methods: Case of the Ecuadorian energy regulation. *Energy Rep.* **2022**, *8*, 3526–3548. [CrossRef]
19. Mirzaei, N. A Multicriteria Decision Framework for Solar Power Plant Location Selection Problem with Pythagorean Fuzzy Data: A Case Study on Green Energy in Turkey. *Sustainability* **2022**, *14*, 14921. [CrossRef]
20. Lee, A.H.I.; Kang, H.-Y.; Lin, C.-Y.; Shen, K.-C. An Integrated Decision-Making Model for the Location of a PV Solar Plant. *Sustainability* **2015**, *7*, 13522–13541. [CrossRef]
21. Al-Shammari, S.; Ko, W.; Al Ammar, E.A.; Alotaibi, M.A.; Choi, H.-J. Optimal Decision-Making in Photovoltaic System Selection in Saudi Arabia. *Energies* **2021**, *14*, 357. [CrossRef]
22. Dang, R.; Li, X.; Li, C.; Xu, C. A MCDM framework for site selection of island photovoltaic charging station based on new criteria identification and a hybrid fuzzy approach. *Sustain. Cities Soc.* **2021**, *74*, 103230. [CrossRef]
23. Seker, S.; Kahraman, C. Socio-economic evaluation model for sustainable solar PV panels using a novel integrated MCDM methodology: A case in Turkey. *Soc. Econ. Plan. Sci.* **2021**, *77*, 100998. [CrossRef]
24. Wang, T.-C.; Tsai, S.-Y. Solar Panel Supplier Selection for the Photovoltaic System Design by Using Fuzzy Multi-Criteria Decision Making (MCDM) Approaches. *Energies* **2018**, *11*, 1989. [CrossRef]
25. Qiu, W.H. Group eigenvalue method. *Appl. Math. Mech.* **1997**, *18*, 1027–1031.
26. Zadeh, L.A. Fuzzy sets. *Inf. Control.* **1965**, *8*, 338–353. [CrossRef]
27. Jasna, P.; Zoran, S.; Maja Levi, J.; Sanja, M. Application of fuzzy AHP method for choosing a technology within service company. *Tech. Technol. Educ. Manag.* **2012**, *7*, 332–341.
28. van Laarhoven, P.J.M.; Pedrycz, W. A fuzzy extension of Saaty's priority theory. *Fuzzy Sets Syst.* **1983**, *11*, 229–241. [CrossRef]
29. Buckley, J.J. Fuzzy hierarchical analysis. *Fuzzy Sets Syst.* **1985**, *17*, 233–247. [CrossRef]
30. Chang, D.-Y. Applications of the extent analysis method on fuzzy AHP. *Eur. J. Oper. Res.* **1996**, *95*, 649–655. [CrossRef]
31. Princy, S.; Dhanakaran, S. Comparison of Triangular and Trapezoidal Fuzzy Membership Function. *J. Comput. Sci. Eng.* **2016**, *2*, 6.
32. Savitha, M.T.; Mary, G. New Methods for Ranking of Trapezoidal Fuzzy Numbers, Advances in Fuzzy Mathematics. *Adv. Fuzzy Math.* **2017**, *12*, 1159–1170.
33. Akbaş, S.; Dalkiliç, T.E. Multi-Criteria Supplier Selection Based on Fuzzy Pairwise Comparison in AHP. *Gazi Univ. J. Sci.* **2018**, *31*, 296–308.
34. Naghadehi, M.Z.; Mikaeil, R.; Ataei, M. The application of fuzzy analytic hierarchy process (FAHP) approach to selection of optimum underground mining method for Jajarm Bauxite Mine, Iran. *Expert Syst. Appl.* **2009**, *36*, 8218–8226. [CrossRef]
35. Chen, Y.-H.; Wang, T.-C.; Wu, C.-Y. Multi-criteria decision making with fuzzy linguistic preference relations. *Appl. Math. Model.* **2011**, *35*, 1322–1330. [CrossRef]
36. Li, X.; Wang, K.; Liu, L.; Xin, J.; Yang, H.; Gao, C. Application of the Entropy Weight and TOPSIS Method in Safety Evaluation of Coal Mines. *Procedia Eng.* **2011**, *26*, 2085–2091. [CrossRef]
37. Lotfi, F.H.; Fallahnejad, R. Imprecise Shannon's Entropy and Multi Attribute Decision Making. *Entropy* **2010**, *12*, 53–62. [CrossRef]
38. Opricovic, S.; Tzeng, G.-H. Compromise solution by MCDM methods: A comparative analysis of VIKOR and TOPSIS. *Eur. J. Oper. Res.* **2004**, *156*, 445–455. [CrossRef]
39. Acuña-Soto, C.M.; Liern, V.; Pérez-Gladish, B. A VIKOR-based approach for the ranking of mathematical instructional videos. *Manag. Decis.* **2019**, *57*, 501–522. [CrossRef]
40. Ju-Long, D. Control problems of grey systems. *Syst. Control. Lett.* **1982**, *1*, 288–294. [CrossRef]
41. Lu, H.S.; Chang, C.K.; Hwang, N.C.; Chung, C.T. Grey relational analysis coupled with principal component analysis for optimization design of the cutting parameters in high-speed end milling. *J. Mater. Process. Technol.* **2009**, *209*, 3808–3817. [CrossRef]
42. Lee, W.-S.; Lin, Y.-C. Evaluating and ranking energy performance of office buildings using Grey relational analysis. *Energy* **2011**, *36*, 2551–2556. [CrossRef]
43. Triantaphyllou, E.; Sánchez, A. A Sensitivity Analysis Approach for Some Deterministic Multi-Criteria Decision-Making Methods. *Decis. Sci.* **1997**, *28*, 151–194. [CrossRef]
44. Simanaviciene, R.; Ustinovichius, L. Sensitivity Analysis for Multiple Criteria Decision Making Methods: TOPSIS and SAW. *Procedia Soc. Behav. Sci.* **2010**, *2*, 7743–7744. [CrossRef]
45. Mukhametzyanov, I.; Pamucar, D. A sensitivity analysis in MCDM problems: A statistical approach. *Decis. Mak. Appl. Manag. Eng.* **2018**, *1*, 51–80. [CrossRef]
46. Syamsuddin, I. Multicriteria Evaluation and Sensitivity Analysis on Information Security. *IJCA* **2013**, *69*, 22–25. [CrossRef]

47. Pamucar, D.; Chatterjee, K.; Zavadskas, E.K. Assessment of third-party logistics provider using multi-criteria decision-making approach based on interval rough numbers. *Comput. Ind. Eng.* **2019**, *127*, 383–407. [CrossRef]
48. Leoneti, A.B. Considerations Regarding the Choice of Ranking Multiple Criteria Decision Making Methods. *Pesqui. Oper.* **2016**, *36*, 259–277. [CrossRef]
49. Hajkowicz, S.; Higgins, A. A comparison of multiple criteria analysis techniques for water resource management. *Eur. J. Oper. Res.* **2008**, *184*, 255–265. [CrossRef]
50. Athawale, V.M.; Chakraborty, S. A comparative study on the ranking performance of some multi-criteria decision-making methods for industrial robot selection. *IJIEC* **2011**, *2*, 831–850. [CrossRef]
51. General Authority for Statistics. *General Information about The Kingdom of Saudi Arabia*; General Authority for Statistics: Riyadh, Saudi Arabia, 23 November 2022.
52. World Bank Group. *Climatology—Saudi Arabia. Climate Change Knowledge Portal for Development Practitioners and Policy Makers*; World Bank Group: Washington, DC, USA, 23 November 2022.
53. Hosseini Dehshiri, S.S.; Firoozabadi, B. A new application of measurement of alternatives and ranking according to compromise solution (MARCOS) in solar site location for electricity and hydrogen production: A case study in the southern climate of Iran. *Energy* **2022**, *261*, 125376. [CrossRef]
54. Al Garni, H.Z.; Awasthi, A. A fuzzy AHP and GIS-based approach to prioritize utility-scale solar PV sites in Saudi Arabia. In Proceedings of the 2017 IEEE International Conference on Systems, Man, and Cybernetics (SMC), Banff, AB, Canada, 5–8 October 2017; pp. 1244–1249.
55. Li, T.; Li, A.; Song, Y. Development and Utilization of Renewable Energy Based on Carbon Emission Reduction—Evaluation of Multiple MCDM Methods. *Sustainability* **2021**, *13*, 9822. [CrossRef]
56. Khemiri, W.; Yaagoubi, R.; Miky, Y. Optimal placement of solar photovoltaic farms using analytical hierarchical process and geographic information system in Mekkah, Saudi Arabia. *AIP Conf. Proc.* **2018**, *2056*, 020025. [CrossRef]
57. Human, S. What Is Global Horizontal Irradiance? Available online: https://solarpoweredblog.com/what-is-global-horizontal-irradiance (accessed on 15 November 2022).
58. Firouzjah, K.G. Assessment of small-scale solar PV systems in Iran: Regions priority, potentials and financial feasibility. *Renew. Sustain. Energy Rev.* **2018**, *94*, 267–274. [CrossRef]
59. Butt, M.J.; Mashat, A.S. MODIS satellite data evaluation for sand and dust storm monitoring in Saudi Arabia. *Int. J. Remote Sens.* **2018**, *39*, 8627–8645. [CrossRef]
60. Imteyaz, B.; Lawal, D.U.; Tahir, F.; Rehman, S. Prospects of large-scale photovoltaic-based power plants in the Kingdom of Saudi Arabia. *Eng. Rep.* **2021**, *3*, e12398. [CrossRef]
61. General Authority for Statistics. *Indicators of Renewable Energy in Saudi Arabia 2017*; General Authority for Statistics: Riyadh, Saudi Arabia, 2017.
62. Climate and Weather Averages—Time and Data. Available online: https://www.timeanddate.com/weather/@108512/climate. (accessed on 23 November 2022).
63. Weather Spark—Climate and Average Weather. Available online: https://weatherspark.com/y/101928/Average-Weather-in-Arar-Saudi-Arabia-Year-Round. (accessed on 23 November 2022).
64. Population of Cities in Saudi Arabia 2022. Available online: https://worldpopulationreview.com/countries/cities/saudi-arabia. (accessed on 23 November 2022).
65. Arishi, A.A. Classification of Sandstorms in Saudi Arabia. *Atmos. Clim. Sci.* **2020**, *11*, 177–193. [CrossRef]
66. Saaty, R.W. The analytic hierarchy process—What it is and how it is used. *Math. Model.* **1987**, *9*, 161–176. [CrossRef]
67. Mian, S.H.; Al-Ahmari, A. Comparative analysis of different digitization systems and selection of best alternative. *J. Intell. Manuf.* **2019**, *30*, 2039–2067. [CrossRef]
68. Awan, A.B.; Zubair, M.; Praveen, R.P.; Bhatti, A.R. Design and comparative analysis of photovoltaic and parabolic trough based CSP plants. *Sol. Energy* **2019**, *183*, 551–565. [CrossRef]
69. Brumana, G.; Franchini, G.; Ghirardi, E.; Perdichizzi, A. Techno-economic optimization of hybrid power generation systems: A renewables community case study. *Energy* **2022**, *246*, 123427. [CrossRef]
70. Mohammed, A.; Ghaithan, A.; Al-Hanbali, A.; Attia, A.M.; Saleh, H.; Alsawafy, O. Performance evaluation and feasibility analysis of 10 kWp PV system for residential buildings in Saudi Arabia. *Sustain. Energy Technol. Assess.* **2022**, *51*, 101920. [CrossRef]
71. Rehman, S.; Bader, M.A.; Al-Moallem, S.A. Cost of solar energy generated using PV panels. *Renew. Sustain. Energy Rev.* **2007**, *11*, 1843–1857. [CrossRef]

sustainability

Article

Real-Time Simulation and Energy Management Attainment of Microgrids

Hani Muhsen [1,2,*], Asma Alkhraibat [2] and Ala'aldeen Al-Halhouli [1,2]

[1] Department of Mechatronics Engineering, School of Applied Technical Sciences, German Jordanian University, Madaba 11180, Jordan
[2] Smart Grid Lab, German Jordanian University, Madaba 11180, Jordan
* Correspondence: hani.mohsen@gju.edu.jo

Abstract: The rapid spread of Microgrid systems has led to the need for an intensive analysis of the system to avoid several challenges such as stability, reliability, power balance, and other aspects. In this context, real-time simulation plays a vital role in the overall system study before the actual implementation stage. This helps avoid many on-site problems of the Microgrid by simulating the system and studying different operation scenarios. This paper uses the OPAL-RT simulator to perform a real-time simulation of an MG case study. Furthermore, it examines the implementation of the Fault Ride Through technique to overcome the total disconnection of the PV system following unpredictable faults. Moreover, a Load curtailment solution method is proposed in this study, to meet the balance and stable operation of the MG. The results prove the effectiveness of both techniques, with FRT implementation reducing the losses by about 62%, and the Load curtailment algorithm maintaining the balance of the MG.

Keywords: real-time simulation; microgrid; OPAL-RT; fault ride through; load curtailment

Citation: Muhsen, H.; Alkhraibat, A.; Al-Halhouli, A. Real-Time Simulation and Energy Management Attainment of Microgrids. *Sustainability* **2023**, *15*, 2696. https://doi.org/10.3390/su15032696

Academic Editor: George Kyriakarakos

Received: 29 December 2022
Revised: 26 January 2023
Accepted: 31 January 2023
Published: 2 February 2023

1. Introduction

Microgrid (MG) systems' proliferation has significantly increased in recent years. The penetration of renewable energy resources, the limitation of carbon dioxide emissions, and the broad deployment of smart power systems have contributed to the wide spread of MG applications.

An MG is a smart small-scale power system comprising a mix of generating resources, controllable loads, energy storage units, transmission lines, transformers, and a point of common coupling (PCC) [1]. The PCC is the main circuit breaker installed in the MG system to control the operation mode of the network. The MG system's operation modes can be grid-connected or grid-islanded [2]. Various research projects have been directed toward the MG topic as a promising field in the electrical energy world [3,4].

The main and critical feature of the MG system is the ability to work in two operation modes: connected or islanded. The potential complexity of such a network is caused by the transition between operation modes while maintaining a stable operation [5].

MG systems have several benefits to different sides of the community, including environmental aspects [6], economic aspects [7], and trading aspects [8]. In addition, the control part of the MG includes energy management [9], stability and resiliency [2], decentralized monitoring [10], real-time monitoring [11], and system protection [12].

The control side of power systems has countless articles highlighting diverse fields and techniques that help to maintain a stable operation. Advanced transactive control strategies have been developed to optimally utilize the installed distributed energy resources and storage units in the energy community [13]. The proposed techniques have been built based on the game theoretical control scheme. In addition, two other innovative control techniques have been established, the whale optimization algorithm (WOA) and the artificial neural network (ANN). They are utilized to control the power flow of power systems comprised of a mix of renewable energy resources.

The availability of software simulations of the MG systems have contributed to its broad deployment worldwide. However, it still needs more accurate and intensive simulators to help in monitoring, and to analyze several cases and scenarios. In this context, Real-Time Simulation (RTS) is considered a valuable and vital tool to test and configure the proposed models before their actual implementation.

The RTS has been revolutionized in recent years; it returns valuable and significant results that are highly considered before the implementation phase of the studied systems. Generally, RTS is a promising technique to test and validate the complex and latest control approaches applied in MGs. Furthermore, it also helps specify the control and management parameter values for the best implementation of the model.

A complete RTS of an MG case study has been introduced in [11]. The authors proposed and studied a reactive power coordination control scheme in a real-time state. Also, they discussed the transition between the operation modes of the MG case study. The voltage and frequency evaluation of an MG system using RTS during islanded mode has been studied in [14].

In MGs, and power systems in general, electrical faults can cause unstable operation of the systems. These faults could occur by failures in the devices or due to human errors [15]. Moreover, the operation of the MG in islanded mode may lead to voltage sag occurrence and hence unstable operation [16]. One of the vital and effective techniques that can be utilized in abnormal operation and voltage sag cases is Fault Ride Through (FRT) [17]. FRT is one of the effective techniques that is applied to the generation units during fault cases. It can be defined as "the ability of the generation units to stay connected and available in case of voltage sag occurrences without interruptions" [18].

Reference [19] has utilized the OPAL-RT simulator to validate dynamic test cases of two test systems. These dynamic test cases have been examined and validated under several disturbances, such as a single line-to-ground fault, line shortage, load changes, and other disturbances. Furthermore, a four-level fault current hierarchical limitation has been proposed in [20] to analyze and resolve the problem of high fault current during FRT. Furthermore, an enhancement of MG models has been proposed in [21], using droop control virtual impedance and secondary power reference generation. It mainly aims to improve the FRT capability of the inverter-based MG system.

A centralized MG system has been designed in [22] to apply the controller Hardware in the Loop (HIL) test. This test is applied for the proposed control algorithm, showing the system's functionality on the power management side and reducing the needed load shedding.

FRT implementation in MG during the islanded mode of operation has been discussed in [23]. Islanded mode operation of the MG refers to the operation of the MG as an independent power system. This can be effected by opening the PCC breaker. Intensive research and efforts are going on to study the different and most important factors of MG operation to support a stable system operation during islanded mode. However, maintaining reliable, stable, and robust operation of the MG, regardless of the current operation mode, is one of the highest priorities of the MG network. Hence, an algorithm is utilized in this paper to perform the load-shedding, which aims to maintain the stability and the power balance factor.

A complete review of load-shedding implementation in MG networks is introduced in [24]. This reference covered a large number of applied intelligent methods, as well as different load-shedding classes. The authors of [25] proposed a new load-shedding algorithm for Inverter-based MGs. The proposed algorithm is based on frequency changes to specify the power shortage in the MG. Compensation for the power shortage is applied in the MG using the proposed load-shedding scheme. The generation side of the studied MG in this reference is generally based on several renewable energy units, including PV systems, solid oxide fuel cells, and energy storage systems.

The research presented so far focuses on MG's characteristics, control systems, protections, management, and optimization issues. However, implementing an RTS of MG is an attractive and promising research avenue that has been given attention recently. Moreover, there is still a significant gap in studying MG control techniques in real-time, especially during the islanding mode operation. This paper emphasizes the implementation of many algorithms and scenarios in a real-time state. Two innovative algorithms are demonstrated for power loss reduction and power balance.

The main objective of this work is the implementation of an RTS, using the OPAL-RT simulator, to study and analyze an MG case study with different cases. Furthermore, the detailed steps of the model transformation into a real-time model in the RT-LAB environment are introduced in this paper. Even though many researchers have worked on the RTS of MG, very few pieces of research were reported on implementing two control techniques that covered both operation modes (connected and islanded modes) of the MG in a real-time state. Hence, as an energy management process, the two operation modes of the MG have been covered in this work by implementing FRT and Load curtailment algorithms.

The rest of the paper is organized as follows: Section 2 presents the proposed MG model that has been created in this work. The detailed RTS part using the OPAL-RT simulator, and the MG modeling in the RT-LAB platform, are addressed in Section 3. Simulation and results analysis are presented in Section 4. Finally, Section 5 summarizes the findings and conclusion of this study.

2. Microgrid Model Understudy

An MG, as defined previously, is represented by a small-scale power network consisting of different generating units, loads (residential, institutions, commercial), transmission lines, transformers, controllers, and a PCC. Figure 1 represents a general MG diagram.

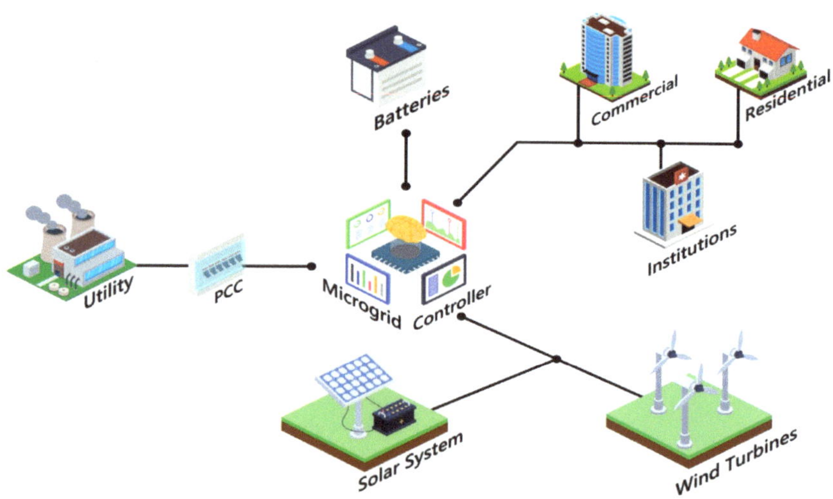

Figure 1. MG diagram.

This paper studies different scenarios of an MG case study in real-time mode. The MG model studied in this work comprises several components: PV systems, load units, battery energy storage systems (BESS), diesel generators (DGs), and the main utility grid. Figure 2 shows the single-line diagram of the MG case study.

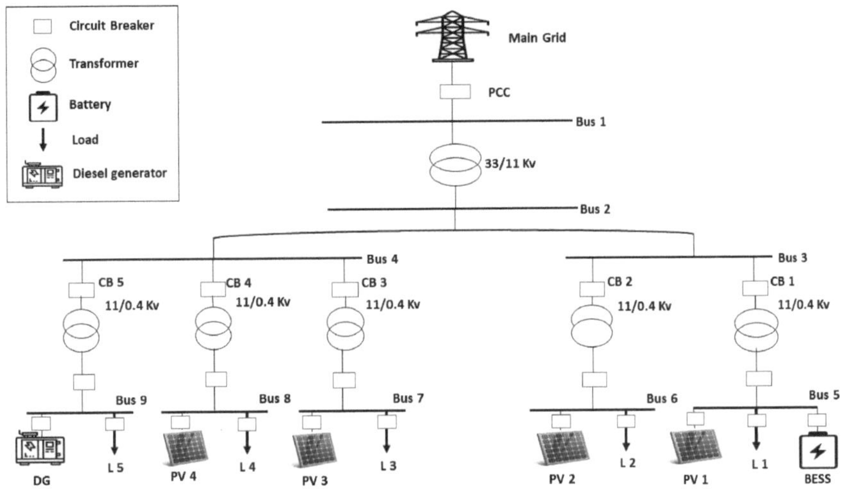

Figure 2. Single line diagram of MG.

The installed PV system consists of four PV areas with a total capacity of around 1.8 MW. The five installed loads are varied in their capacity and priority, which have been considered in performing several algorithms in case of blackout occurrence. The load priority of the loads has been defined as 25%, 21%, 11%, 13%, and 30% for the installed loads 1, 2, 3, 4, and 5, respectively.

The utility grid usually controls the MG operation during the connected mode phase. The main grid follows the voltage and frequency parameters required to maintain the MG's stable and reliable operation. The control scheme of the MG during the islanded mode depends on providing and absorbing the active and reactive power difference between the generation and load units. Hence, the voltage and frequency parameters can be maintained at stable values during the islanded operating mode.

Regarding the operation of the MG during the islanded mode, or any fault cases, the installed resources must have voltage or frequency ride-through capabilities. These capabilities help maintain the MG's continuous operation during voltage anomalies.

The modeling of the system has been performed on MATLAB SIMULINK, as shown in Figure 3. The model's left side represents the utility grid and its connection with step-down transformers and a PCC. While the right side of the MG model illustrates the model classification into five areas, as explained previously in the single-line diagram.

The energy storage system is built with a battery control system that manages its operation. It helps provide power to the MG system in case of any power shortage. Moreover, inside the PV and BESS subsystems, a control system has been included to manage and organize the system's operation. As both systems are DC power supplies, inverters have been inserted to convert their output into AC power.

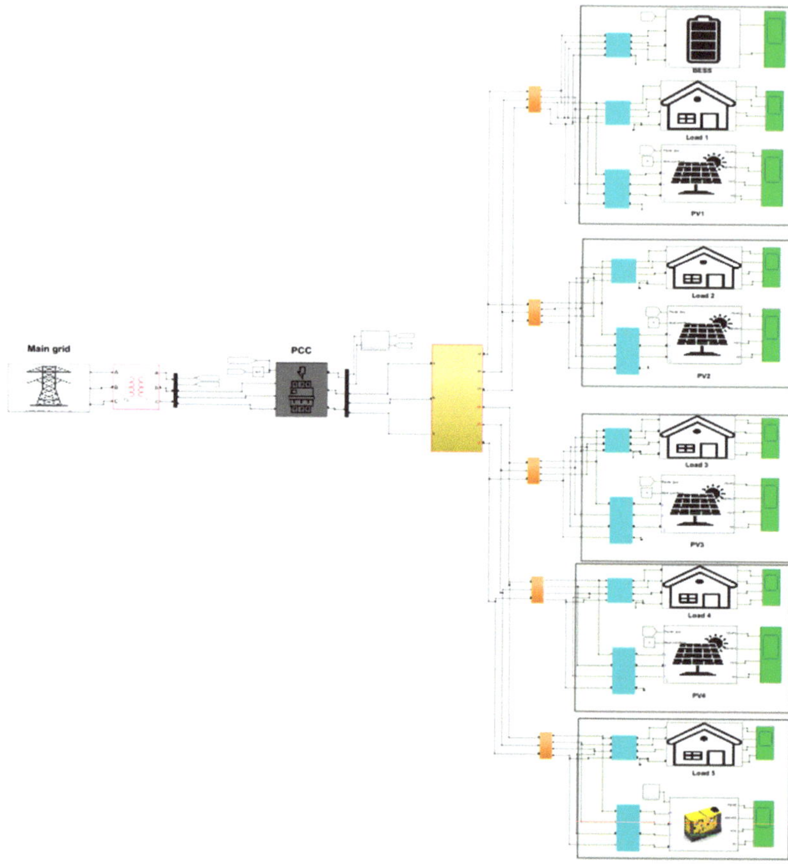

Figure 3. MG Model in MATLAB/SIMULNK.

3. Developed Microgrid Model for RTS

RTS is a promising step in testing several scenarios and technologies. It has contributed considerably to the evolution of several research communities, especially in the power systems field. RTS platforms have significantly contributed to analyzing and studying power system performance issues.

The simulation part in an RTS is strictly based on a discrete-time model with a fixed-step size. The RT-LAB software is an OPAL-RT real-time simulation platform that meets the RTS objectives and improves the user's access experience. It enables the model to perform its real-world implementation, allowing Hardware-In-Loop (HIL) simulations, for instance.

The OPAL-RT is a cutting-edge technology that combines distributed processing software and hardware to obtain a high parallel speed and RTS. Figure 4 represents the OPAL-RT simulator (OP1420 PHIL Microgrid test bench) used in this study. The RT-LAB software is used in the second stage of the simulation part. The use of the fixed-step solver in RT-LAB software is mandatory. The system structure and interaction between the Real-time target and the software platform (Host PC) are represented in Figure 5.

Generally, and regarding the RTS part, the applied process of the RTS can be briefly summarized into four main steps. In the first step, the Simulink model has to be opened through the RT-LAB software, where RT-LAB version 2021.3.4.320 has been used in this stage. Then, the base model will be reconstructed, to be composed of three subsystems called master, slave, and console subsystems. The master subsystem is the main subsystem in the real-time model, and its prefix is "SM_". Therefore, it cannot be accessed after starting the simulation.

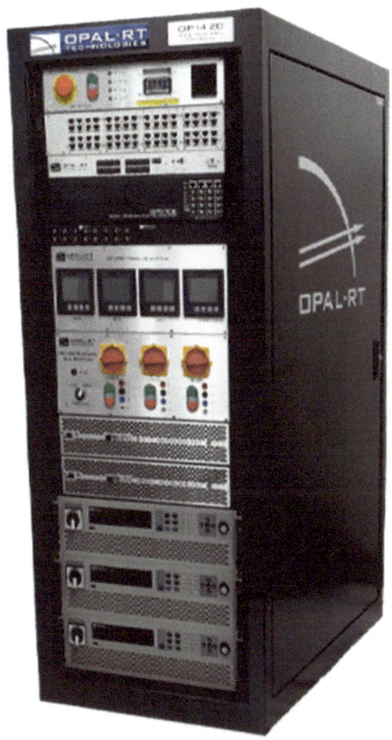

Figure 4. OPAL-RT Simulator.

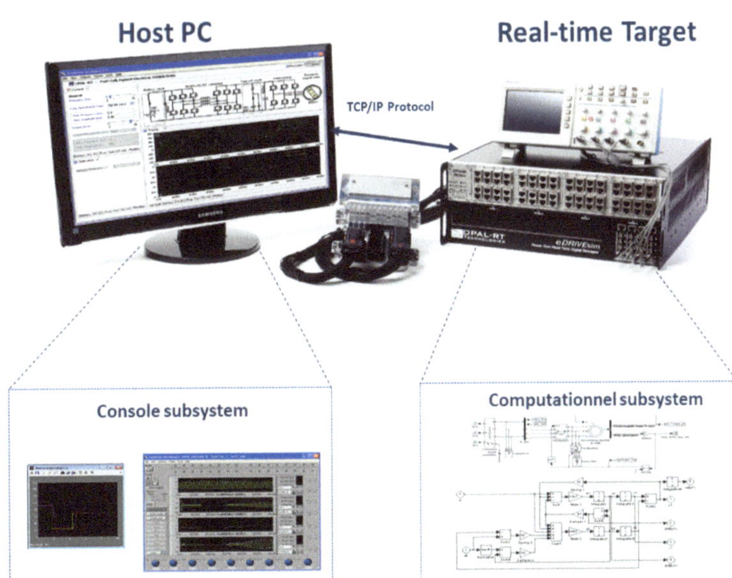

Figure 5. Interaction between Real-time Target and Host PC.

On the other hand, users can access the console subsystem during the simulation running time. The console subsystem contains all the scope blocks and other variables that need to be modified during the running time. Finally, the slave subsystem is an optional component in the RT-LAB model, which can be inserted in the case of large and complex models. Real-time models can include many slave subsystems based on the model structure.

The second step of the RTS is the compilation of the real-time model. This stage involves the model's conversion into a real-time application based on the C language. The time utilized in this stage depends on the model's complexity and the predefined configurations of the real-time process. The compiling step involves six stages: separate model, generate code, clean target directory, transfer files to target, compile and link generated code, and retrieve files from the target.

The third stage of the RTS process is "Execute." As shown in Figure 5, the host contains the model construction and the platform where the results will be illustrated. The number of cores used has to be specified before the execution stage, to perform the simulation part in less computational time and enhance the allocation of the resources. The interact step is the final step of the RTS. This stage refers to the utilization of the graphical interface in order to perform the available features, such as data aggregation and control implementation. The flow chart shown in Figure 6 represents the main steps in the RTS phase.

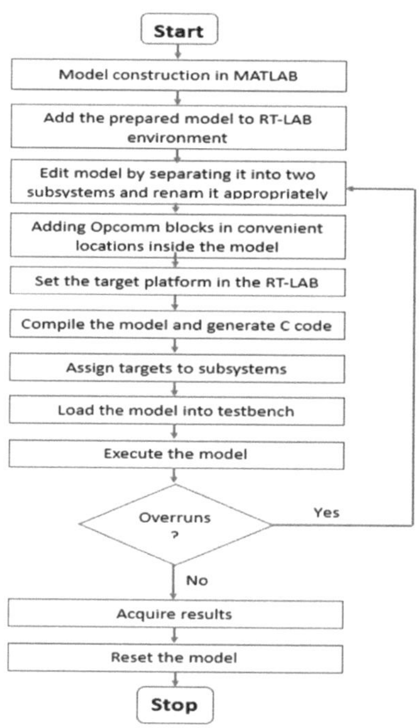

Figure 6. RTS Flowchart.

ARTEMIS and RT-Events are modeling tools available in the RT-LAB platform. These tools help in simulating complex power systems with high-performance improvement. ARTEMIS is a fixed-time solver applied in the real-time simulation of Simscape power systems. It allows users to benefit from the available algorithms that help provide reliable, stable, and accurate simulation. The utilization of the ARTEMIS tool allows the users to perform parallel execution of the circuits on different cores. Additionally, it enables the real-time simulation of models that consist of a large number of switches.

RT-Event block-set enables the simulation of systems that contain both continuous and discrete systems. Moreover, it supports fixed-step size simulation models and works with RT-LAB to perform real-time simulations.

In the initial stage of the RTS, the MG model has to follow specific modifications to meet the RTS considerations. One of these conditions is that the model's upper view must be composed of two or three subsystems, as mentioned in the previous section. After the first stage modifications, the MG model is represented in Figure 7, where the red subsystem is the master subsystem, while the blue one is the console subsystem. The 'powergui' block is the same as the conventional models. It performs its functions represented by choosing the simulation solver type, obtaining steady-state values, adjusting the initial state, FFT (Fast Fourier Transform) analysis, etc. Moreover, the "ARTEMIS" block is a fixed step solver utilized as an RTS optimization tool. It is fully compatible with MATLAB and the "SimPowerSystems" library.

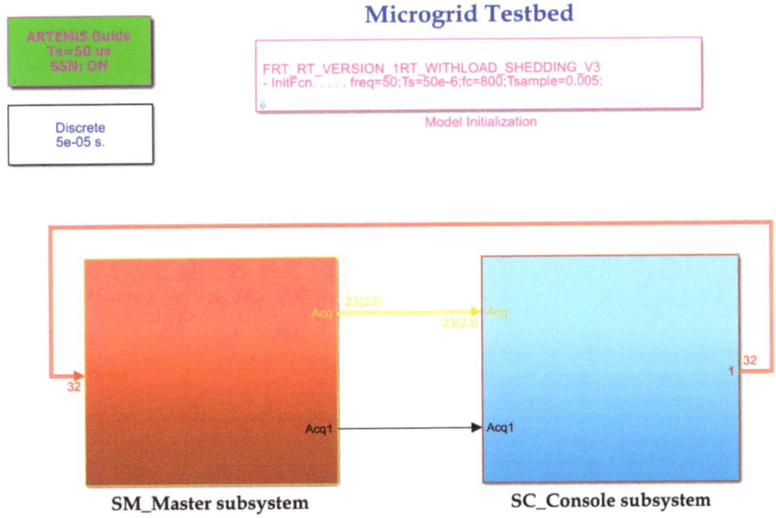

Figure 7. MG model in RT-LAB.

The "OpComm" block is a critical component in the model during the RTS part. It can be installed from the RT-LAB library. It is responsible for providing the communication infrastructure between the subsystems and the communication between the GUI subsystem and other computational subsystems. The essential condition in the RTS process is that the model has to be run well in the MATLAB platform without any errors. After performing the required modifications, we can open the MG model through the RT-LAB platform and start the RTS by building the model, loading the model on the target, executing the model, and acquiring the needed results from the console window. The specifications applied through this simulation are shown in Table 1.

Table 1. Specification of the RTS.

RT-LAB version	v2021.3.4.320
Host platform	Windows
Target platform	OPAL-RT Linux (x86-based)
Architecture	I686
Target OS version	Red Hat 5.2 (2.6.29.6-opalrt-6.2.1)
Matlab version	v9.9 (64 Bit) (R2020B)

4. Simulation and Results

This section presents the proposed mechanisms: FRT and Load curtailment algorithms, and their implementation in the MG case study.

4.1. Fault Ride through

The first technique that has been studied in the real-time state through this paper is FRT. The operational principle of the FRT technique is represented in the flowchart in Figure 8. It is mainly based on the injection of the reactive current. The FRT algorithm starts with sensing the voltage sag. Based on the regulations, the voltage limits measured at the PCC must be in the range of 0.9–1.1 pu. This indicates the normal operation of the MG system. However, if the voltage value differs from the acceptable limits, the FRT will start its operation based on the specified K factor. This factor can be calculated using the following equations:

$$K = \frac{I_Q}{\Delta V} = \frac{I_Q}{V_N - V_{PCC}}, K \geq 2 \quad (1)$$

$$K = -\tan\delta \quad (2)$$

where K is the FRT gain, I_Q is the reactive current, ΔV is the voltage difference, V_N is the nominal voltage (1 pu), and V_{PCC} is the voltage value at the PCC point.

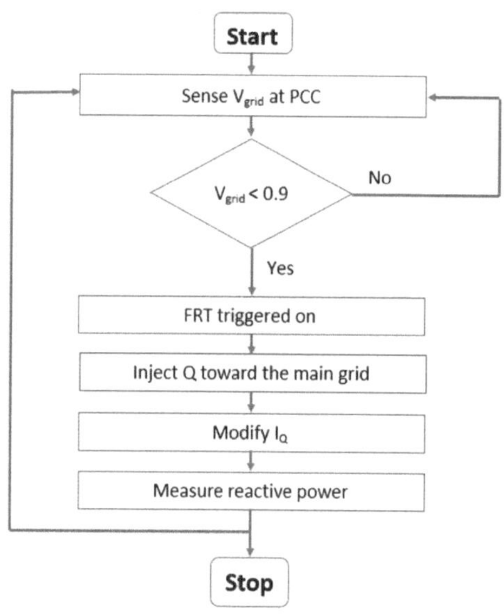

Figure 8. FRT flowchart.

If the voltage parameter exceeds acceptable limits, FRT will start operating by injecting the reactive current from the Grid Side Converter (GSC) to the PCC side to oppose the voltage sag. Generally, the operational process of the applied FRT is composed of three stages:

1. The first stage involves the sensing operation of the voltage parameter, in which a comparison between the measured voltage and the nominal voltage occurs. Referring to the output of this comparison, the model continually monitors any voltage sag cases and recovers the system by the proposed FRT.
2. The second step indicates the moving state of the reactive current I_Q parameter to be 1 pu as the system faced a voltage sag.
3. The third phase is responsible for adjusting the reactive power reference factor fed to the PV area block, to control the reactive power injected into the grid side.

The FRT technique will be studied on the MG model during the connected mode by inserting a fault at t = 1 s. As the fault occurs, the system will monitor a voltage sag case, and the FRT algorithm will start to be triggered. It is worth noting that the running time of the MG simulation is 10 s.

Figure 9 shows the FRT triggering signal in a real-time state, representing the FRT's starting operation at t = 1 s. The reactive current (I_Q) signal and the changing state of its value at the FRT operation are shown in Figure 10. The PV block's reactive power setpoint, considered a reference factor, will be adjusted during the FRT operation. Figure 11 displays the MG system's reactive power signal at the GSC and PCC. The first signal involves a negative value of the reactive power, showing the flow of the reactive power in the opposite direction, toward the main grid. Further, the second signal represents the increasing change of the reactive power value injected from the GSC towards the grid side, to oppose the voltage sag.

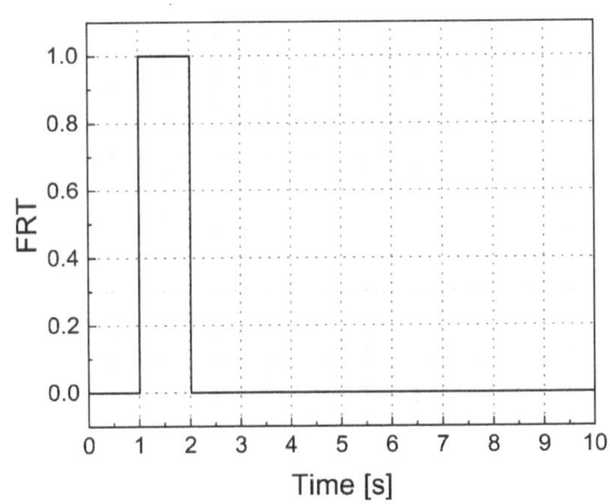

Figure 9. FRT triggering signal.

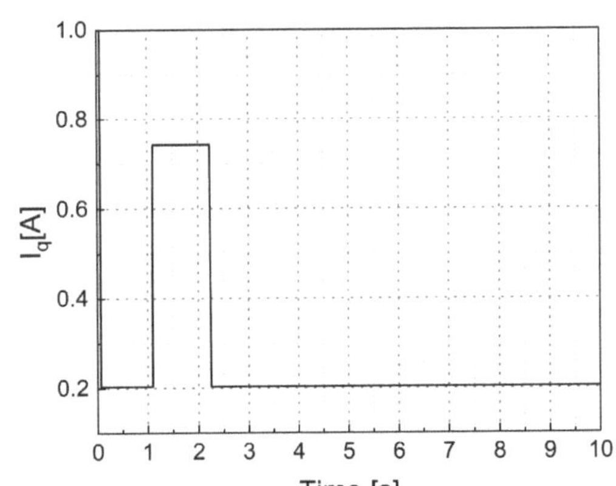

Figure 10. Reactive current signal in real-time.

Figure 12 shows the Active power signal at the first PV area. The spikes generated in the figures above are due to the loads associated with the PV systems. Furthermore, the applied maximum point power tracking (MPPT) technique in the PV subsystem can generate oscillations in some cases, especially in inductive load cases. Load switching and fault clearness can also create transients, due to the interaction between the mechanical energy stored in rotating machines and the energy stored in the inductance and capacitance when connecting and disconnecting the generation units.

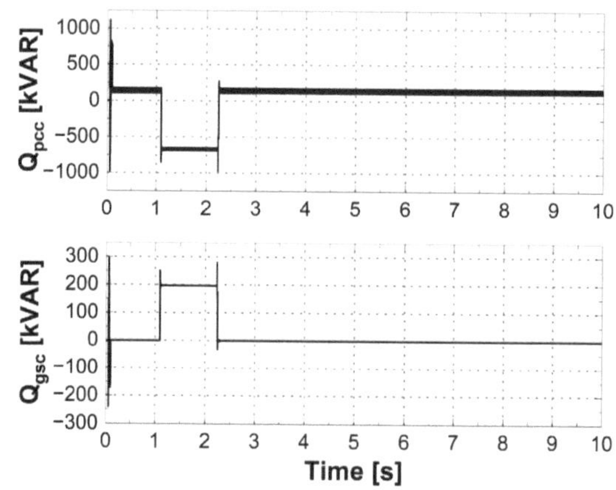

Figure 11. Reactive power signal in real-time.

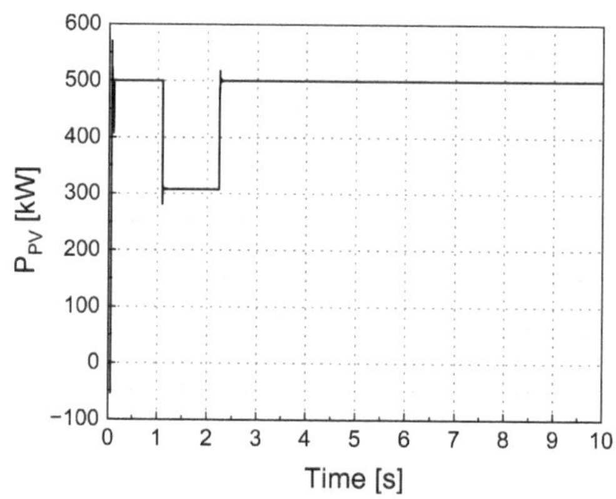

Figure 12. Active power signal at PV area 1 in real-time.

Equation (3) represents the power losses of the PV system in the two cases: utilizing the FRT technique, and without FRT. The losses in the output power of the PV system at area "A" reached about 190 kW. While in the case of the MG, without applying the FRT technique, the PV system will be disconnected as the fault occurs, leading to a cut of the output power from the PV side.

$$P_{pv\_losses} = \frac{P_{normal} - P_{fault}}{P_{normal}} \times 100\% \quad (3)$$

where, P_{pv_losses} is the losses percentage in the PV output power, P_{normal} is the measured output PV power in ordinary cases, P_{fault} is the measured output PV power in fault cases.

Based on the results, the output power of the PV is decreased by about 190 kW, resulting in power losses of about 38%. However, comparing this result with the losses in case of the total disconnection of the PV system in case of voltage sag, the output power will reach zero. The achievement of power loss reduction by the FRT technique leads to a significant reduction in the economic losses and the operational cost of the system. A minimization function (f_1) of the cost factor of the studied MG is represented below by implementing the FRT algorithm.

$$\min.f_1 = \min(C_{losses}) \quad (4)$$

$$\min.f_1 = \min\left(P_{losses} \cdot C_p \cdot 365\right) \quad (5)$$

C_{losses} is the cost of the losses, and Cp is the price of the energy use in kWh in Jordan, which is USD 0.123. The MG's power losses factor with FRT implementation is 192 kW, while it is 500 kW in the case of the absence of the FRT implementation. These results lead to a cost reduction of USD 22,447.5 to 8619.84 per year by applying FRT. So a minimization factor of about USD 13,827.66 is achieved out of the f_1 function.

4.2. Load Curtailment

Power losses and power balance keys are considered vital factors in the energy management of power systems. They have a direct impact on the system's performance and economical operation. Power balance relies on the generation and demand levels. It is based on the power difference parameter. The generation side of the MG is classified based on the installed generation units. RES (PV, wind, etc.), DGs, grid sources, and other units are examples of the generation side. In addition to the case of BESS installation, Equations (6) and (7) present the power balance condition:

$$P_{generation\_side} = P_{consumption\_side} \quad (6)$$

$$P_{Grid} + P_{PV} + P_{SS\_discharge} + P_{DG} = P_{Demand} + P_{SS\_charge} \quad (7)$$

The problem formulation of this scenario started in the MG's islanded mode operation. During the islanded mode of the MG, the generation side relies on the installed types of generation, including intermittent generation units such as PV systems and dispatchable generation units such as diesel generators. Moreover, the installed intermittent generation units and load demand units vary depending on several factors, such as weather forecasts and human activities. This leads to the need for continuous monitoring of the generation/demand balance in the MG system.

The power difference parameter (P_{Diff}) is the key to achieving power balance. It has to be within a specified range, as shown in Equation (8). A zero value of P_{Diff} indicates the power balance state. Several strategies have to be exploited in the case of P_{Diff} values that are more or less than zero. Values greater than zero indicate cases of surplus power. The surplus power can be stored in an energy storage system, or injected into the main grid, or sold to other utilities. Values below zero represent the cases of generation failures covering the installed load. Load curtailment techniques can be exploited in these cases. They rely on load minimization through a gradual disconnection process of the load.

$$\begin{gathered} P_{Diff} = P_{generation\_side} - P_{consumption\_side} \\ \text{Where } P_{Diff}^{min} < P_{Diff} < P_{Diff}^{max} \end{gathered} \quad (8)$$

where

$$\begin{gathered} {P_{generation\_side}}^{min} \le P_{generation\_side} \le {P_{generation\_side}}^{max} \\ {P_{consumption\_side}}^{min} \le P_{consumption\_side} \le {P_{consumption\_side}}^{max} \\ P_{diff} \ge 0, \text{(typical case)} \end{gathered} \quad (9)$$

A second minimization function (f_2) of the power difference factor of the studied MG is represented below by implementing the Load curtailment algorithm.

$$\min.f_2 = \min(P_{Diff})\Big|_{P_{grid}=0},\ P_{Difff} \in R \quad (10)$$

$$\min.f_2 = \min\left(P_{generated} - P_{demand}\right)\Big|_{P_{grid}=0} \quad (11)$$

$$P_{generated} = P_{PV} + P_{SS} + P_{DG};\ P_{grid} = 0 \quad (12)$$

The f_2 mainly manages the balance state of the system by decreasing the P_{diff} parameter in the case of islanded mode operation. The main constraints of this function are the generated and demand power, which together specify the P_{diff} parameter. Based on equation 9, P_{diff} is considered an inequality constraint with several active, inactive, and violated cases at the design point (zero).

As the $P_{generated}$ depends on the installed generation units, f_2 decreases the demand side as the P_{grid} is zero (islanded mode operation). The shedding process of the installed load depends mainly on the priority of the load. Therefore, the load with the lowest priority is considered the first choice in the shedding stage. The disconnecting loads decrease the overall demand, and the algorithm recalculates the P_{Diff} parameter. In the MG case study, the load priority of the installed loads 1, 2, 3, 4, and 5 are ordered as follows: 25%, 21%, 11%, 13%, and 30%, respectively.

In this case study, the MG will operate in islanded mode at t = 3 s. Figure 13 represents the power difference signal of the MG. The power significantly decreases when the MG operates in islanded mode.

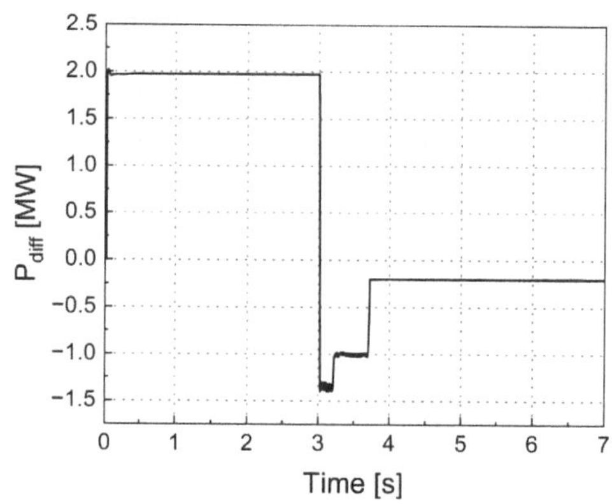

Figure 13. Power difference signal in real-time.

The Load curtailment algorithm starts its operation as it senses the value of the grid current I_{grid}. The zero value of I_{grid} indicates the MG is operating in islanded mode. Figure 14 represents the operation principle of the proposed mechanism. As the value of I_{grid} reaches zero, and the P_{diff} factor is negative, Load curtailment will start its operation automatically. Load 3 (L3) will be the first choice to be disconnected since it has the lowest priority. At about t = 3.2, L3 has disconnected, increasing the P_{diff} factor, as shown in Figure 13. However, the P_{diff} signal is still a negative value, which means that the demand side still exceeds the generation side. Hence, the Load curtailment algorithm continues its operation by disconnecting load 4. This process proceeds automatically by the Load curtailment model, based on a sensing system and logical blocks to send ON/OFF signals to the demand side. Moreover, in RTS, RT-LAB allows the users to manually send the ON/OFF signal during the running simulation, using control signals from the console subsystem. After the disconnection of load 4, the P_{diff} signal becomes positive, and this indicates the elimination of the unbalanced case.

Generally, as the implementation of FRT and Load shedding techniques leads to cost reduction, and demand scheduling maintaining a stable and reliable operation, both techniques can be considered as a simplified approach of the power flow scheme. One of the future directions of this study is to develop an Optimal power flow of the MG system to enhance several factors. An example of these factors is the enhancement of the voltage stability using the demand response management system, referring to [26].

The monitoring view window in Figure 15 presents the CPU usage for the tasks implemented in a real-time state. It helps to monitor overrun cases, in which our simulation shows that the number of overruns is zero. Furthermore, this tab can specify each task's duration by identifying the task's start and stop times.

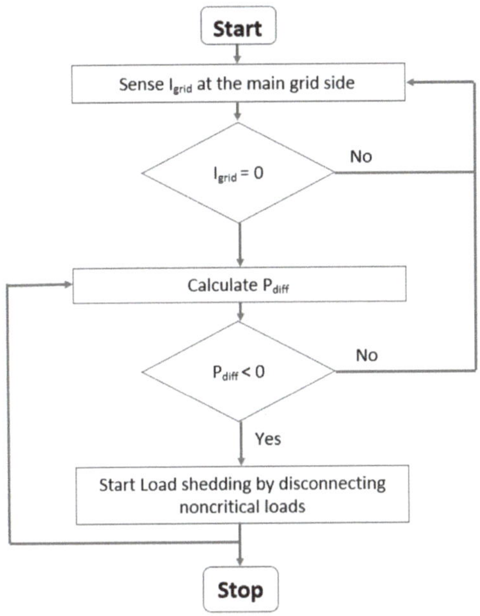

Figure 14. Load curtailment flowchart.

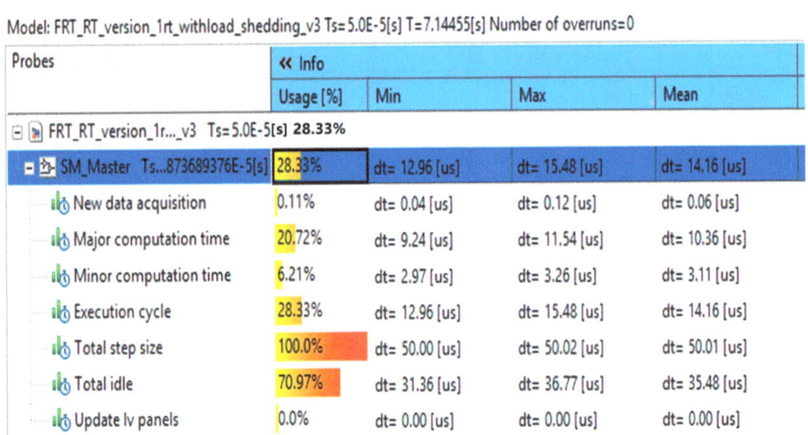

Model: FRT_RT_version_1rt_withload_shedding_v3 Ts=5.0E-5[s] T=7.14455[s] Number of overruns=0

Probes	« Info			
	Usage [%]	Min	Max	Mean
FRT_RT_version_1r..._v3 Ts=5.0E-5[s] 28.33%				
SM_Master Ts...873689376E-5[s]	28.33%	dt= 12.96 [us]	dt= 15.48 [us]	dt= 14.16 [us]
New data acquisition	0.11%	dt= 0.04 [us]	dt= 0.12 [us]	dt= 0.06 [us]
Major computation time	20.72%	dt= 9.24 [us]	dt= 11.54 [us]	dt= 10.36 [us]
Minor computation time	6.21%	dt= 2.97 [us]	dt= 3.26 [us]	dt= 3.11 [us]
Execution cycle	28.33%	dt= 12.96 [us]	dt= 15.48 [us]	dt= 14.16 [us]
Total step size	100.0%	dt= 50.00 [us]	dt= 50.02 [us]	dt= 50.01 [us]
Total idle	70.97%	dt= 31.36 [us]	dt= 36.77 [us]	dt= 35.48 [us]
Update lv panels	0.0%	dt= 0.00 [us]	dt= 0.00 [us]	dt= 0.00 [us]

Figure 15. Real-time monitoring view window for the proposed model.

5. Conclusions

In this paper, an MG case study has been built in MATLAB/SIMULINK and studied in a real-time state. The real-time simulator used in this study is the OP1420 PHIL Microgrid testbench simulator. The MG model has been edited in the RT-LAB platform to implement the required modifications in a real-time state. Two scenarios have been proposed and modeled through this work. The first scenario is the FRT, which is implemented to prevent the total outage of the PV system that may occur following unpredictable faults. The results show the effectiveness of this technique in decreasing the losses of the system and hence decreasing the economic losses. The FRT implementation achieved a minimization in losses of about 60%. The second scenario is performed to obtain the load-generation balance by applying the proposed Load curtailment algorithm. The results show that the

Load curtailment algorithm contributes to maintaining the balance and stable operation of the MG.

Author Contributions: Conceptualization, H.M. and A.A.; methodology, H.M.; software, A.A.; validation, H.M., A.A. and A.A.-H.; formal analysis, H.M.; investigation, A.A.; resources, H.M.; data curation, H.M. and A.A.; writing—original draft preparation, H.M., A.A. and A.A.-H.; writing—review and editing, H.M, A.A. and A.A.-H.; visualization, H.M.; supervision, H.M. and A.A.-H.; project administration, H.M.; funding acquisition, H.M. All authors have read and agreed to the published version of the manuscript.

Funding: This research was funded by Deanship of Scientific Research at the German Jordanian University under seed grant SATS 07/2021.

Institutional Review Board Statement: Not applicable.

Informed Consent Statement: Not applicable.

Data Availability Statement: Not applicable.

Acknowledgments: The authors acknowledge the financial support of the Deanship of Scientific Research at the German Jordanian University under seed grant SATS 07/2021 for the presented work in this paper.

Conflicts of Interest: The authors declare no conflict of interest.

References

1. Caruso, M.; Di Tommaso, A.O.; Miceli, R.; Nevoloso, C.; Pellitteri, F.; Puccio, C.; Schettino, G. Experimental Prototyping of a Microgrid with Mechanical Point of Common Coupling. In Proceedings of the 2020 8th International Conference on Smart Grid (icSmartGrid), Paris, France, 17–19 June 2020; pp. 214–219.
2. Ibrahim, M.; Alkhraibat, A. Resiliency Assessment of Microgrid Systems. *Appl. Sci.* **2020**, *10*, 1824. [CrossRef]
3. Cagnano, A.; De Tuglie, E.; Mancarella, P. Microgrids: Overview and guidelines for practical implementations and operation. *Appl. Energy* **2020**, *258*, 114039. [CrossRef]
4. Al-Agtash, S.; Al-Mutlaq, N.; Elabbas, M.; Alkhraibat, A.; Al Hashem, M. Multi-Agents for Microgrids. *Energy Power Eng.* **2021**, *13*, 293–305. [CrossRef]
5. Talapur, G.G.; Suryawanshi, H.M.; Xu, L.; Shitole, A.B. A reliable microgrid with seamless transition between grid connected and islanded mode for residential community with enhanced power quality. *IEEE Trans. Ind. Appl.* **2018**, *54*, 5246–5255. [CrossRef]
6. Adefarati, T.; Bansal, R.C. Bansal. Reliability, economic and environmental analysis of a microgrid system in the presence of renewable energy resources. *Appl. Energy* **2019**, *236*, 1089–1114. [CrossRef]
7. Xin-gang, Z.; Ze-qi, Z.; Yi-min, X.; Jin, M. Economic-environmental dispatch of Microgrid based on improved quantum particle swarm optimization. *Energy* **2020**, *195*, 117014. [CrossRef]
8. Muhsen, H.; Allahham, A.; Al-Halhouli, A.A.; Al-Mahmodi, M.; Alkhraibat, A.; Hamdan, M. Business Model of Peer-to-Peer Energy Trading: A Review of Literature. *Sustainability* **2022**, *14*, 1616. [CrossRef]
9. Nejabatkhah, F.; Li, Y.W. Overview of power management strategies of hybrid AC/DC microgrid. *IEEE Trans. Power Electron.* **2014**, *30*, 7072–7089. [CrossRef]
10. Samadi, E.; Badri, A.; Ebrahimpour, R. Decentralized multi-agent based energy management of microgrid using reinforcement learning. *Int. J. Electr. Power Energy Syst.* **2020**, *122*, 106211. [CrossRef]
11. Al-Agtash, S.; Alkhraibat, A.; Al Hashem, M.; Al-Mutlaq, N. Real-Time Operation of Microgrids. *Energy Power Eng.* **2021**, *13*, 51. [CrossRef]
12. Dagar, A.; Gupta, P.; Niranjan, V. Microgrid protection: A comprehensive review. *Renew. Sustain. Energy Rev.* **2021**, *149*, 111401. [CrossRef]
13. Mignoni, N.; Scarabaggio, P.; Carli, R.; Dotoli, M. Control frameworks for transactive energy storage services in energy communities. *Control Eng. Pract.* **2023**, *130*, 105364. [CrossRef]
14. Farzinfar, M.; Jazaeri, M.; Nair, N.K.C.; Razavi, F. Stability evaluation of Microgrid using real-time simulation. In Proceedings of the 2014 Australasian Universities Power Engineering Conference (AUPEC), Perth, WA, Australia, 28 September–1 October 2014; pp. 1–6.
15. Kundur, P.S.; Balu, N.J.; Lauby, M.G. Power system dynamics and stability. *Power Syst. Stab. Control* **2017**, *3*, 827–950.
16. Leake Enquay Weldemariam. *Monitoring and Regulation of Voltage Dips in the Distribution Network*; LAP LAMBERT Academic Publishing: Sunnyvale, CA, USA, 2018.
17. Brahma, S.; Pragallapati, N.; Nagpal, M. Protection of islanded microgrid fed by inverters. In Proceedings of the 2018 IEEE Power & Energy Society General Meeting (PESGM), Portland, OR, USA, 5–10 August 2018.

18. Muhsen, H.; Alkhraibat, A.; ElMoaqet, H. Competencies of Fault Ride through in Microgrid Modeling: A case study. In Proceedings of the 2021 IEEE Jordan International Joint Conference on Electrical Engineering and Information Technology (JEEIT), Amman, Jordan, 16–18 November 2021.
19. Singh, S.K.; Padhy, B.P.; Chakrabarti, S.; Singh, S.N.; Kolwalkar, A.; Kelapure, S.M. Development of dynamic test cases in OPAL-RT real-time power system simulator. In Proceedings of the 2014 Eighteenth National Power Systems Conference (NPSC), Guwahati, India, 18–20 December 2014; pp. 1–6.
20. Liu, X.; Li, C.; Shahidehpour, M.; Gao, Y.; Zhou, B.; Zhang, Y.; Yi, J.; Cao, Y. Fault current hierarchical limitation strategy for fault ride-through scheme of Microgrid. *IEEE Trans. Smart Grid* **2019**, *10*, 6566–6579. [CrossRef]
21. Buraimoh, E.; Davidson, I.E. Fault ride-through analysis of current-and voltage-source models of grid supporting inverter-based microgrid. *IEEE Can. J. Electr. Comput. Eng.* **2021**, *44*, 189–198. [CrossRef]
22. Sun, C.; Joos, G.; Ali, S.Q.; Paquin, J.N.; Rangel, C.M.; Al Jajeh, F.; Novickij, I.; Bouffard, F. Design and real-time implementation of a centralized microgrid control system with rule-based dispatch and seamless transition function. *IEEE Trans. Ind. Appl.* **2020**, *56*, 3168–3177. [CrossRef]
23. Shen, X.; Huang, W.; Shen, C.; Shen, Y.; Shuai, Z. Fault-Ride through Strategy for Islanded Microgrids Via Dynamically Reconfigurable Voltage Reference. In Proceedings of the 2022 IEEE Energy Conversion Congress and Exposition (ECCE), Detroit, MI, USA, 9–13 October 2022; pp. 1–6.
24. Madiba, T.; Bansal, R.C.; Mbungu, N.T.; Bettayeb, M.; Naidoo, R.M.; Siti, M.W. Under-frequency load shedding of microgrid systems: A review. *Int. J. Model. Simul.* **2022**, *42*, 653–679. [CrossRef]
25. Dehghanpour, E.; Karegar, H.K.; Kheirollahi, R. Under frequency load shedding in inverter based microgrids by using droop characteristic. *IEEE Trans. Power Deliv.* **2020**, *36*, 1097–1106. [CrossRef]
26. Yao, M.; Molzahn, D.K.; Mathieu, J.L. An optimal power-flow approach to improve power system voltage stability using demand response. *IEEE Trans. Control Netw. Syst.* **2019**, *6*, 1015–1025. [CrossRef]

sustainability

MDPI

Article

Hybrid Tripping Characteristic-Based Protection Coordination Scheme for Photovoltaic Power Systems

Feras Alasali [1,*], Abdelaziz Salah Saidi [2,3], Naser El-Naily [4], Mahmoud A. Smadi [1] and William Holderbaum [5,*]

1. Department of Electrical Engineering, Faculty of Engineering, The Hashemite University, P.O. Box 330127, Zarqa 13133, Jordan
2. Department of Electrical Engineering, King Khalid University, Abha. P.O. Box 394, Abha City 61411, Saudi Arabia
3. Laboratoire des Systèmes Électriques, École Nationale d'Ingénieurs de Tunis, Université de Tunis El Manar, Tunis 1068, Tunisia
4. College of Electrical and Electronics Technology-Benghazi, Benghazi 23P7F49, Libya
5. School of Biomedical Engineering, University of Reading, Whiteknights, Reading RG6 6AY, UK

* Correspondence: ferasasali@hu.edu.jo (F.A.); w.holderbaum@reading.ac.uk (W.H.)

Abstract: Due to the high penetration of renewable energy sources into the electrical power network, overcurrent relays coordination with highly sensitive and selective protection systems are now two of the most important power protection concerns. In this research, an optimal coordination strategy utilising a new hybrid tripping scheme based on current–voltage characteristics has been devised for overcurrent relays in a power network coupled to a photovoltaic system. This research develops and proves a new optimal coordination scheme based on two optimisation methods, the vibrating particles system and particle swarm optimisation algorithms, in consideration of the impact of renewable sources on fault characteristics. The new optimal coordination approach aims to improve the sensitivity and dependability of the protection system by reducing the tripping time of the overcurrent relays by employing a new hybrid tripping scheme. A specific case study, Conseil International des Grands Réseaux Electriques (CIGRE) distribution network connected to two photovoltaic systems is constructed and presented utilising Industrial software (namely ETAP), and the outcomes of the proposed optimal coordination scheme are compared with standard and recent characteristics from the literature. The hybrid tripping scheme and optimisation techniques are evaluated using different fault and power network model scenarios. The results show that the optimal hybrid tripping scheme provided successfully decreases the overall operating time of the overcurrent relays and increases the sensitivity of the relay during all fault scenarios. The reduction in overall time for the proposed hybrid tripping scheme was 35% compared to the literature for the scenario of a power grid with and without photovoltaic systems.

Keywords: overcurrent relays; optimal coordination; renewable energy; nonstandard tripping characteristics

Citation: Alasali, F.; Saidi, A.S.; El-Naily, N.; Smadi, M.A.; Holderbaum, W. Hybrid Tripping Characteristic-Based Protection Coordination Scheme for Photovoltaic Power Systems. *Sustainability* **2023**, *15*, 1540. https://doi.org/10.3390/su15021540

Academic Editors: Ayman Al-Quraan and Ahmad M.A. Malkawi

Received: 17 December 2022
Revised: 7 January 2023
Accepted: 11 January 2023
Published: 13 January 2023

1. Introduction

Photovoltaic (PV) arrays, wind turbines, and fuel cells are all examples of Distributed Generators (DGs) that can be integrated into an existing Distribution Network (DN) to save energy costs, improve reliability, and satisfy environmental regulations. Furthermore, utilities benefit from higher power quality with such DG systems in place. However, DGs cause a change in the level and characteristics of the fault current and bidirectional power flow [1–3]. As a result, the conventional method of protecting DN with DGs based on the use of Overcurrent Relays (OCRs) is becoming more challenging. In addition, when designing a protective system to deal with varying fault levels and different grid operation modes (with and without DG), the DN's ability to operate in these modes is an important concept to take into account [4]. Redesigning or replacing the protection system in a DN due to the integration of the DGs can be costly and technically challenging. By increasing the capacity of DG penetration in DN, the OCR settings must be modified to cope with the changing power flow and fault. Adaptive protection systems in radial DNs often adjust the relay settings according to the role of the communication system. However, in many DNs, establishing a communication infrastructure for power protection systems may be an expensive choice. The recent and main OCR coordination approaches for DN with DG are divided into six categories, which are shown in Figure 1, as follows [4,5]:

- Developing a new objective function to address the OCR coordination problem.
- Applying and developing a dual-setting approach in OCR.
- Designing and developing a non-standard characteristic for OCR.
- Using a new optimisation algorithm to solve the complex OCR coordination problem for a DN with DG.
- Applying and developing new constraints to the objective function.
- Designing and developing a hybrid tripping characteristic for OCR.

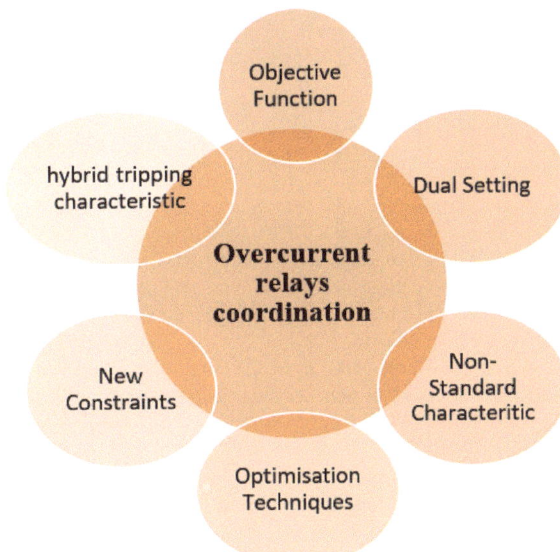

Figure 1. The recent and major OCR coordination approaches for DN with DGs.

The primary purpose of all of the aforementioned approaches is to obtain the appropriate setting for protection schemes to maintain the reliability of protection scheme performance for interconnected DNs with DGs. Therefore, this research presents a new non-standard current–voltage characteristic for programmable OCRs as a hybrid tripping characteristic approach. The magnitudes of the faulty phase voltage and current are used by the suggested method to calculate the relay's operation time by employing a new optimisation method. Plus, it can safeguard DNs with extensive penetration of DGs in the grid-connected mode of DN operation, and it does not rely on any sort of communication infrastructure. Academics and industry professionals have focused significantly on the problem of DN protection due to the stochastic behaviour of DGs. Several approaches have been proposed to provide an adequate method of protecting the microgrid in both modes (with and without DGs). Some of them concentrate on communication channel protection [6,7]. Slemaisar-doo et al. [8] suggested a differential protection approach employing a non-nominal frequency-current during a microgrid fault that is superior to traditional overcurrent protection in detecting the microgrid fault. Aghdam et al. [9] proposed a differential protection method based on variable tripping times, and a multiagent protection scheme was designed to improve the coordination of adjacent relays. Communication-based strategies are realistic microgrid protection options. However, the reliability of this type of protection highly depends on the communication facilities and performance; it is also not an economically viable solution. In addition [6,10], these schemes are affected by a communication failure, imbalanced loads, and transients' events during the connection and disconnection of DGs. These days, it is common practice to employ programmable relays (microprocessor relays) to apply non-standard characteristics. The literature proposed several concepts, including a logarithm characteristic for OCRs [5,10], a combination of standard characteristics and a non-standard term based on voltage [5], and a standard characteristic under new constants [5,10] for DNs with DG to reduce the total amount of time spent operating the OCR. However, these approaches are utilised in radial DNs with DG, and all of these techniques require a communication infrastructure. Therefore, the purpose of this study is to introduce a new hybrid optimal coordination strategy that does not require a communication link between the OCRs. This is intended to reduce the demand for communications infrastructure while improving the coordination approach of the OCR. Furthermore, the suggested method will reduce computing costs and the necessity to access voluminous PV and network data.

Because of phase OCR's inapplicability for handling the complexity of DN-integrated PV systems, the voltage term is being explored as a potential term for solving the OCRs coordination issue [11]. In [12], a voltage-restricted overcurrent relay is presented using phase voltage and current to set the necessary threshold. Nevertheless, the PV plant's control method may cause phase currents to be larger in a healthy phase than in a faulty phase, leading to the relay in miscoordination events. Few studies have looked into the use of voltage terms in the OCRs coordination problem [13]. The use of voltage-restricted OCRs coordination schemes for network protection was discussed, for example, in [14]. However, no voltage limitation was presented with the OCR algorithm or result [14]. The voltage–current–time inverse model presented by Singh et al. [15] is based on variations in currents and voltages during fault events. The suggested OCR coordination model improves operating time and maintains protection coordination for power networks with PV systems without considering different modes of the grid. El-Sayed et al. [16] introduced a current–voltage scheme for directional OCR based on measuring harmonic currents and voltages at the relay position to guarantee optimal protection coordination for the islanded microgrid without taking into account bidirectional power flow and grid-connected operation mode. Protection schemes that use conventional inverse time–current characteristics are becoming increasingly unsuitable as the penetration of DGs in the DN rises. Authors [17,18] presented a voltage-based protection scheme to minimise the operating time of the relay compared to the traditional inverse time current scheme. Another study [19] proposes a strategy for protecting the DN using the superconducting fault limiter based on the voltage parameter. There is a limited number of research that used the non-standard logarithmic function for developing a current–voltage protection coordination scheme for DN with DGs under different grid operation modes and fault scenarios.

The foregoing challenges and evidence point to the necessity of having a flexible protection mechanism for more dynamic power networks with DG in the future. OCR protection schemes must be compatible with dynamic power systems in terms of their ability to overcome and accommodate these emerging features, raising questions about the appropriateness of standard characteristics of OCR protection schemes, stability, and protection selectivity for DN with DG. In this paper, we introduce a novel hybrid tripping scheme based on non-standard current–voltage characteristics for fast response OCR prevention in DNs with PV farm systems without using a communication protocol. The following are some of the study's contributions that aim to bridge this research area gap:

1. To enhance the performance of the protection system, a new non-standard logarithmic and hybrid tripping coordination scheme based on current–voltage characteristics is established for DN with DGs. In this article, a significant reduction in total operational time is achieved, with no instances of miscoordination compared to typical

characteristics of the OCR scheme. This work compares the proposed new hybrid tripping OCR scheme (HOC) with the common inverse time–current characteristic (SIC) and the time–current–voltage characteristic (CVC) from the literature [16,20].

2. In the literature [20], the use of modern optimisation techniques in solving protection problems, such as the particle swarm algorithm [21,22] assists in achieving the optimal OCR settings. To solve the OCR coordination problem based on the non-standard current–voltage characteristic and reduce the total operational time of OCRs, a new optimisation technique, the Vibrating Particles System (VPS) approach, has been designed and employed.
3. Since the proposed hybrid optimal coordination scheme in this work only uses locally obtained measurements, no medium of communication between the OCRs is necessary. This eliminates the demand for communications infrastructure and reduces computational costs and the requirement for access to extensive PV and network data.
4. The sensitivity and selectivity of the proposed hybrid optimal coordination scheme have been investigated for DN (benchmark power network, CIGRE) with DGs under different fault scenarios and operation modes. This aims to provide network operators with a preliminary indicator regarding the potential impact of DGs on the fault contribution and relay setting.

The following sections are ordered as follows: The optimum OCRs coordination problem formulation is introduced in Section 2. In addition, the proposed hybrid tripping scheme is developed. The results and comparison are discussed in Section 3. In Section 4, the summary and conclusion of this work are presented.

2. Problem Formulation of Optimum OCRs Coordination

The coordination problem of OCRs in a DN with DGs can be formulated as an optimisation problem [1,5]. This formulation's purpose is to identify the OCRs settings that reduce the overall operational times of OCRs without sacrificing the selectivity between primary and backup relays. Figure 2 describes the structure and workflow taken in this paper.

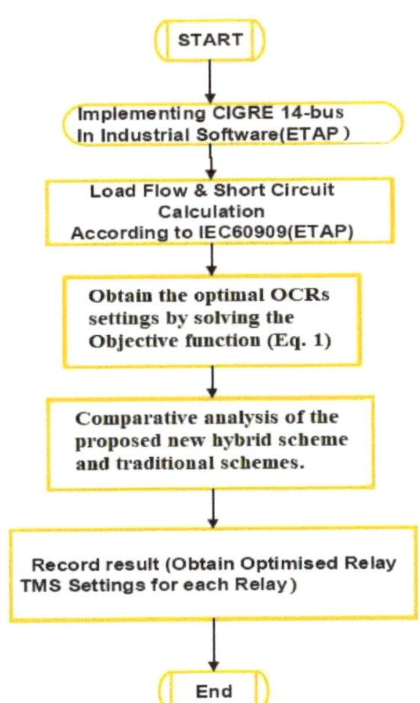

Figure 2. The structure and general workflow for optimum OCRs coordination.

This section offers a mathematical formulation of the proposed optimisation approach to solve the coordination problem and enhance the performance of the optimisation strategies for OCRs. The objective function (OT) is, therefore, utilised to minimise the overall operating time of the primary and backup OCRs. OT can be expressed as described in Equation (1) [1,5].

$$\mathrm{OT} = \sum_{r=1}^{R} \sum_{f=1}^{F} t_{r,f} \tag{1}$$

where $t_{r,f}$ is the operational time of the relay, r, at the fault location, f. R is the number of OCRs and F is the total location of the fault in the DN. The objective function, OT, is subject to the following constraints.

2.1. Selectivity Constraints

The objective of the selectivity or coordination restriction is to extend the lag time between the primary and backup OCRs in terms of propositional operations. This producer will help in reducing the impact of a power loss on the network by pinpointing the precise location of the defect; in this case, the backup OCRs should be inoperable unless the primary OCRs fail to do so. As a collection of inequality constraints, selectivity and coordination criteria can be stated using the Coordination Time Interval (CTI) [1,4].

$$t_{\mathrm{backup}} - t_{\mathrm{primary}} \geq \mathrm{CTI} \tag{2}$$

where the t_{primary} is the operational time for the primary OCR and t_{backup} is the operational time of the backup OCR. The CTI value is based on several criteria, including relay type and circuit breaker speed. To achieve selectivity, the CTI is often set between 0.2 and 0.5 s according to IEEE-242. Therefore, the CTI is equal to 0.3 in this study, similar to [1,4].

2.2. Relay Settings Constraints

To preserve the OCRs operational time constraints, minimum and maximum OCR operational time requirements must be presented. However, protective relays should run with the shortest operational time possible; if they take longer to work, equipment damage and power system instability will result. Here are the minimum and maximum operational time limits:

$$t_{min} \leq t_r \leq t_{max} \tag{3}$$

$$TMS_{min} \leq TMS_r \leq TMS_{max} \tag{4}$$

where t_{min} and t_{max} are the minimum and maximum OCR operational time, t_r is the operational time of the relay r. TMS_{min} and TMS_{max} are the minimum and maximum Time Multiplier Settings (TMS). TMS_r is the TMS value for relay r. In this study, TMS is handled as a continuous variable, and OCRs need to work within the operating period of the protection schemes. Therefore, the TMS should be set within the maximum and minimum values under different fault conditions, or even during a light overload. TMS limits range from 0.01 to 3 [5], and the TMS for the majority of industrial and microgrid OCRs falling within this range.

2.3. Characteristics of the Relay

In general, the operation time of an overcurrent relay (OCR), t, is dictated by a standard inverse function of the fault current and operational time. The characteristic equation governing the relay's working time varies based on the OCR's manufacturer and type of the relay. In this paper, standard OCRs are employed in conjunction with the IEC255-3 standard characteristic equation as described in (5) as a reference scheme to compare our proposed scheme with it [4,5].

$$t = \left[\frac{A}{\left(\frac{I_f}{I_{pi}}\right)^B - 1} \right] TMS \quad (5)$$

where I_f is the fault current, I_{pi} is the pickup current, A and B are constants defined based on the OCR standard such as IEEE, IEC, and AREVA. Numerical and programmable relays, such as OCRs (microprocessor relays), can typically have their time-operating properties modified and updated via a network connection and real-time data. In this article, the numerical relay (OCR) is used in this article based on industry standard (IEC) where A and B are equal to 014 and 0.02, respectively [2,5]. The normal standard inverse time-current characteristic is shown in Figure 3 for the relation between operational time, t, and $\frac{I_f}{I_{pi}}$.

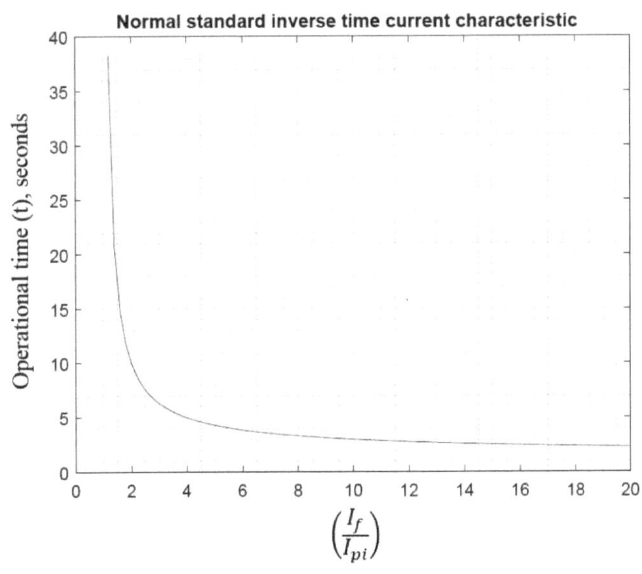

Figure 3. The normal standard inverse time–current characteristic for the OCR.

In addition to the standard inverse characteristic, in [23] a time–current–voltage characteristic for OCR based on the phase fault voltage, V, is proposed as follows:

$$t = \left(\frac{1}{e^{1-V}}\right)^K TMS \left(\frac{A}{\left(\frac{I_f}{I_{pi}}\right)^B - 1} \right) \quad (6)$$

The time–current–voltage characteristic of [19], as described in Equation (6) and shown in Figure 4, is also used as a reference OCR scheme, which will be compared to the hybrid tripping scheme in this work.

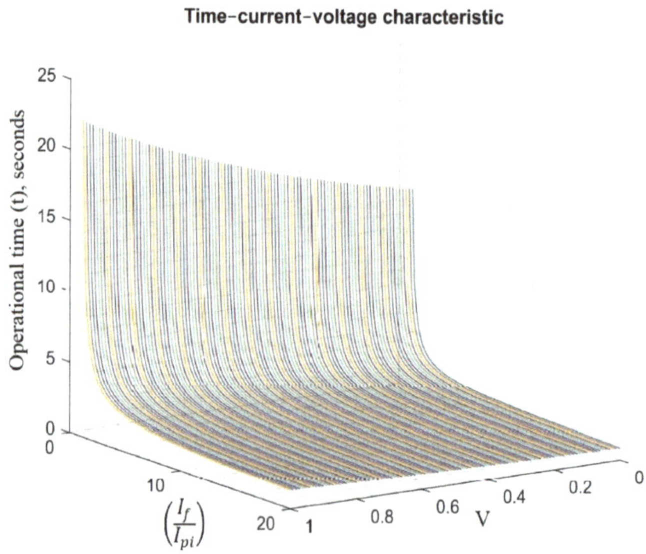

Figure 4. Time–current–voltage characteristic for OCR.

2.4. Proposed Hybrid Tripping OCR Characteristic

In the literature, the typically inverse time–current characteristic [1,2] and time–current–voltage characteristic [23] are restricted to certain values of the setting of low and high multipliers. As previously explained, the fault characteristics of a power system network incorporating renewable energy sources are challenging and distinct. The limitation of using characteristics from the literature reduces OCR's sensitivity in the event of maximum and minimum fault currents at different operation modes for the DN [5]. This work proposes an advanced hybrid OCRs characteristic and coordination approach based on a non-standard current–voltage characteristic based on a logarithmic function to alleviate the limitations in the standard characteristics and the protection difficulties for DG systems. Furthermore, the tripping characteristics of OCRs as described in Equations (5) and (6) [5] are significantly longer for DGs with a weak fault infeed, making them unsuitable for use in power networks with low fault currents. This limitation is alleviated in this work by the introduction of a new hybrid tripping OCR characteristic based on a logarithmic function for the currents ($\frac{I_f}{I_{pi}}$) and the phase voltage (V_P). The new hybrid tripping OCR characteristic is presented in Equation (7):

$$t = \left(5.8 - 1.35 * \log_e\left(\frac{I_f}{I_{pi}}\right)\right) \log_e(9V_P + 1)\ TMS \tag{7}$$

To achieve selectivity in OCR coordination, the grading time must be independent of the network fault location and fault current level. Equation (7) presents the proposed hybrid tripping OCR characteristic utilizing logarithmic and constant coefficients for all relays. This helps the classification time to be independent of the fault level and location. To establish the best *TMS* in Equation (7) that minimises the operation tripping time, the next section describes the optimisation algorithm that has been used to find the optimal OCR setting and *TMS*. Figure 5 presents the current–voltage characteristic curve based on a logarithmic function.

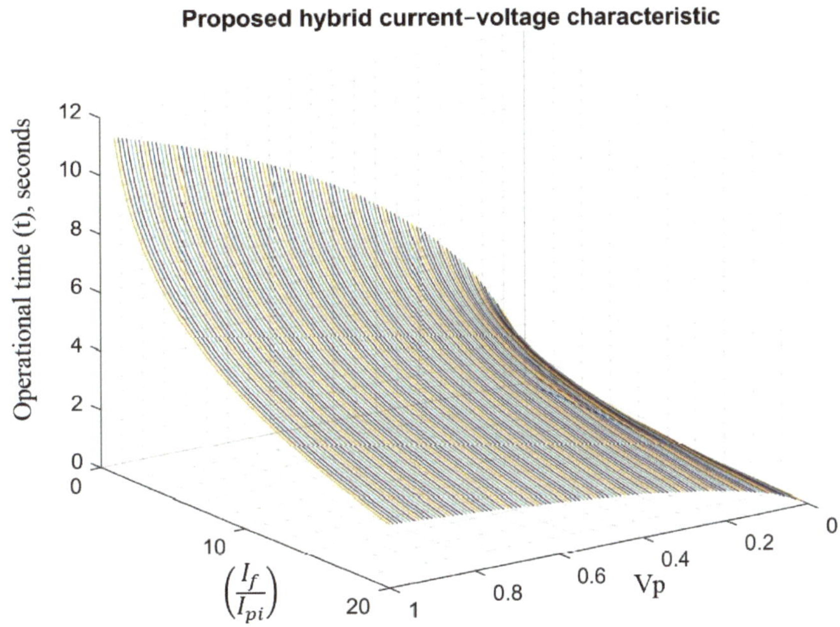

Figure 5. Proposed hybrid current–voltage characteristic for OCR.

2.5. *Optimisation Methods for Solving the OCRs Coordination Problem*

In this section, the OCRs coordination problem represented by Equation (1) for a DN with connected DGs is treated as an optimisation problem. This section presents two optimisation algorithms, Particle Swarm Optimisation (PSO) and the Vibrating Particle System (VPS) approach, as common and new powerful optimisation algorithms for solving OCR coordination problems using the standard inverse time–current characteristic [2,4], the time–current–voltage characteristic [20,23], and the hybrid current–voltage characteristic.

2.5.1. PSO Algorithm

This research introduces and employs the PSO algorithm, as depicted in Figure 6 [24,25], as a common optimisation technique to solve OCR coordination problems. Kennedy and Eberhart established PSO as one of the most effective approaches for solving complex engineering optimisation problems [2,26]. Furthermore, the PSO is a cutting-edge heuristic optimisation technique, and its inherent simplicity means that it uses fewer CPU resources. Consequently, the PSO's global solution finding, rapid convergence, and easy implementation have made it popular in a variety of contexts, including energy, power flow, and protection system problems [5,26]. Based on the benefits of this strategy, the PSO method is used here to provide a common and reference algorithm for solving the proposed coordination problem, as described in Equation (1). In general, PSO mimics and is inspired by human social behaviour and the behaviour of swarming animals [5]. The objective of the PSO algorithm is to maintain and govern the population of agents or particles (called a "swarm"), where each swarm of particles represents a potential solution. In PSO, the population of the swarm represents the solution, and each particle represents an individual result. The particles memorize their current and optimal population position relative to the objective function under consideration (Equation (1)). The trajectory of the particles will be modified based on their position and the optimal position of the swarm. In each iteration, improved placements and optimal solutions are discovered, which influence particle movement in the swarms. Each particle moves inside the search space with a variable velocity. In the end, the PSO model will identify the optimal answer among all conceivable alternatives, which is known as the "global solution".

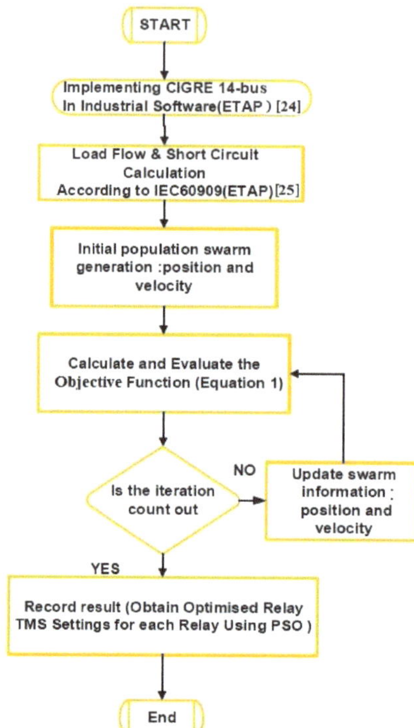

Figure 6. Outline of the optimisation technique (PSO) for solving the OCRs coordination problem [24,25].

Figure 5 depicts the outline of the optimisation technique for the proposed coordination of OCRs problem. First, using the measured network information, load flow and short circuits are computed. Then, the PSO model is used to generate an initial swarm population based on all of the input data. In this article, we describe the PSO optimisation model through the minimal cost function, Equation (1), and the protection and network constraints, Equations (2) to (4). After the swarm's data have been updated, the process will repeat itself based on the fitness values of the goal function. Finally, the OCRs problem's global solution will be found by the PSO model at the end of each iteration, and the TMS will be determined as a result.

2.5.2. VPS Algorithm

In 2017, an evolutionary metaheuristic search strategy, the Vibrating Particles System (VPS) approach [27], was created by Kaveh and Ilchi Ghazaan. It encourages natural vibration of systems with a single degree of freedom subject to viscous damping. VPS has been used to solve a variety of structural and engineering optimisation problems, with positive results in terms of convergence and accuracy [28]. The newly presented optimisation algorithm (VPS) is intended to tackle complicated optimisation problems with minimal computational effort and expense. Therefore, the proposed VPS algorithm can be an extremely efficient and potent algorithm to tackle complex protection coordination problems for a power network with integrated DGs. Similarly, to other population-based metaheuristics, VPS starts with a random pool of initial solutions and analyses them as single degrees of freedom in vibrating systems. When a freely vibrating system is subjected to damping conditions, it oscillates and eventually finds its equilibrium position according to a predetermined formula. Differential equations are used to easily demonstrate this case. VPS improves the quality of the particles on a regular basis during the optimisation process by oscillating them forward to the equilibrium position using a combination of randomness and exploitation of the data collected [27,28]. Consider that the equilibrium position of each particle includes three parts: the highest optimal position (HOP), a good optimal particle (GOP), and a bad optimal particle (BOP). Therefore, there are essentially three concepts for the movement towards the global optimal solution:

1. Self-adaptation: In a process of self-adaptation, the particle shifts its trajectory to approach HOP.
2. Collaborations: The new particle's location can be influenced by the GOP and BOP.
3. Competitions: In terms of competition, GOP is more influential than BOP.

For each particle, VPS modifies the weights in HP, GP, and BP. Following a population-level application of a penalised objective function sort, values are then ranked in ascending order. Both the GOP and the BOP for a given particle are drawn at random from the first and second halves of the population, respectively. Good particles, bad particles, and the algorithm's best particle up to this point are compared using different constant weights, and the rand is a uniformly produced random number between 0 and 1. It should be mentioned that VPS adjusts the location of particles using memory based on the harmony search approach [27]. The best solution (TMS) will be chosen at the end of the iteration process, when the number of objective function evaluation iterations (NFEs) exceeds the maximum number of NEFs (maxNFES), as shown in Figure 7.

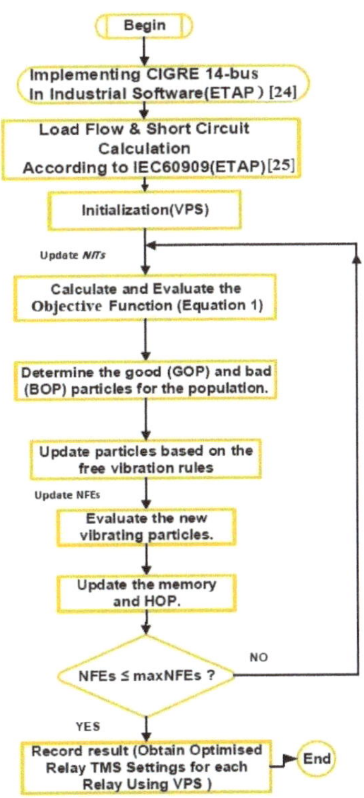

Figure 7. Outline of the optimisation technique (VPS) for solving the OCRs coordination problem [24,25].

3. Results

The formulation described in Section 2 for the OCRs coordination problem is evaluated using a CIGRE distribution network as a common and standard DN. Figure 7 shows the CIGRE distribution network, and a detailed description of the CIGRE grid is discussed in [29]. The proposed hybrid tripping OCR scheme suggested in this paper was first applied to protect the CIGRE network with or without DGs (two PV systems). This section will describe the results of the hybrid tripping OCR characteristic provided under different network operation modes (with and without DG) and different fault scenarios. Furthermore, this section compares the proposed hybrid tripping OCR scheme (HOC) to the common inverse time–current characteristic (SIC) and time–current–voltage characteristic (CVC) [20,23]. Presented is a comparison of the total operational time for HOC, SIC, and CVC in several power network scenarios. Using the three tripping characteristics, the VPS and PSO optimisation algorithms are applied to solve the OCR protection coordination problem. The outcomes of both the VPS and PSO models are also evaluated and contrasted. The HOC approach has been evaluated using industrial software (ETAP), and the findings are provided and compared to the SIC and CVC schemes.

3.1. Description of the CIGRE Distribution Network under Study

The proposed HOC scheme is tested on a CIGRE distribution network, as shown in Figure 8, to determine optimal OCRs coordination and achieve the minimum tripping time. The CIGRE grid is developed based on a 14-bus feeder, and all details are described in [29]. In general, the CIGRE grid is fed by a utility HV/M source and protected by 12 OCRs. Furthermore, this grid is connected to two PV units (each rated at 5MVA) through a 1/20 kV set-up transformer, as described in more detail in [2,5]. On each line, three-phase faults are done at nodes (F1–F12) that represent the near-end and far-end fault locations. Each fault location is assigned two primary OCRs, with one backup OCR for each primary OCR. The Plug Setting (PS), pickup current (I_{pi}), and Current Transformer Ratio (CTR) for each OCR are displayed in Table 1. The load flow was generated to determine the CTR and PS for each OCR initially. In addition, a three-phase short-current determine according to IEC-60909 for different locations. Consequently, the HOC, SIC, and CVC schemes utilised in this investigation have been detailed. The three-phase short-circuit was examined using the ETAP software and the relevant data. The OCR data required to simulate the power network model are listed in Table 1.

Table 1. The PS, I_{pi}, and CTR for each OCR.

Relay	CTR	PS	I_{pi}(A)
HOC1	200/1	60	120
HOC2	200/1	60	120
HOC3	100/1	50	50
HOC4	100/1	50	50
HOC5	100/1	20	20
HOC6	100/1	20	20
HOC7	100/1	60	60
HOC8	100/1	50	50
HOC9	100/1	30	30
HOC10	100/1	20	20
HOC11	100/1	30	30
HOC12	100/1	50	50

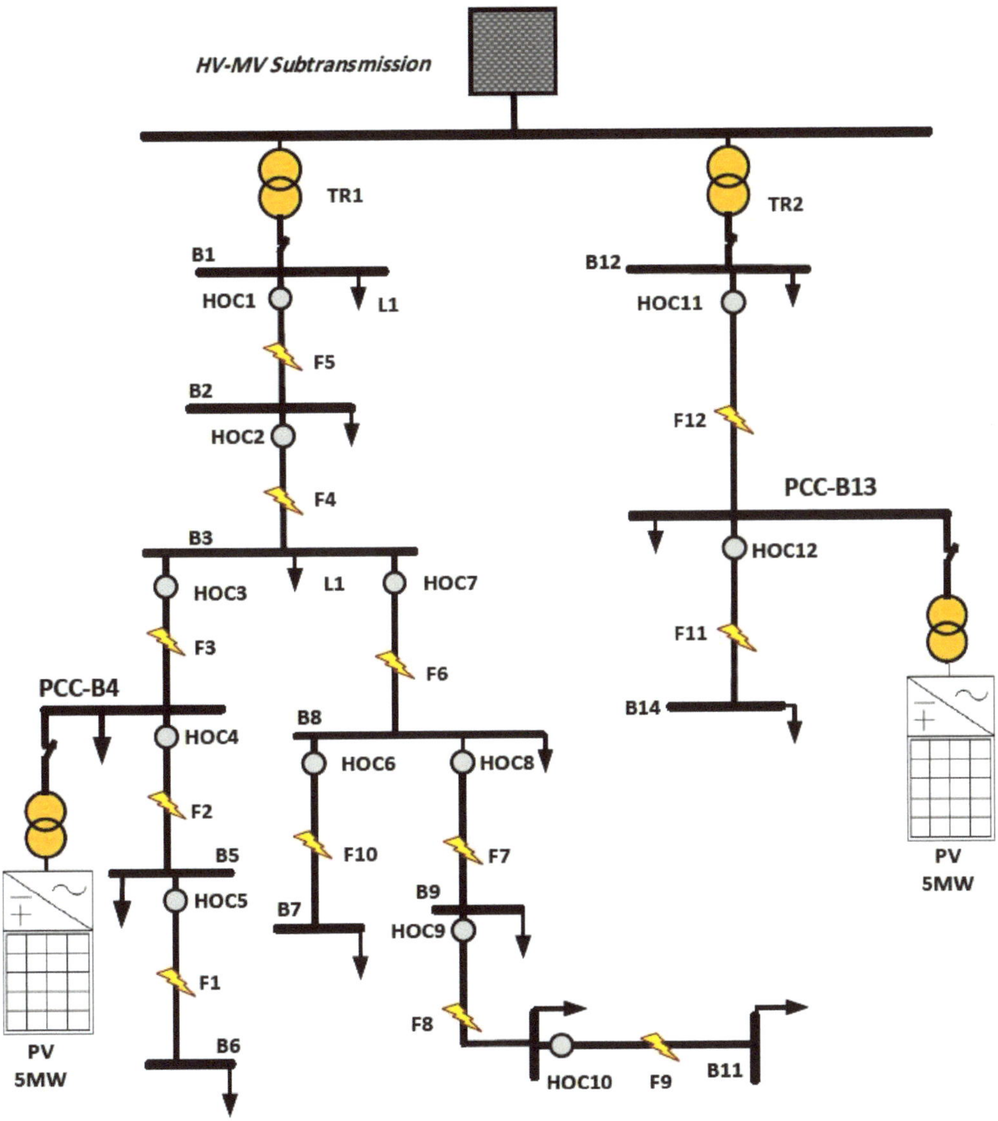

Figure 8. CIGRE distribution network.

3.2. Test Results for CIGRE Distribution Network without PVs

To demonstrate the performance advantage of the proposed hybrid tripping scheme (HOC) in terms of providing the shortest tripping time while preserving suitable CTI across OCR pairs, Table 1 displays the tripping times of all OCR pairs (primary and backup) at various fault sites for the distribution network without PVs. In addition, Table 2 displays the number of recorded delays in trip time when the SIC and CVC approaches were employed in comparison to the HOC approach. For instance, 0.27 s in comparison to 0.59 and 0.56 s for OCR 4 when utilizing the HOC compared to SIC and CVC, respectively, in the event of a three-phase fault at F3.

Table 2. The operating time values in seconds (S.) of OCR for the SIC, CVC, and HOC approaches for the CIGRE distribution network without PVs.

Fault	Fault Current (A)	OCR	SIC (S.)	CVC (S.)	HOC (S.)
F1	1223.0	5	0.02	0.01	0.00
	1223.0	4	0.32	0.31	0.31
F2	1415.0	4	0.31	0.26	0.12
	1415.0	3	0.61	0.56	0.42
F3	1500.0	3	0.59	0.52	0.27
	1500.0	2	0.89	0.82	0.57
F4	1603.0	2	0.87	0.77	0.54
	1603.0	1	1.17	1.07	0.84
F5	3239.0	1	0.91	0.69	0.54
	3239.0	—	0.00	0.01	0.00
F6	1394.0	7	0.90	0.89	0.40
	1394.0	2	0.92	0.87	0.60
F7	1352.0	8	0.60	0.52	0.15
	1352.0	7	1.00	0.94	0.46
F8	1258.0	9	0.32	0.29	0.21
	1258.0	8	0.61	0.59	0.51
F9	1205.0	10	0.02	0.01	0.00
	1205.0	9	0.32	0.31	0.30
F10	1195.0	6	0.03	0.01	0.01
	1195.0	7	0.98	1.01	0.60
F11	2039.0	11	0.02	0.01	0.005
	2039.0	12	0.32	0.54	0.48
F12	2943.0	12	0.29	0.45	0.34
	2943.0	—	0.02	0.00	0.00

Observably, the HOC approach reduces the trip time of all OCRs compared to the SIC and CVC approaches. The use of the HOC method as mentioned in Table 2 reveals the tendency of all OCRs to create an exact CTI among all fault conditions. In addition, the OCR times satisfy all optimisation constraints when the HOC approach is applied. As shown in Table 2, the sensitivity task is adequately provided using the HOC, demonstrating the superiority of the suggested method.

3.3. *Test Results for CIGRE Distribution Network with PVs*

To evaluate the effects of the novel HOC approach, the ideal OCR configuration for the CIGRE distribution network fed by a utility feeder and two PV systems was found, as illustrated in Figure 7. Table 3 displays the trip time for the primary and backup OCR pairs for three-phase faults for the HOC, SIC, and CVC characteristics under grid-connected PV mode. The HOC strategy outperforms both the standard OCRs method (SIC) and CVC. The operational time of all OCRs, as provided in Table 3, indicates that the HOC decreased with the maintaining of sufficient CTI between primary and backup OCR pairs, as compared to SIC and CVC approaches. For example, the HOC approach achieved a trip time of 0.535 s for OCR 2 during F4, while the SIC and CVC approach achieved times of 0.894 and 0.817 s, respectively. By decreasing the time it takes for relays to trip, the HOC technique has the potential to shorten the amount of time relays need to be in operation, which in turn would improve the reliability of the power grid.

Table 3. The operating time values in seconds (S.) of OCR for the SIC, CVC and HOC approaches for the CIGRE distribution network with PVs.

Fault	Fault Current (A)	OCR	SIC (S.)	CVC (S.)	HOC (S.)
F1	1223.0	5	0.021	0.009	0.004
	1223.0	4	0.322	0.309	0.305
F2	1415.0	4	0.309	0.259	0.102
	1415.0	3	0.604	0.554	0.411
F3	1500.0	3	0.593	0.518	0.249
	1500.0	2	0.894	0.817	0.564
F4	1603.0	2	0.871	0.772	0.535
	1603.0	1	1.171	1.072	0.820
F5	3239.0	1	0.914	0.692	0.533
	3239.0	—	0.000	0.000	0.000
F6	1394.0	7	0.898	0.796	0.406
	1394.0	2	0.925	0.868	0.599
F7	1352.0	8	1.610	0.519	0.192
	1352.0	7	0.904	0.821	0.466
F8	1258.0	9	0.317	0.293	0.213
	1258.0	8	1.647	0.593	0.516
F9	1205.0	10	0.021	0.008	0.002
	1205.0	9	0.321	0.310	0.300
F10	1195.0	6	0.029	0.013	0.006
	1195.0	7	0.886	0.899	0.610
F11	2039.0	11	0.018	0.009	0.005
	2039.0	12	0.318	0.541	0.476
F12	2943.0	12	0.292	0.449	0.347
	2943.0	—	0.000	0.000	0.000

3.4. Discussion and Comparison

This section contrasts the efficiency of the suggested HOC approach compared to traditional SIC and CVC strategies in the two power grid scenarios (with and without PV). In general, Table 4 displays the overall OCRs' tripping times for the two power grid scenarios. The TMS and overall operational time in each power grid scenario are reported in Table 4 and were computed using the VPS optimisation technique. To analyse the effectiveness of these acquired settings in terms of total operational time and CTI, we used ETAP software to simulate fault scenarios and then investigated their performance using HOC, SIC, and CVC techniques. In Table 4, we can see that the comparison results in shorter operating times for all OCRs in all cases. The overall operational times for the power grid connected to PVs were 11.86, 11.14, and 7.68 s for the SIC, CVC, and HOC approaches, respectively. Furthermore, the reduction in the overall time for HOC was 35.3% and 33% compared to SIC and CVC, respectively, for the scenario of a power grid without PV.

Table 4. The overall operating time in seconds (S.) and TMS values of OCRs for the SIC, CVC, and HOC approaches for the CIGRE distribution network with and without PVs.

OCR	Without PVs			With PVs		
	TMS					
	SIC	CVC	HOC	SIC	CVC	HOC
1	0.445	0.464	0.597	0.445	0.464	0.584
2	0.330	0.465	0.474	0.331	0.465	0.472
3	0.299	0.666	1.28	0.298	0.66	1.19
4	0.151	0.328	0.62	0.155	0.328	0.524
5	0.01	0.01	0.01	0.01	0.01	0.01
6	0.01	0.01	0.01	0.01	0.01	0.01
7	0.416	0.977	0.995	0.424	0.873	1
8	0.293	0.667	1.31	0.3	0.667	1.61
9	0.150	0.354	0.88	0.154	0.354	0.862
10	0.01	0.01	0.01	0.01	0.01	0.01
11	0.01	0.01	0.01	0.01	0.01	0.01
12	0.175	0.339	0.615	0.177	0.339	0.627
Overall operational time (S.)	11.84	11.44	7.66	11.86	11.14	7.68

3.5. Results of the Proposed Coordination Schemes Based on Different Optimisation Algorithms

Two optimisation methods (PSO and VPS), as common and new powerful optimisation algorithms, respectively, were utilised to determine the optimal coordination setting for all OCRs, and their results were compared to those of the SIC, CVC, and HOC approaches in each of the two power grid scenarios (with and without PV). It is clear from Table 5 that the suggested HOC approach is the best option for reducing the overall operating time of OCRs for both optimisation algorithms (PSO and VPS). In addition, for the SIC, CVC, and HOC approaches, the coordination performance is slightly enhanced by the use of the VPS optimisation algorithm compared to the PSO algorithm. For the power grid connected to PVs, the VPS algorithm achieved an overall tripping time of 7.68, 11.14, and 11.86 s for HOC, CVC, and SIC, respectively, compared to 7.73, 11.25, and 12 s for the PSO algorithm. However, the main improvement in the OCRs' coordination and time-tripping performance was shown through the use of the novel protection scheme in this work (HOC) compared to the SIC and CVC protection schemes.

Table 5. The overall operating time in seconds (S.) of OCRs for the SIC, CVC, and HOC approaches using PSO and VPS optimisation algorithms for the CIGRE network with and without PVs.

Coordination Scheme	Without PVs		With PVs	
	Optimisation Algorithms			
	PSO	VPS	PSO	VPS
SIC	11.9	11.84	12	11.86
CVC	11.6	11.44	11.25	11.14
HOC	7.7	7.66	7.73	7.68

It is intriguing to examine the convergence rate of the proposed optimisation technique (VPS) under different protection schemes. Figures 9 and 10 depict the convergence curves for the VPS technique under the different protection schemes (SIC, CVC, and HOC) and both power network scenarios (with and without PVs). First, Figure 9 demonstrates and compares the performance of VPS for the SIC, CVC, and HOC schemes for power networks without PV scenarios in terms of convergence. In general, the HOC algorithm has a faster and smoother convergence curve, achieving optimal results compared to SIC and CVC. This shows that the HOC has reduced computation costs while increasing CPU utilisation efficiency. Second, the convergence rate results for a PV power network with PVs are presented in Figure 10. The convergence curve of the HOC is smoother, especially when the number of iterations exceeds 400 while the curves of CVC and SIC were stochastic and volatile over all iterations.

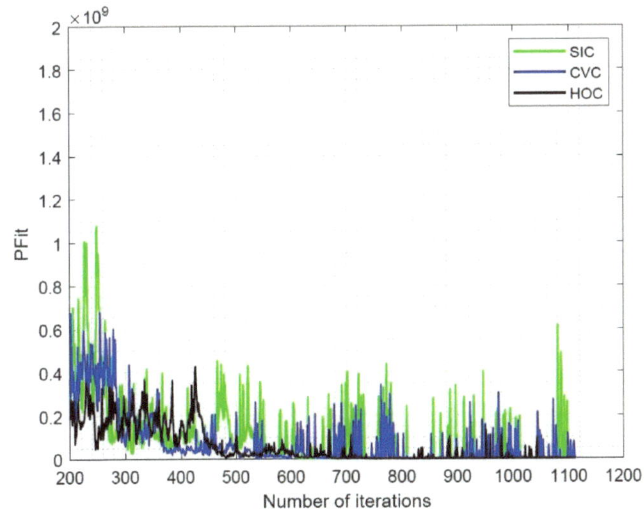

Figure 9. The convergence rate of the VPS technique for power networks without PVs for SIC, CVC, and HOC protection schemes.

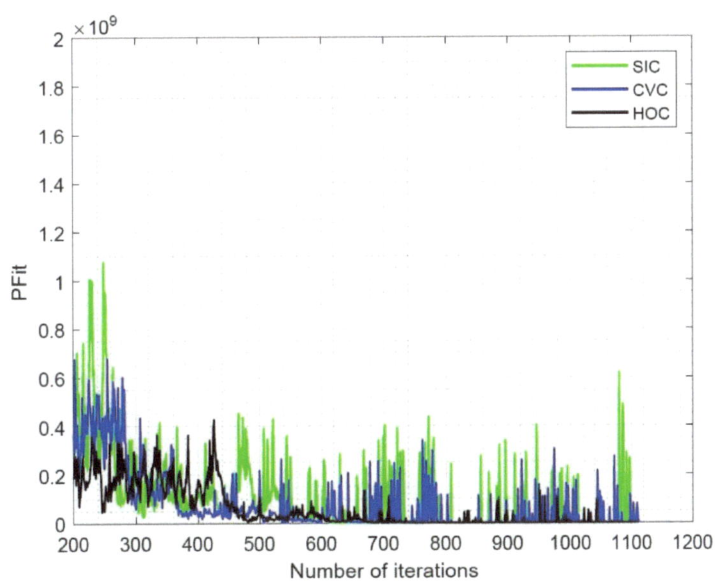

Figure 10. The convergence rate of the VPS technique for power networks with PVs for SIC, CVC, and HOC protection schemes.

3.6. Evaluation Using Industrial Software (ETAP)

ETAP is a typical simulation tool and industrial software that is widely used to show the model of power networks with various types of relays. ETAP is a user-friendly interface for analysing the effectiveness, performance, and viability of protection and power operating systems. In this study, ETAP was used to investigate the performance of OCR settings under different protection schemes (SIC, CVC, and HOC) for the given coordination problem, without introducing misoperation. In this section, a simulation model example was developed to demonstrate optimal OCR coordination utilising ETAP for the power network with PVs using the VPS algorithm, as depicted in Figures 11 and 12. The purpose of these figures is to illustrate the Time Current Curve (TCC) and the OCR coordination between primary and backup OCRs. Figure 11 presents the OCRS coordination between relays 3 and 4 at F2. The proposed HOC scheme outperformed SIC and CVC for the primary relay (4) and the backup relay (3). The primary relay (4) will operate in 0.109, 0.259, and 0.318 s for HOC, CVC, and SIC, respectively. The OCRs coordination at F6 (1.481 Ka at 20 kV) is presented and compared in Figure 12. The proposed HOC scheme recorded the minimum tripping time for primary and backup relays compared to the SIC and CVC schemes. The backup relay at F6 will operate in 0.597, 0.828, and 0.931 s for HOC, CVC, and SIC, respectively.

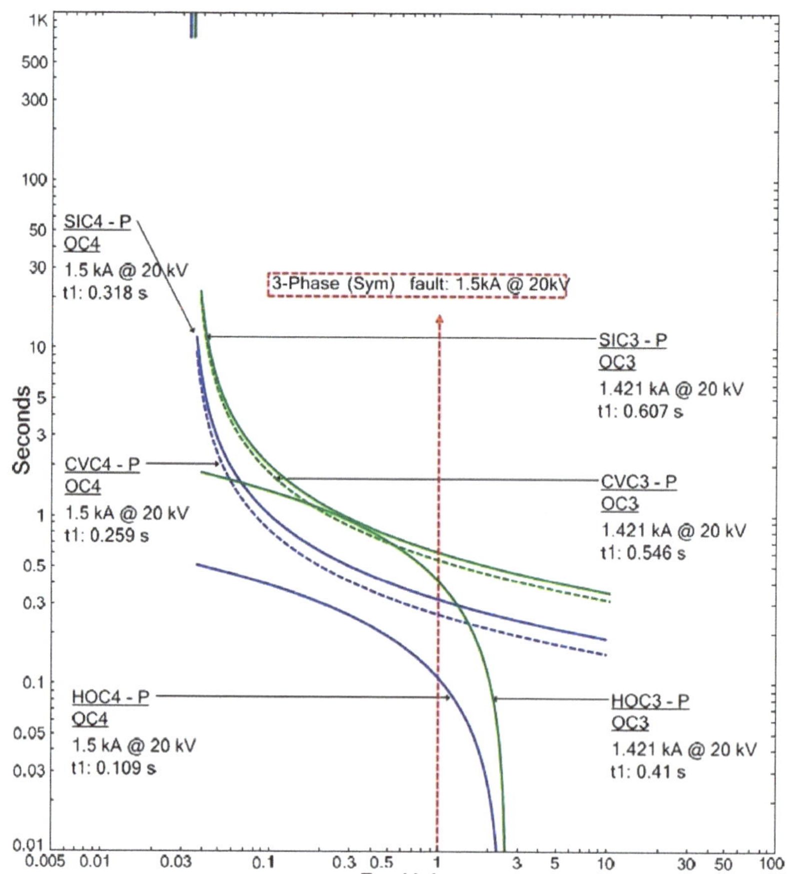

Figure 11. TCC graphs of OCRs for power network with PVs at F2.

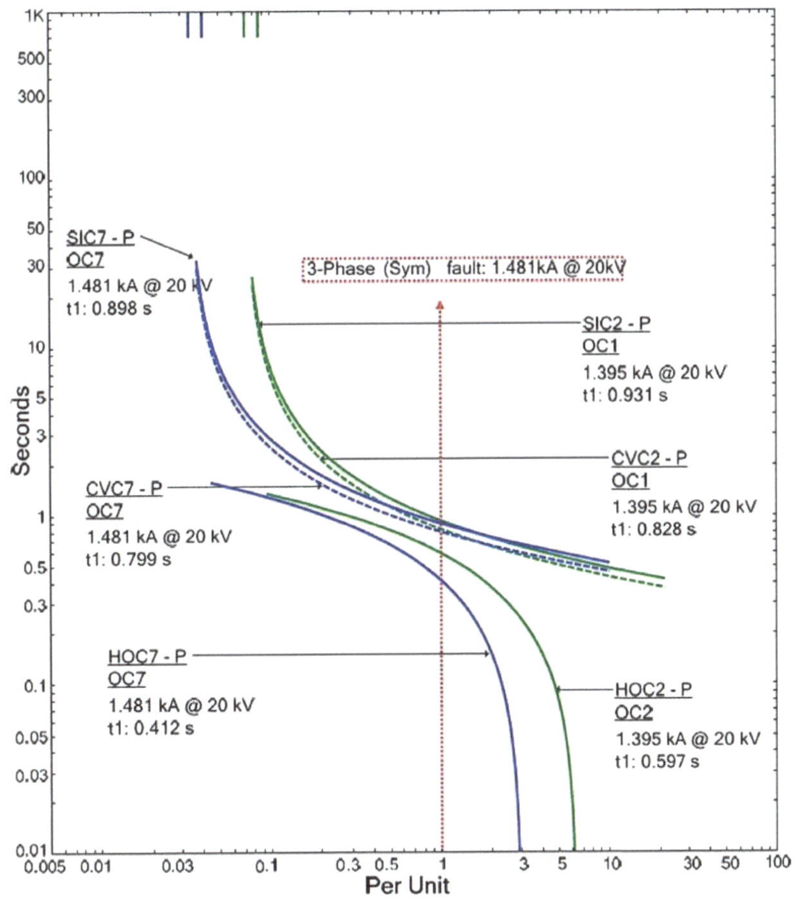

Figure 12. TCC graphs of OCRs for power network with PVs at F6.

4. Conclusions

The goal of this work was to implement a rapid response hybrid tripping scheme for the OCR protection system in modern power network architecture (with and without PVs). The suggested HOC scheme controls and minimises the total operational time for all OCRs by utilizing a new current–voltage characteristic. The optimal OCRs coordination problem under grid constraints was established using the HOC method. The suggested HOC was formulated and solved to provide optimal OCR settings under various fault conditions and different network topologies. The VPS and PSO algorithms were employed to fine-tune and resolve the optimal configuration of all OCRs. The consequences of the suggested HOC approach were not only the fulfilment of the coordination assignment but also a significant reduction in the overall operating time compared to the common SIC and CVC. For example, the overall operational times for the power grid connected to PVs were 11.86, 11.14, and 7.68 s for SIC, CVC, and HOC approaches, respectively. In addition, the VPS and PSO optimisation methods were evaluated to identify a solution with a rapid trip time. The high and complex protection challenges will, in the future, require using machine learning, artificial intelligence, and different optimisation algorithms to minimise the tripping time and improve the protection selectivity performance.

Author Contributions: Conceptualization, F.A., N.E.-N. and A.S.S.; methodology, F.A., M.A.S., N.E.-N. and. W.H; software, F.A., W.H. and N.E; validation, F.A., A.S.S. and N.E.-N.; formal analysis, F.A, W.H. and M.A.S.; investigation, F.A., A.S.S. and W.H.; resources, all authors; data curation, all authors; writing—original draft preparation, F.A., M.A.S. and A.S.S.; writing—review and editing, all authors; visualization, all authors; supervision, all authors; project administration, F.A. All authors have read and agreed to the published version of the manuscript.

Funding: The authors extend their appreciation to the Deanship of Scientific Research at King Khalid University for funding this work through Research Groups Program under grant number RGP.2/81/43.

Institutional Review Board Statement: Not applicable.

Informed Consent Statement: Not applicable.

Data Availability Statement: Not applicable.

Acknowledgments: We would like to thank University of Reading and The Hashemite University (Renewable Energy Center) for their support and funding this article.

Conflicts of Interest: The authors declare no conflict of interest.

Nomenclature

CIGRE	Conseil International des Grands Réseaux Electriques	$t_{r,f}$	The operational time of the relay, r, at the fault location, f
PV	Photovoltaic	t_{backup}	Backup OCR
DGs	Distributed generators	$t_{primary}$	Primary OCR
DN	Distribution network	TMS_{min}	Minimum TMS
OCRs	Overcurrent relays	TMS_{max}	Maximum TMS
HOC	The proposed new hybrid tripping OCR scheme	I_f	Fault current
SIC	The common inverse time–current characteristic	I_{pi}	Pickup current
CVC	The time–current–voltage characteristic	A, B	A and B are constants
VPS	Vibrating particles system	V	The phase fault voltage
PSO	Particle swarm optimisation	HOP	Highest optimal position
CTI	Coordination Time Interval	GOP	Good optimal particle
OT	Objective function	BOP	Bad optimal particle
TMS	Time Multiplier Settings	maxNFES	OF evaluation iterations OF evaluation iterations

References

1. Singh, P.; Pradhan, A. A Local measurement based protection technique for distribution system with photovoltaic plants. *IET Renew. Power Gener.* **2020**, *4*, 996–1003. [CrossRef]
2. Alasali, F.; Zarour, E.; Holderbaum, W.; Nusair, K. Highly Fast Innovative Overcurrent Protection Scheme for Microgrid Using Metaheuristic Optimisation Algorithms and Nonstandard Tripping Characteristics. *IEEE Access* **2022**, *10*, 42208–42231. [CrossRef]
3. Malik, P.; Sinha, S.; Awasthi, M.; Aggarwal, K. Hybrid operational approach for PV/DG microgrid without storage device. In Proceedings of the IEEE 2nd International Conference on Smart Technologies for Power, Energy and Control (STPEC), Bilaspur, India, 19–22 December 2021. [CrossRef]
4. Jamali, S.; Bahabadi, H. Non-communication protection method for meshed and radial distribution networks with synchronous-based DG. *Int. Electr. Power Energy Syst.* **2017**, *93*, 468–478. [CrossRef]
5. Alasali, F.; El-Naily, N.; Zarour, E.; Saad, S. Highly sensitive and fast microgrid protection using optimal coordination scheme and nonstandard tripping characteristics. *Int. J. Electr. Power AND Energy Syst.* **2021**, *128*, 106756. [CrossRef]
6. Chakraborty, S.; Das, S. Communication-less protection scheme for AC microgrids using hybrid tripping characteristic. *Electr. Power Syst. Res.* **2020**, *187*, 106453. [CrossRef]
7. Ji, L.; Cao, Z.; Hong, Q.; Chang, X.; Fu, Y.; Shi, J.; Mi, Y.; Li, Z. An Improved Inverse-Time Over-Current Protection Method for a Microgrid with Optimized Acceleration and Coordination. *Energies* **2020**, *13*, 5726. [CrossRef]
8. Soleimanisardoo, A.; Karegar, H.K.; Zeineldin, H.H. Di_erential frequency protection scheme based on off-nominal frequency injections for inverter-based islanded microgrids. *IEEE Trans. Smart Grid* **2019**, *10*, 2107–2114. [CrossRef]

9. Aghdam, T.S.; Karegar, H.K.; Zeineldin, H.H. Variable tripping time differential protection for microgrids considering DG stability. *IEEE Trans. Smart Grid* **2019**, *10*, 2407–2415. [CrossRef]
10. Rajput, V.N.; Pandya, K.S.; Hong, J.; Geem, Z.W. A Novel Protection Scheme for Solar Photovoltaic Generator Connected Networks Using Hybrid Harmony Search Algorithm-Bollinger Bands Approach. *Energies* **2020**, *13*, 2439. [CrossRef]
11. Mishra, P.; Prahan, A.; Bajpai, P. A Positive Sequence Relaying Method for Solar Photovoltaic Integrated Distribution System. *IEEE Trans. Power Deliv.* **2021**, *36*, 3519–3528. [CrossRef]
12. Shih, M.; Conde, A.; Leonowicz, Z.; Martirano, L. An adaptive overcurrent coordination scheme to improve relay sensitivity and overcome drawbacks due to distributed generation in smart grids. *IEEE Trans. Ind. Appl.* **2017**, *53*, 5217–5228. [CrossRef]
13. Jamali, S.; Bahabadi, H. Recloser time–current–voltage characteristic for fuse saving in distribution networks with DG. *IET Gener. Transm. Distrib.* **2017**, *11*, 272–279. [CrossRef]
14. Tumilty, R.; Brucoli, M.; Burt, G.M.; Green, T. Approaches to Network Protection for Inverter Dominated Electrical Distribution Systems. In Proceedings of the 3rd IET International Conference on Power Electronics, Machines and Drives, Dublin, Ireland, 4–6 April 2006.
15. Singh, M.; Agrawal, A. Voltage–current–time inverse-based protection coordination of photovoltaic power systems. *IET Gener. Transm. Distrib.* **2019**, *13*, 794–804. [CrossRef]
16. El-sayed, W.; Azzouz, M.; Zeineldin, H.; El-Saaadany, E. Harmonic Time-Current-Voltage Directional Relay for Optimal Protection Coordination of Inverter-Based Islanded Microgrids. *IEEE Trans. Smart Grid* **2021**, *12*, 1904–1917. [CrossRef]
17. Momesso, A.; Bernardes, W.; Asada, E. Fuzzy adaptive setting for time-current-voltage based overcurrent relays in distribution systems. *Int. Electr. Power Energy Syst.* **2019**, *108*, 135–144. [CrossRef]
18. Marcolino, M.; Leite, J.; Mantovani, S. Optimal coordination of overcurrent directional and distance relays in meshed networks using Genetic Algorithm. *IEEE Lat. Am. Trans.* **2015**, *13*, 2975–2982. [CrossRef]
19. Bernardes, W.; Santos, F.; Asada, E.; Ramos, M. Application of discrete PSO and evolutionary PSO to the coordination of directional overcurrent relays. In Proceedings of the IEEE XVII International Conference on Intelligent System Applications to Power Systems, Tokyo, Japan, 1–4 July 2013; pp. 1–6.
20. Saleh, K.; Zeineldin, H.; Al-Hinai, A.; El-Saadany, E. Optimal Coordination of Directional Overcurrent Relays Using a New Time–Current–Voltage Characteristic. *IEEE Trans. Power Deliv.* **2014**, *30*, 537–544. [CrossRef]
21. Niranjan, P.; Anand, A.; Singh, N.; Choudhary, N. Comparative Study of Optimal Relay Coordination with Current and Voltage-based Relay in Distributed Generation Systems. In Proceedings of the IEEE Students Conference on Engineering and Systems (SCES), Prayagraj, India, 1–3 July 2022. [CrossRef]
22. Santos, G.; Vieira, J.; Piardi, A. Coordination and Selectivity Analysis of Voltage-based Relays for Enhancing Microgrid Protection. In Proceedings of the IEEE PES Innovative Smart Grid Technologies Conference—Latin America (ISGT Latin America), Lima, Peru, 15–17 September 2021. [CrossRef]
23. Lim, S.-H.; Lim, S.-T. Analysis on Coordination of Over-Current Relay Using Voltage Component in a Power Distribution System with a SFCL. *IEEE Trans. Appl. Supercond.* **2019**, *29*, 1–5. [CrossRef]
24. ETAP. Electrical Power System Analysis & Operation Software. Available online: https://etap.com/ (accessed on 6 January 2023).
25. *IEC 60909*; Short-Circuit Currents in Three-Phase a.c. Systems—Part 0: Calculation of Currents. IEC: Geneva, Switzerland, 2016. Available online: https://webstore.iec.ch/publication/24100 (accessed on 6 January 2023).
26. Tiwari, S.; Kumar, A. Advances and bibliographic analysis of particle swarm optimisation applications in electrical power system: Concepts and variants. *Evol. Intell.* **2021**, *1*. [CrossRef]
27. Kaveh, A.; Ilchi Ghazaan, M. A new meta-heuristic algorithm: Vibrating particles system. *Int. J. Sci. Technol.* **2017**, *24*, 551–556. [CrossRef]
28. Talatahari, S.; Jalili, S.; Azizi, M. Optimum design of steel building structures using migration-based vibrating particles system. *Structures* **2021**, *33*, 1394–1413. [CrossRef]
29. e-cigre. Benchmark Systems for Network Integration of Renewable and Distributed Energy Resources. 2014. Available online: https://e-cigre.org/publication/ELT_273_8-benchmark-systems-for-network-integration-of-renewable-and-distributed-energy-resources (accessed on 6 January 2023).

MDPI AG
Grosspeteranlage 5
4052 Basel
Switzerland
Tel.: +41 61 683 77 34

Sustainability Editorial Office
E-mail: sustainability@mdpi.com
www.mdpi.com/journal/sustainability

www.ingramcontent.com/pod-product-compliance
Lightning Source LLC
Chambersburg PA
CBHW041417210926
51850CB00013B/3294
9783725827787